NEUROMETHODS

Series Editor
Wolfgang Walz
University of Saskatchewan
Saskatoon, SK, Canada

For further volumes:
http://www.springer.com/series/7657

For further volumes: http://www.springer.com/series/7657

Neuromethods publishes cutting-edge methods and protocols in all areas of neuroscience as well as translational neurological and mental research. Each volume in the series offers tested laboratory protocols, step-by-step methods for reproducible lab experiments and addresses methodological controversies and pitfalls in order to aid neuroscientists in experimentation. *Neuromethods* focuses on traditional and emerging topics with wide-ranging implications to brain function, such as electrophysiology, neuroimaging, behavioral analysis, genomics, neurodegeneration, translational research and clinical trials. *Neuromethods* provides investigators and trainees with highly useful compendiums of key strategies and approaches for successful research in animal and human brain function including translational "bench to bedside" approaches to mental and neurological diseases.

Computational Neuroscience

Edited by

Drozdstoy Stoyanov

Division of Translational Neuroscience, Medical University of Plovdiv, Plovdiv, Bulgaria

Bogdan Draganski

University of Lausanne, CHUV, Lausanne, Switzerland

Paolo Brambilla

Pathophysiology and Transplantation, University of Milan, Milan, Italy

Claus Lamm

Cognitive & Affective Neuroscience Unit, University of Vienna, Vienna, Austria

Humana Press

Editors
Drozdstoy Stoyanov
Division of Translational Neuroscience
Medical University of Plovdiv
Plovdiv, Bulgaria

Bogdan Draganski
University of Lausanne, CHUV
Lausanne, Switzerland

Paolo Brambilla
Pathophysiology and Transplantation
University of Milan
Milan, Italy

Claus Lamm
Cognitive & Affective Neuroscience Unit
University of Vienna
Vienna, Austria

ISSN 0893-2336 ISSN 1940-6045 (electronic)
Neuromethods
ISBN 978-1-0716-3229-1 ISBN 978-1-0716-3230-7 (eBook)
https://doi.org/10.1007/978-1-0716-3230-7

This Humana imprint is published by the registered company Springer Science+Business Media, LLC, part of Springer Nature.
The registered company address is: 1 New York Plaza, New York, NY 10004, U.S.A.

Preface to the Series

Experimental life sciences have two basic foundations: concepts and tools. The Neuromethods series focuses on the tools and techniques unique to the investigation of the nervous system and excitable cells. It will not, however, shortchange the concept side of things as care has been taken to integrate these tools within the context of the concepts and questions under investigation. In this way, the series is unique in that it not only collects protocols but also includes theoretical background information and critiques which led to the methods and their development. Thus it gives the reader a better understanding of the origin of the techniques and their potential future development. The Neuromethods publishing program strikes a balance between recent and exciting developments like those concerning new animal models of disease, imaging, in vivo methods, and more established techniques, including, immunocytochemistry and electrophysiological technologies. New trainees in neurosciences still need a sound footing in these older methods in order to apply a critical approach to their results.

Under the guidance of its founders, Alan Boulton and Glen Baker, the Neuromethods series has been a success since its first volume published through Humana Press in 1985. The series continues to flourish through many changes over the years. It is now published under the umbrella of Springer Protocols. While methods involving brain research have changed a lot since the series started, the publishing environment and technology have changed even more radically. Neuromethods has the distinct layout and style of the Springer Protocols program, designed specifically for readability and ease of reference in a laboratory setting.

The careful application of methods is potentially the most important step in the process of scientific inquiry. In the past, new methodologies led the way in developing new disciplines in the biological and medical sciences. For example, Physiology emerged out of Anatomy in the nineteenth century by harnessing new methods based on the newly discovered phenomenon of electricity. Nowadays, the relationships between disciplines and methods are more complex. Methods are now widely shared between disciplines and research areas. New developments in electronic publishing make it possible for scientists that encounter new methods to quickly find sources of information electronically. The design of individual volumes and chapters in this series takes this new access technology into account. Springer Protocols makes it possible to download single protocols separately. In addition, Springer makes its print-on-demand technology available globally. A print copy can therefore be acquired quickly and for a competitive price anywhere in the world.

Wolfgang Walz

Wolfgang Walz

Preface

This volume presents the advances in imaging neuroscience methods using magnetic resonance imaging (MRI) and electroencephalography (EEG) to study the healthy and diseased brain.

One of the great challenges in the field appears to be the integration across levels of observation, spanning from molecular to neurophysiological and imaging methods in vivo.

Despite the accumulated evidence in the past few decades, imaging neuroscience methods are still confined to a single level of observation rather than spanning data acquisition or analysis over multiple levels. In a similar vein, there is major controversy with reference to a causal inference stemming from one single imaging modality and level of observation.

This notion is introduced in the opening chapter which sets the stage of the following chapters by proposing an innovative concept for translational imaging neuroscience with the potential to overcome those challenges.

The first section of the edited volume is then dedicated to molecular methods. The chapter by Ivanovska and associates delivers an overview of the different applied methods for data acquisition, while the chapter by Maes provides a machine learning approaches to process and interpret them in the clinical context. According to this approach, computational models can re-shape psychiatric classifications and taxonomy.

The second section deals with neurophysiological methods for assessment, such as quantitative EEG and event-related potentials. Event-related oscillations are the focus of the contribution by Yordanova and associates, while the other three chapters by Panov, Kandilarova, and Riecansky address in a consolidated fashion the principles and the use of QEEG in mood disorders and schizophrenia research.

The third section of the book summarizes the advances and innovations in computational anatomy. In the field of structural imaging, Draganski and co-authors present the potential benefits of quantitative MRI over the traditional T1-weighted imaging that carries the promise of "in vivo histology."

Korda and co-authors deliver innovative perspectives on structural MRI data analysis, adopting approaches from chaos theory and non-linear systems. In the field of functional MRI, Vaisvilaite and colleagues analyze the confounding effect of various factors that may influence the BOLD signal. Molecular neuroimaging methods (magnetic-resonance spectroscopy) are the focus of the contribution by Brambilla and Squarcina. This group of techniques are bridge neurochemical and functional levels of explanation in psychopathology and psychopharmacology. Disorder-specific alteration of amygdala and anterior insula are discussed in the chapters by Kandilarova, and Sladky, Todeva-Radneva, and Lamm, which also illustrate caveats of the existing technologies and analysis approaches. Another disorder-focused group led by Vassileva explores the possibilities for use of computational approaches to addictions research. It is outlined in this chapter the difference between data-driven, agnostic (theory-driven), and hybrid applications.

The fourth section directly addresses the challenge before the computational neuroscience to find wider translational applications in the field of psychiatry and mental health. Namely the translation and integration across levels of explanation. Maggioni and colleagues set the background for multimodal imaging, while Kherif and co-workers illustrate the premises for multivariate linear methodology as possible resolutions for integration of diverse data sets.

In summary, the edited volume stimulates an effort to follow the epistemological pathways of computational neuroscience in psychiatry and mental health by tracing its various applications to data from molecular, neurophysiological, and imaging level methods and their potential integration into non-linear complex models. The traditional bio-medical science is actually based by default on such models of disease since early twenty-first century, recognized as nomothetic networks.

Our volume offers an exciting opportunity for psychiatry and mental health to incorporate into broader medical knowledge by iterating similar computational approaches, beyond research purposes, for translation into utility and validity of clinical methods.

Plovdiv, Bulgaria — *Drozdstoy Stoyanov*
Lausanne, Switzerland — *Bogdan Draganski*
Milan, Italy — *Paolo Brambilla*
Vienna, Austria — *Claus Lamm*

Contents

Contributors

WOO-YOUNG AHN • *Department of Psychology, Seoul National University, Seoul, South Korea*
MICAEL ANDERSSON • *Umeå Center for Functional Brain Imaging, Umeå University, Umeå, Sweden; Department of Integrative Medical Biology, Umeå University, Umeå, Sweden*
CHRISTINA ANDREOU • *Translational Psychiatry group, Department of Psychiatry and Psychotherapy, University Hospital Lübeck (UKSH), Lübeck, Germany*
ALEXANDRA BALDZHIEVA • *Department of Medical Microbiology and Immunology "Prof. Dr. Elissay Yanev", Faculty of Pharmacy, Research Institute at Medical University-Plovdiv (RIMU); Center of Competence – Personalized Innovative Medicine (PERIMED), Medical University, Plovdiv, Bulgaria; Laboratory of Clinical Immunology, University Hospital "St. George", Plovdiv, Bulgaria;* НУКБПИ-*BBMRI.BG,* Д*01-395/18.12.2020, Plovdiv, Bulgaria*
ANNA M. BIANCHI • *Department of Electronics, Information and Bioengineering, Politecnico di Milano, Milan, Italy*
ELENA BONDI • *Department of Electronics, Information and Bioengineering, Politecnico di Milano, Milan, Italy; Department of Pathophysiology and Transplantation, University of Milan, Milan, Italy*
STEFAN BORGWARDT • *Translational Psychiatry group, Department of Psychiatry and Psychotherapy, University Hospital Lübeck (UKSH), Lübeck, Germany*
MARTINA BOZHKOVA • *Department of Medical Microbiology and Immunology "Prof. Dr. Elissay Yanev", Faculty of Pharmacy, Research Institute at Medical University-Plovdiv (RIMU); Center of Competence – Personalized Innovative Medicine (PERIMED), Medical University, Plovdiv, Bulgaria; Laboratory of Clinical Immunology, University Hospital "St. George", Plovdiv, Bulgaria;* НУКБПИ-*BBMRI.BG,* Д*01-395/18.12.2020, Plovdiv, Bulgaria*
PAOLO BRAMBILLA • *Department of Pathophysiology and Transplantation, University of Milan, Milan, Italy; Department of Neurosciences and Mental Health, Fondazione IRCCS Ca' Granda, Ospedale Maggiore Policlinico, Milan, Italy*
BOGDAN DRAGANSKI • *LREN—Département des neurosciences cliniques, CHUV, Université de Lausanne, Lausanne, Switzerland; Neurology Department, Max-Planck-Institute for Human Cognitive and Brain Sciences, Leipzig, Germany*
MARINA FRISMAN • *Translational Psychiatry group, Department of Psychiatry and Psychotherapy, University Hospital Lübeck (UKSH), Lübeck, Germany*
MARIYA IVANOVSKA • *Department of Medical microbiology and Immunology "Prof. Dr. Elissay Yanev", Faculty of Pharmacy, Research Institute at Medical University-Plovdiv (RIMU); Center of Competence – Personalized Innovative Medicine (PERIMED), Medical University, Plovdiv, Bulgaria; Laboratory of Clinical Immunology, University Hospital "St. George", Plovdiv, Bulgaria*
TEODORA KALFOVA • *Department of Medical Microbiology and Immunology "Prof. Dr. Elissay Yanev", Faculty of Pharmacy, Research Institute at Medical University-Plovdiv (RIMU); Center of Competence – Personalized Innovative Medicine (PERIMED),*

Medical University, Plovdiv, Bulgaria; НУКБПИ*-BBMRI.BG,* Д*01-395/18.12.2020, Plovdiv, Bulgaria*

SEVDALINA KANDILAROVA • *Department of Psychiatry and Medical Psychology, Research Institute at Medical University of Plovdiv, Plovdiv, Bulgaria; Department of Psychiatry and Medical Psychology and Research Institute at Medical University of Plovdiv, Plovdiv, Bulgaria*

FERATH KHERIF • *Department of Clinical Neurosciences, Laboratory for Research in Neuroimaging (LREN), Centre for Research in Neuroscience, Lausanne University Hospital, University of Lausanne, Lausanne, Switzerland; LREN—Département des neurosciences cliniques, CHUV, Université de Lausanne, Lausanne, Switzerland; Department of Clinical Neurosciences, Laboratory for Research in Neuroimaging LREN, Centre for Research in Neuroscience, Lausanne University Hospital, University of Lausanne, Lausanne, Switzerland*

ROUMEN KIROV • *Institute of Neurobiology, Bulgarian Academy of Sciences, Sofia, Bulgaria*

VASIL KOLEV • *Institute of Neurobiology, Bulgarian Academy of Sciences, Sofia, Bulgaria*

ALEXANDRA KORDA • *Translational Psychiatry group, Department of Psychiatry and Psychotherapy, University Hospital Lübeck (UKSH), Lübeck, Germany*

CLAUS LAMM • *Social, Cognitive and Affective Neuroscience (SCAN) Unit, Department of Cognition, Emotion, and Methods in Psychology, Faculty of Psychology, University of Vienna, Vienna, Austria*

ADELIYA LATYPOVA • *LREN—Département des neurosciences cliniques, CHUV, Université de Lausanne, Lausanne, Switzerland; Department of Clinical Neurosciences, Laboratory for Research in Neuroimaging LREN, Centre for Research in Neuroscience, Lausanne University Hospital, University of Lausanne, Lausanne, Switzerland*

JEUNG-HYUN LEE • *Department of Psychology, Seoul National University, Seoul, South Korea*

MICHAEL MAES • *Department of Psychiatry, Faculty of Medicine, Chulalongkorn University, Bangkok, Thailand; Deakin University, IMPACT, the Institute for Mental and Physical Health and Clinical Translation, School of Medicine, Barwon Health, Geelong, Australia; Department of Psychiatry, Medical University of Plovdiv, Plovdiv, Bulgaria; Research Institute, Medical University Plovdiv, Plovdiv, Bulgaria*

ELEONORA MAGGIONI • *Department of Electronics, Information and Bioengineering, Politecnico di Milano, Milan, Italy; Department of Neurosciences and Mental Health, Fondazione IRCCS Ca'Granda Ospedale Maggiore Policlinico, Milan, Italy*

MARIANNA MURDJEVA • *Department of Medical Microbiology and Immunology "Prof. Dr. Elissay Yanev", Faculty of Pharmacy, Research Institute at Medical University-Plovdiv (RIMU); Center of Competence – Personalized Innovative Medicine (PERIMED), Medical University, Plovdiv, Bulgaria; Laboratory of Clinical Immunology, University Hospital "St. George", Plovdiv, Bulgaria;* НУКБПИ*-BBMRI.BG,* Д*01-395/18.12.2020, Plovdiv, Bulgaria*

GEORGI PANOV • *Psychiatric Clinic, University Hospital for Active Treatment "Prof. Dr. Stoyan Kirkovich" Trakia University, Stara Zagora, Bulgaria; Medical Faculty, University "Prof. Dr. Asem Zlatarov", Burgas, Bulgaria*

ROSITSA PAUNOVA • *Department of Psychiatry and Medical Psychology, Research Institute Plovdiv, Medical University Plovdiv, Plovdiv, Bulgaria; Department of Psychiatry and Medical Psychology, Medical University Plovdiv, Plovdiv, Bulgaria*

Steliyan Petrov • *Department of Medical Microbiology and Immunology "Prof. Dr. Elissay Yanev", Faculty of Pharmacy, Research Institute at Medical University-Plovdiv (RIMU); Center of Competence – Personalized Innovative Medicine (PERIMED), Medical University, Plovdiv, Bulgaria;* НУКБПИ-*BBMRI.BG, Д01-395/18.12.2020, Plovdiv, Bulgaria*

Maria Chiara Piani • *Translational Research Center, University Hospital of Psychiatry and Psychotherapy, Bern, UPD, Switzerland; Graduate School of Health Sciences, University of Bern, Bern, Switzerland*

Elena Psederska • *Bulgarian Addictions Institute, Sofia, Bulgaria; Department of Cognitive Science and Psychology, New Bulgarian University, Sofia, Bulgaria*

Cristina Ramponi • *Department of Clinical Neurosciences, Laboratory for Research in Neuroimaging LREN, Centre for Research in Neuroscience, Lausanne University Hospital, University of Lausanne, Lausanne, Switzerland*

Igor Riečanský • *Social, Cognitive and Affective Neuroscience Unit, Department of Cognition, Emotion, and Methods in Psychology, Faculty of Psychology, University of Vienna, Vienna, Austria; Department of Behavioural Neuroscience, Centre of Experimental Medicine, Slovak Academy of Sciences, Bratislava, Slovakia; Department of Psychiatry, Faculty of Medicine, Slovak Medical University in Bratislava, Bratislava, Slovakia*

Ronald Sladky • *Social, Cognitive and Affective Neuroscience (SCAN) Unit, Department of Cognition, Emotion, and Methods in Psychology, Faculty of Psychology, University of Vienna, Vienna, Austria*

Karsten Specht • *Department of Biological and Medical Psychology, University of Bergen, Bergen, Norway; Mohn Medical and Imaging Visualization Centre, Haukeland University Hospital, Bergen, Norway; Department of Education, UiT/The Arctic University of Norway, Tromsø, Norway*

Letizia Squarcina • *Department of Pathophysiology and Transplantation, University of Milan, Milan, Italy*

Drozdstoy Stoyanov • *Department of Psychiatry and Medical Psychology, Faculty of Medicine, Translational Neuroscience Division; Research Institute at Medical University-Plovdiv (RIMU), Plovdiv, Bulgaria; Department of Psychiatry and Medical Psychology, Faculty of Medicine, Research Institute at Medical University – Plovdiv (RIMU), Plovdiv, Bulgaria*

Hristo Taskov • *Department of Medical Microbiology and Immunology "Prof. Dr. Elissay Yanev", Faculty of Pharmacy, Research Institute at Medical University-Plovdiv (RIMU); Center of Competence – Personalized Innovative Medicine (PERIMED), Medical University, Plovdiv, Bulgaria; Laboratory of Clinical Immunology, University Hospital "St. George", Plovdiv, Bulgaria;* НУКБПИ-*BBMRI.BG, Д01-395/18.12.2020, Plovdiv, Bulgaria*

Anna Todeva-Radneva • *Department of Psychiatry and Medical Psychology, Faculty of Medicine, and Research Institute, Medical University of Plovdiv, Plovdiv, Bulgaria*

Liucija Vaisvilaite • *Department of Biological and Medical Psychology, University of Bergen, Bergen, Norway; Mohn Medical and Imaging Visualization Centre, Haukeland University Hospital, Bergen, Norway*

JASMIN VASSILEVA • *Institute for Drug and Alcohol Studies, Virginia Commonwealth University, Richmond, VA, USA; Department of Psychiatry, Virginia Commonwealth University, Richmond, VA, USA*
MENG-YUN WANG • *Department of Biological and Medical Psychology, University of Bergen, Bergen, Norway; Mohn Medical and Imaging Visualization Centre, Haukeland University Hospital, Bergen, Norway*
JULIANA YORDANOVA • *Institute of Neurobiology, Bulgarian Academy of Sciences, Sofia, Bulgaria*
DORA ZLATAREVA • *Department of Diagnostic Imaging, Medical University of Sofia, Sofia, Bulgaria*

Part I

Introduction

Chapter 1

Toward Methodology for Strategic Innovations in Translational and Computational Neuroscience in Psychiatry

Drozdstoy Stoyanov, Sevdalina Kandilarova, and Ferath Kherif

Abstract

This chapter provides an outline of the conceptual rationale behind the agenda of the present book. We argue here that the task-related functional MRI should be translated into clinical reasoning by more intensive and extensive use of clinical tests as stimuli in fMRI paradigms. Further the integration of multimodal imaging data (resting state connectivity, structural studies, magnetic resonance spectroscopy, etc.) into machine learning models, which can further be incorporated into nomothetic networks by including molecular biomarkers, is discussed. Nomothetic networks psychiatry provides trajectory for application of computational and translational methodology platform for clinical reasoning and decision-making in psychiatry.

Key words Computational methods, Translational psychiatry, Machine learning, Neuroimaging, Neurobiology, Nomothetic networks

1 Background

The aim of translational neuroscience is to integrate basic research on brain morphology and functional activity in vivo, with the needs of patients suffering from disorders of the central nervous system (CNS). A translational approach is needed in a number of specialties such as neurology, psychiatry, and neurosurgery, as it will help them treat disorders more effectively than they do today. This is particularly true in psychiatry where, in contrast to other medical specialties, the diagnostic process relies solely on the clinical judgment of the practitioner assessing the patient's behavior and the subjective experiences (as reported by the patient and their relatives). A reliable biomarker for mental illness has been found despite decades of research in various medical domains (genetics, immunology, molecular biology, imaging, etc.). The recent

Drozdstoy Stoyanov et al. (eds.), *Computational Neuroscience*, Neuromethods, vol. 199,
https://doi.org/10.1007/978-1-0716-3230-7_1,

discovery and expanding use of the functional MRI raised hopes for a change in this status quo.

Mainstream task-related fMRI uses diagnostically irrelevant stimulus material (auditory, visual, tactile) that is suitable for exploring specific aspects of cognitive and emotional process in health as well as in different pathological states but cannot easily be translated into clinical practice. To address this gap, we have developed our paradigm using diagnostically relevant and widely used clinical instruments such as depression and paranoid self-assessment scales in contrast to diagnostically irrelevant statements or in contrast between the two scales [1, 2]. Furthermore, for translational approaches, it is crucial to integrate a large and rich set of informational data that captures the relationships between behavior and biology, which is why multivariate linear methods (MLM) are particularly useful for translational approaches. The methods allow for the identification of latent abstractions and eigencomponents that are most effective in capturing redundant information and maximizing the variances explained when decomposing the data. The development of the translational approach and its supporting methods resulted from individual medical needs, coupled with precise measurements and integration of information collected at different levels of brain observation, from molecular differences to anatomical differences to population differences (Fig. 1).

2 Current Advancements

Using this novel paradigm of translational cross-validation of clinical scales and fMRI, together with machine learning-based analysis approaches (multivariate linear method, MLM), we were able to find discriminative value of the specific brain activation signatures related to the diagnostically relevant stimuli [3]. The objective of our first study [4] was to develop a bottom-up unsupervised machine learning approach for classification of psychiatric diagnoses. Multivariate linear model is applied to the high-dimensional data, such as fMRI or structural MRI brain imaging, in combination with additional data, such as experimental, behavioral, and others, to explain the variance of the dataset with minimal loss of information, by combining all possible features. Using MLM approach, we were able to derive the specific brain signatures that are able to distinguish different diagnostic groups based on their different brain signatures. In our study we used principal component analysis (PCA), as a statistical method, in order to construct those brain signatures that yield the different conditions in our paradigm (depression (DS) and paranoid (PS) scale scores, particularly incorporated within PD-S, and diagnostically neutral (DN) items from the scale of interests). This served for the purpose

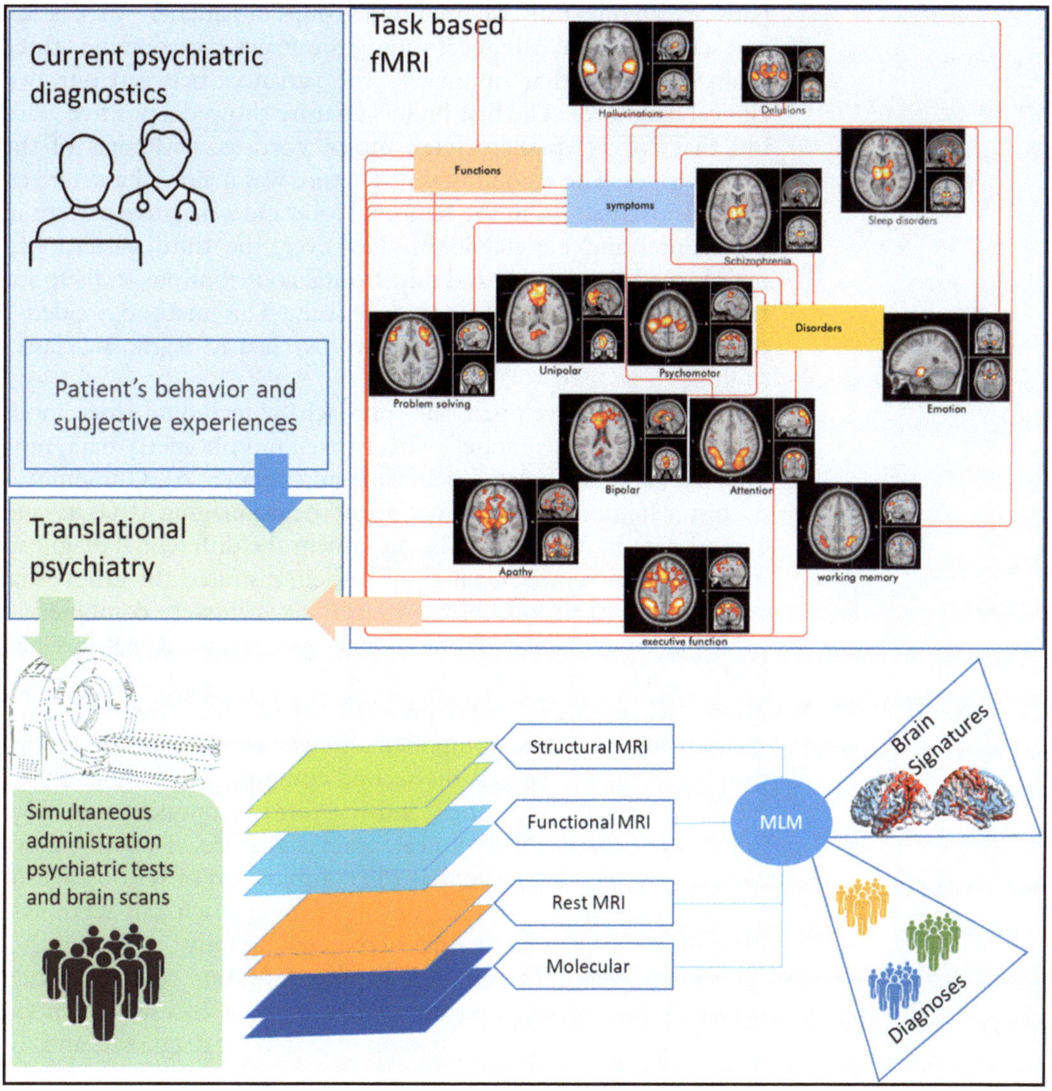

Fig. 1 For translational approaches, it is crucial to integrate a large and rich set of informational data that captures the relationships between behaviour and biology, which is why multivariate methods, MLM, are particularly useful for translational approaches. The methods allow for the identification of latent abstractions and eigencomponents that are most effective in capturing redundant information and maximising the variances explained when decomposing the data. The development of the translational approach and its supporting methods resulted from individual medical needs, coupled with precise measurements and integration of information collected at different levels of brain observation, from molecular differences to anatomical differences to population differences

of creating cross-validation markers that have predictive value for the variance at the level of clinical populations and ultimately delineate diagnostic and classification groups. Using machine learning methodology, we were able to identify three brain patterns that summarized all the individual variabilities by their brain patterns. In our study we used two groups of patients—depression and schizophrenia. After performing the analysis, we obtained three brain signatures that summarize the variances between our two clinical diagnoses. The first brain signature showed a positive loading that covers visual parietal, motor cortices, and parts of the frontal lobes. The second brain signature was mainly characterized by a positive pattern in the temporal lobe and a negative pattern in the frontal and parietal lobes. However, the third pattern had mainly medial temporal and mid-frontal contributions to the positive and negative signature, respectively. This method yielded a discrimination accuracy between two psychiatric diagnostic classes of 90%.

We further developed our approach by including three modalities in the analysis model—voxel-based morphometry data, resting state, and task-related neuroimaging data [5, 6]. Our aim was to test whether the combination of neuroimaging datasets and clinical data in MLM analysis can inform the differential diagnosis in terms of incremental validity. We demonstrated the differential contribution of various MRI modalities which were combined in principal components of brain signatures. Those signatures had high capacity for discrimination of the two observational diagnoses—schizophrenia and depression. We identified three principal components—the first component showed almost equal contribution for all the modalities, the second component was more driven by the functional modalities, and the third component was more informative for the anatomical brain signature. The model we created provided evidence that the combination of various neuroimaging data and clinical data in MLM approach can serve as a diagnostic tool demonstrating incremental validity.

Other areas of translation explored by our research group involve a number of transdisciplinary projects focused on mood disorders and especially depression in search for diagnostic and/or differential diagnostic biomarkers or constellations by combining fMRI and immunological (specific proteins such as zonulin) or molecular markers (expression of lncRNA). It is well known that in major depressive disorder (MDD) increased IgM-mediated autoimmune responses to oxidation-specific epitopes (OSEs) and nitric oxide (NO) adducts are observed. However, these immunological reactions are not explored in bipolar disorder type 1 (BP1) and BP2. Therefore, we have investigated IgM responses to malondialdehyde (MDA), phosphatidylinositol, oleic acid, azelaic acid, and NO adducts in healthy controls and patients with MDD, BP1, and BP2. In addition, levels of serum peroxides, IgG to oxidized LDL (oxLDL), and IgM/IgA directed to lipopolysaccharides (LPS) were explored. We found that both MDD and BP1 were characterized by increased IgM responses to OSEs and NO adducts (OSENO) as compared to healthy individuals. Moreover, MDD group compared to BP2 patients had higher IgM to OSEs. Using partial least squares (PLS) analysis, we demonstrated that 57.7% of the variance in the clinical phenome of mood disorders was explained by the number of episodes, a latent vector extracted from IgM to OSENO, IgG to oxLDL, and peroxides. Significant specific indirect effects of IgA/IgM to LPS on the clinical phenome mediated by peroxides, IgM OSENO, and IgG oxLDL were identified. Moreover, a data-driven nomothetic networks was constructed using PLS. This nomothetic networks encompassed the causome (increased plasma LPS load), the adverse outcome pathways (viz., neuro-affective toxicity), and the clinical phenome features of mood disorders in a data-driven model. Using cluster analysis on those features, a new diagnostic class was identified. Increased plasma LPS load, peroxides, autoimmune responses to OSENO, and increased phenome scores were characteristic of this new class. In summary, the new nomothetic networks approach yielded a transdiagnostic class indicating neuro-affective toxicity in 74.3% of the mood disorder patients [7]. Translational research line of our group has been focused on schizophrenia, exploring the phenomenon of auditory verbal hallucinations (AVHs) [8, 9]. AVHs have been long considered a defining feature of schizophrenia but are also observed in other psychiatric conditions as well as in the general population. They have been extensively researched by various neuroimaging methods, including electroencephalography, positron emission tomography, and structural and functional MRI. However, their neural mechanisms remain poorly understood. Usually, the investigation of AVHs is based on measurements during the hallucinatory episode with focus on the identification of brain regions and networks that are activated. But AVHs are not constant; they fluctuate over time (with a duration of second or minutes), and they are spontaneous (the patient cannot control them). An interesting and still unanswered question is what neuronal events reflect the initiation and cessation of a hallucinatory episode. To answer this question, we recruited 66 individuals experiencing AVHs and investigated the changes in the blood oxygenation level dependent (BOLD) signal during fMRI. The subjects indicated when the hallucinations started and stopped by a button press. Each participant had multiple AVH episodes during the scanning session. For the subsequent analysis, we used the BOLD time series from 10 s before to 15 s after participants indicated the onset and offset of an episode. We observed the changes in brain activations, and we identified a specific region in the ventromedial prefrontal cortex (VMPFC) that demonstrated significantly increased signal starting just a few seconds before the indication of the beginning of an episode. Moreover, this same region showed decreased activation few seconds before the button press indication the cessation of the AVH. We suggest that the finding of corresponding VMPFC activation change prior to awareness of the presence or absence of the hallucinations suggests that this region is acting as a switch for turning the episodes on and off. Potential clinical implications of this finding may be found especially in the field of brain stimulation used for treatment.

Translational cross-validation thus seems to be a way to combine clinical state-dependent measures and functional MRI. It is based on the hypothesis that simultaneous administration of the previously described methods could create dataset with enhanced synchronization and compliance. The main idea of translational validation is to bridge the explanatory gap by using clinical psychometric tools during fMRI procedure and then translating the results back to clinical reality. With our studies, we can identify common diagnostic task-specific measures in terms of brain activations and network modulations. On the other hand, our results need more investigation in order to be confirmed as a diagnostic tool for specific disorders or syndromes. All these statements raise the important question of why there are currently no conventional psychiatric classification criteria. Based on our exploratory findings, we propose here a novel algorithm for translational validation. The design itself involves pre-selection of a clinically validated instrument (test), based on psychometric characteristics, which is already adapted for an fMRI protocol, exploring the brain neural correlates in healthy individuals, compared to a patient population with defined diagnosis and, in the end, analysis of the differences between the selected chosen diagnostic classes [10].

Thus, we have further developed our paradigm for translational cross-validation of clinically relevant tools and tasks tailored to measure meaningful responses in the brain MRI signals and integrate them with other relevant imaging and healthcare data (Fig. 2). These data cover all levels of explanation of mental disorders in vivo, including molecular (biochemical) imaging of brain structure and function—at rest and under diagnostic task conditions. We believe that this could be a solution to the existing explanatory gap between subjective clinical assessment in psychiatry (phenomenological level of explanation) and more objective measurement of brain structure and function (anatomical/physiological level).

Our research line is complementary with the *National Strategy for Mental Health 2030*, adopted by the Bulgarian Government and particularly with the Early Intervention Program component. It is designed to impact the levels of healthcare education and training, decision-making, health policy, and economics by means of providing artificial intelligence (AI)/machine learning solutions for risk/resilience assessment in mental health. Those approaches might be summarized as nomothetic networks psychiatry and are exposed in detail elsewhere [11]. The contemporary understanding of major depressive disorder and bipolar disorder has a lot of controversies created by different competing schools. However, the

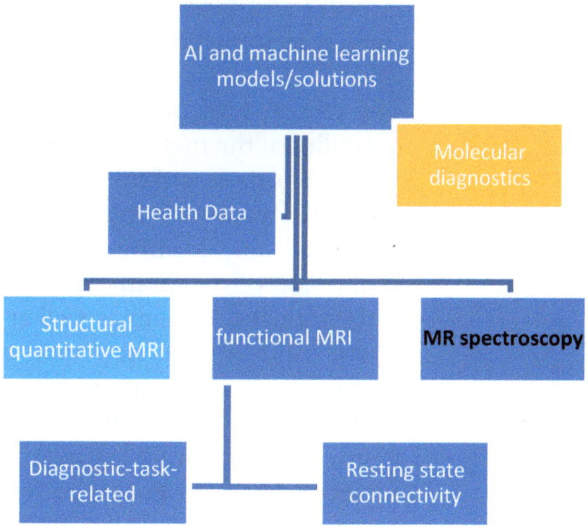

Fig. 2 Model delivery

DSM/ICD taxonomies have solidified their perception about the problem as gold standard for diagnosing patients with major depressive disorder and bipolar disorder. In a recent review,Maes and Stoyanov [12] discussed the dogmas that reign in the research of major depressive disorder (MDD) and bipolar disorder (BD), reflecting the newly implemented data-driven and machine learning-based approaches to the model of psychiatric illness, which is named nomothetic networks psychiatry. The so-called false dogmas which are discussed are the mind–brain interactions/disorders that can be best capitalized using a bio–psycho–social model; mood disorders caused by medical disease are determined to chemical imbalances or psychological stress; DSIM/ICD should be seen as gold standard in order to conclude MDD/BD diagnosis; the severity of the illness should be measured only by using rating scales, as well as the evaluation of the remission should be only defined by thresholding rating scale scores; and there are too many restrictions in order to diagnose an individual with bipolar disorder. On the other hand, the nomothetic networks approach models show that the spectrum of major depressive disorder/bipolar disorders is not psychosocial or mind–brain but are systematic medical disorders. Moreover, there is a common core, called reoccurrence of illness (ROI), which underpins the involvement of depressive and manic episodes and suicidal behaviors. Therefore, mood disorders should be seen as ROI-defined. In addition, the term includes the mediation of nitro-oxidative stress pathways and early lifetime trauma upon mood disorders. Furthermore, the treatment response and severity of illness should be marked out by means of a nomothetic networks approach-derived pathway, ROI, and integrated scores. In conclusion it seems that major depressive disorder and bipolar disorder are the same illness, and both can convert at some point of a lifetime.

2.1 Future Research Goals

The critical goal of our future research program is to prove the reliability of the method for translational cross-validation of clinical assessment tools and fMRI by implementing the already developed task fMRI protocols in a larger sample of patients (paranoid and depressed) and healthy controls. We plan to further develop the application of the multivariate linear method for the analysis of the gathered data. We aim as well at expanding the scope of the diagnostic utility of the new method by developing new paradigms investigating different psychiatric diagnostic entities or syndromes [13]. Moreover, we will be implementing the new paradigms in a sufficiently large sample of patients and healthy control subjects. The final goal is to fill in the gap between neuroscience and clinical practice by developing specific algorithms for diagnosis and differential diagnosis based on the analysis of the imaging data from diagnostic task-related fMRI. The latter may also be complemented with resting state functional (and effective) connectivity data, able to aid diagnostic reasoning and classification, by means of novel approaches for data analysis such as graph theory [14, 15].

2.2 Expected Results

The outcome of our research program is expected to be the development of a reliable innovative algorithm for the diagnosis of psychiatric disorders (psychotic, affective) based on machine learning approach to analysis of multimodal MRI data, including quantitative structural MRI, MR spectroscopy, and functional MRI. The model will involve selecting a set of highly informative brain structural and functional signatures related to the clinical assessment scales used in the task fMRI. These brain signatures would potentially serve as clinically relevant biomarkers. Such an algorithm can be an object of intellectual property protection (IPP) that could further be commercially used by healthcare providers. Research advances include advanced clinical and neurobiological studies into the shared and distinct mechanisms of mental disorders using functional MRI, quantitative MRI, and proton MR spectroscopy to identify new complex fingerprints of diagnosis, treatment prediction, and monitoring of the pharmacological response.

3 Impact

The successful development of such a diagnostic algorithm in psychiatry would represent a scientific breakthrough that could foster a completely new field in translational neuroscience bridging the gap between clinical assessment and brain imaging. Furthermore, the method has the potential to change not only the diagnostic process but the management of the patient as well with the development of

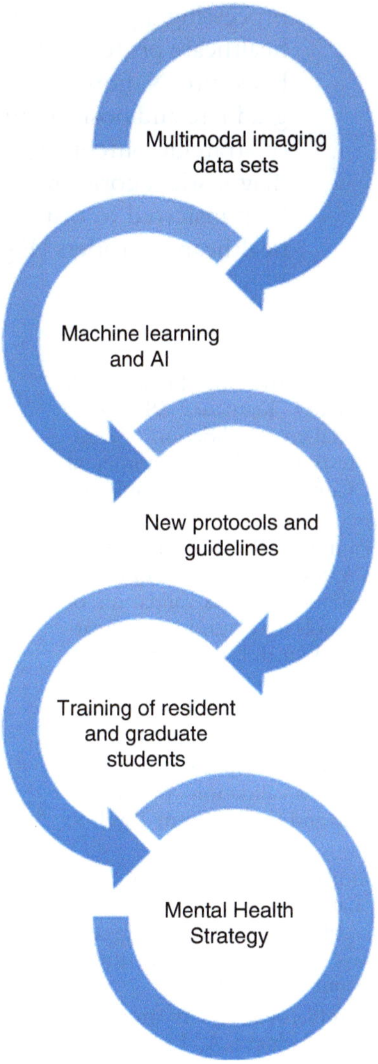

Fig. 3 Implementation strategy

similar algorithm in combination with pharmacological MRI. On the health-economic level, the use of such diagnostic tools may be considered for reimbursement by the National Health Insurance in the process of diagnosis and management of severe mental illness, thereby reducing the social and economic burden. As early diagnosis of major psychiatric disorders would significantly improve patients' long-term survival and quality of life, our approach has the potential to save and change lives. Besides the enormous impact on patients' health, this will also have a positive effect on the economy, as much less funds will be needed for the treatment and more patients will start the treatment earlier especially if the algorithm is adopted in the assessment of high-risk patient groups. The generation of new algorithms for multimodal multivariate data processing by means of machine learning will generate new AI healthcare protocols and guidelines. The implementation of guidelines into National Mental Health Strategy is associated with the graduate and post-graduate professional training (Fig. 2). In terms of strategic outcomes, the results of the program and established diagnostic algorithms for assessment of psychiatric disorders may be transferred to other interested clinical and imaging centers and healthcare providers (Fig. 3).

References

1. Stoyanov D, Kandilarova S, Sirakov N, Stoeva M, Velkova KG, Kostianev SS (2017) Towards translational cross-validation of clinical psychological tests and fMRI: experimental implementation. Compt Rend Acad Bulg Sci 70:879–884
2. Aryutova K, Paunova R, Kandilarova S, Todeva-Radneva A, Stoyanov D (2021) Implications from translational cross-validation of clinical assessment tools for diagnosis and treatment in psychiatry. World J Psychiatr 11(5): 169
3. Stoyanov D, Kandilarova S, Borgwardt S, Stieglitz RD, Hugdahl K, Kostianev S (2018) Psychopathology assessment methods revisited: on translational cross-validation of clinical self-evaluation scale and fMRI. Front Psych 9: 21
4. Stoyanov D, Kandilarova S, Paunova R, Barranco Garcia J, Latypova A, Kherif F (2019) Cross-validation of functional MRI and paranoid-depressive scale: results from multivariate analysis. Front Psych 10:869
5. Stoyanov D, Kandilarova S, Aryutova K, Paunova R, Todeva-Radneva A, Latypova A, Kherif F (2020) Multivariate analysis of structural and functional neuroimaging can inform psychiatric differential diagnosis. Diagnostics 11(1):19
6. Paunova R, Kandilarova S, Todeva-Radneva A, Latypova A, Kherif F, Stoyanov D (2022) Application of mass multivariate analysis on neuroimaging data sets for precision diagnostics of depression. Diagnostics 12(2):469
7. Simeonova D, Stoyanov D, Leunis JC, Murdjeva M, Maes M (2021) Construction of a nitro-oxidative stress-driven, mechanistic model of mood disorders: a nomothetic network approach. Nitric Oxide 106:45–54
8. Hugdahl KA, Craven R, Johnsen E, Ersland L, Stoyanov D, Kandilarova S, Sandøy LB, Kroken RA, Løberg E-M, Sommer IE (2022) Neural activation in the ventromedial prefrontal cortex precedes conscious experience of being in or out of a transient hallucinatory state. Schizophrenia Bulletin:sbac028. https://doi.org/10.1093/schbul/sbac028
9. Weber S, Johnsen E, Kroken RA, Løberg EM, Kandilarova S, Stoyanov D et al (2020) Dynamic functional connectivity patterns in schizophrenia and the relationship with hallucinations. Front Psych 11:227
10. Stoyanov D (2022) Perspectives before incremental trans-disciplinary cross-validation of clinical self-evaluation tools and functional MRI in psychiatry: 10 years later. Front Psych 13
11. Stoyanov D, Maes MH (2021) How to construct neuroscience-informed psychiatric classification? Towards nomothetic networks psychiatry. World J Psychiatr 11(1):1
12. Maes MH, Stoyanov D (2022) False dogmas in mood disorders research: towards a nomothetic network approach. World J Psychiatr 12(5):651
13. Stoyanov DS (2023) Endophenotypes and pathway phenotypes in neuro-psychiatry: crossdisciplinary implications for diagnosis, cns & neurological disorders. Drug Targets 22(2). https://doi.org/10.2174/187152732202220914125530
14. Stoyanov D, Khorev V, Paunova R, Kandilarova S, Simeonova D, Badarin A et al (2022) Resting-state functional connectivity impairment in patients with major depressive episode. Int J Environ Res Public Health 19(21):14045
15. Pitsik E, Kurkin S, Hramov A, Paunova R, Simeonova D, Kandilarova S, Stoyanov D (2022) A graph convolutional network for classification of resting-state fMRI data. In 2022 6th Scientific School Dynamics of Complex Networks and their Applications (DCNA) (pp. 223–225). IEEE

Part II

Molecular Methods

Chapter 2

Molecular Methods in Neuroscience and Psychiatry

Mariya Ivanovska, Teodora Kalfova, Steliyan Petrov, Martina Bozhkova, Alexandra Baldzhieva, Hristo Taskov, Drozdstoy Stoyanov, and Marianna Murdjeva

Abstract

Clinical practice in all fields is working to provide an approach to patients based on their individual characteristics. A personalized approach in the field of psychiatry and neuroscience is mandatory. New technologies, new methods, and interdisciplinary methodologies are a must. Personalized psychiatry will improve the diagnosis, treatment, and prevention of this huge group of psychiatric disorders. We explored relevant literature related to molecular methods in the field of neuroscience and psychiatry at PubMed/MEDLINE, Google Scholar, Scopus, and other scientific online databases. In neuroscience and psychiatry research, molecular mechanisms are linked to cellular and system mechanisms in various brain regions involved in information processing and behavior generation. Neurotranscriptomics, neuroproteomics, terminal differentiation of DNA regions, gene expression, DNA patterning, the interaction between distal chromosomal regions, flow cytometry ELISpot, and fluorometry are our methods of interest and a new wave of the personalized translational diagnostic approach. All of this should be cost-effective and should increase life expectancy by avoiding unnecessary diagnostic methods and treatment.

Key words Molecular methods, Neuroscience, Psychiatry, Personalized medicine, Translational research

1 Introduction

Difficulties in the field of psychiatry and neuroscience include determining which method to use for the best diagnosis and working with big data analyses. Different psychiatric disorders and conditions necessitate a distinct approach to diagnosis. Knowing what method to employ is insufficient. A multidisciplinary approach, a good team, a hospital, and a university are all essential.

As stated by the World Health Organization (WHO), depression affects more than 264 million individuals of all ages worldwide. Experts estimate that 20 million people worldwide suffer from schizophrenia, while 46 million suffer from bipolar disorder [1]. Knowing this, collaboration between psychiatry, neurocience,

Drozdstoy Stoyanov et al. (eds.), *Computational Neuroscience*, Neuromethods, vol. 199,
https://doi.org/10.1007/978-1-0716-3230-7_2,

psychology, and laboratory medicine is needed. Also, to understand the pathogenesis of psychiatric disorders, the research center, hospital, and university should be able to offer an excellent opportunity for educating young researchers.

Novel translational research will be possible if utilizing new educational resources and understanding how and where to combine cutting-edge data analysis technology with multiple data sets in each field. Clinical and basic science in psychiatry and neuroscience should go together, hand in hand. As a result, translational research is required to understand the pathophysiology of psychiatric disorders and to develop new therapeutic and diagnostic approaches [2].

Molecular methods in neuroscience and psychiatry research seek to explain brain processes by connecting molecular mechanisms to cellular and system mechanisms in various brain regions involved in information processing and behavior generation. This article will go over the following techniques: neurotranscriptomics, neuroproteomics, terminal differentiation of DNA regions, gene expression, DNA patterning, interaction between distal chromosomal regions, flow cytometry ELISpot, and fluorometry.

2 Methods

We searched online libraries, including PubMed/MEDLINE, Google Scholar, and Scopus. The main searched terms were "molecular methods" [MeSH], "methods in psychiatry" [MeSH], "methods in neuroscience" [MeSH], "Personalized medicine" [MeSH], "flow cytometry," "neurotranscriptomics" [MeSH], and "ELISpot" [MeSH] with filters activated, namely, publication date from January 1, 1990, to September 30, 2022, and papers written in English.

3 Results and Discussions

3.1 Methods in Neurotranscriptomics

In order to define neurotranscriptomics, we must first understand what lies behind the term transcriptome. Ribonucleic acids (RNAs) are macromolecules made of linear chains of nucleotides that perform a variety of cellular and biological activities. Proteins can be synthesized from them, using their sequence as a template, or they (RNAs) can serve a large number of diverse regulatory functions in the organism [3]. Depending on the nature of the experiment, a total set of messenger RNAs (mRNAs) or all RNA present in a sample is called transcriptome. The study of all RNAs and their abundant roles in gene regulation at specific time and space is referred to as transcriptomics [4]. This provides a comprehensive view of molecular activity in cells, which affects both human physiology and pathology. Major directions of transcriptomics studies [5]:

- Expression patterns can be used to characterize different states of cells (i.e., developmental stages), tissues, or cell cycle phases.
- Explore the molecular mechanisms underlying a specific phenotype.
- Discover biomarkers that differ in expression between the sick and healthy states.
- Distinguish disease stages or subtypes (e.g., cancer stages).
- To better understand illness etiology, establish the causal connection between genetic variations and gene expression patterns.

Neurotranscriptomics is a sub-field which aims to investigate the transcriptome related to neuroscience [6]. Modern transcriptomics analyzes the expression of many transcripts in diverse physiological or pathological settings, which is increasingly improving our understanding of the links between the transcriptome and the phenotype across a wide spectrum of living beings. Next-generation sequencing and microarrays are two important technological platforms for transcriptome studies.

In contrast to the relatively static DNA molecule, whose primary job is to transmit information along cell lineages, RNAs can function as both information carriers and catalytic agents [5]. Ribonucleic acids come in different forms. They are represented by ribosomal RNA (rRNA), transfer RNA (tRNA), small nuclear RNA (snRNA), small nucleolar RNA (snoRNA), micro-RNA (miRNA), short-interfering RNA (siRNAs), long non-coding RNA, and pseudogenes, which perform a wide range of cellular activities [4]. The central dogma of molecular biology is that RNAs code for proteins. The information transfer from DNA to RNA, then to protein, suggests that mRNA and protein expression should follow similar time-dependent and tissue-specific patterns [7]. Due to various layers of post-transcriptional control and variable degradation rates, RNA and protein abundance are not always tightly connected. As a result, the relationship between mRNA and protein expression is not evident [8].

Techniques often used for transcriptome investigation include expressed sequence tag (EST)-based approaches, serial analysis of gene expression (SAGE), hybridization-based microarray, real-time PCR, NGS-based RNA-sequencing (RNA-seq) methods, RNA interference, and bioinformatics tools. RNA isolation, purification, quantification, copy DNA (cDNA) library creation, and high-throughput sequencing are all part of the [7–12]. The following factors should be considered while selecting transcriptomics tools: cost-effectiveness, sensitivity, high throughput, and minimal beginning RNA concentration.

The extraction of RNA is the first step in the transcriptomics process. Depending on the experiment, either total or mRNA is extracted from the tissues. However, the extraction of RNA is a time-consuming operation; therefore specialist assistance is required for transcriptomics research. RNA extraction is difficult because it is weaker than DNA and thus easily fragmented, either by external factors or by RNase. Furthermore, sterile conditions must be meticulously maintained during RNA extraction because RNase can be found anywhere—on our hands, a desk, or even a pipette [7]. Although ready-to-use RNA extraction kits are strongly recommended, they are still not sufficiently standardized [5]. All we can say is that extracting RNA is an art, and the quality of the extract is determined by the researcher's experience. In any case, our pure mRNA is now ready for downstream processing.

One way of detecting a molecule in the sample that is being analyzed is by using RNA probes. RNA probes are stretches of single-stranded RNA used to detect the presence of complementary nucleic acid sequences (target sequences) by hybridization [13]. RNA probes are typically labeled with radioisotopes, epitopes, biotin, or fluorophores to permit their detection. RNA probes are often referred to as riboprobes or cRNA probes [14]. It is traditionally utilized in the northern blot hybridization procedure, in which the RNA probes are produced and then hybridized on nitrocellulose paper. Because our target single-stranded nucleic acid is already immobilized on the nitrocellulose paper, the hybridization signal may be identified using autoradiography [5]. Recently, RNA probes have been employed in microarrays, in which millions of probes are immobilized on a solid surface. Using nucleic acid permits hybridization of just complimentary DNA sequences, allowing for the detection of many different copy number variations in a single test [4]. Microarrays are a powerful tool in understanding the relations between different molecules, their expression levels, and action mechanisms of neurodegenerative conditions, such as Alzheimer's disease, Huntington's disease, and Parkinson's disease, and allow for more refined understanding and treatment of brain tumors [15]. Microarrays will transform our understanding of these and other complicated disorders and will soon have an impact on medical practice [4].

Another modern molecular technique is the single-cell RNA sequencing (scRNA-seq). scRNA-seq counts the number of RNA molecules in each cell of a given sample [16]. This data provides an overview of the transcriptome (the genes being transcribed) at the time the cells were sampled. Since its introduction, scRNA-seq has found a wide range of uses [17–19]. Although the final outcome of a gene's expression is a protein, identifying its messenger RNA (mRNA) indicates that the gene has been activated and hence has the potential to be translated and expressed. Additionally, co-transcribed genes can be utilized to infer the gene regulatory networks that underpin the phenotypes of certain cells. A cellular population's transcriptional differences can aid in the identification of subpopulations, such as malignant cells in a tumor mass. scRNA-seq is also utilized to investigate key gene transcription properties including splicing patterns and monoallelic gene transcription [15].

Isolated cells are lysed, and their polyadenylated mRNA molecules are enriched with poly[T]-primers. This is a key step since the majority of RNA molecules in a cell are ribosomal (rRNA): quite big and typically not the objective of transcriptome sequencing [9]. The poly[T]-primed, single-stranded mRNAs are then converted into complementary DNA (cDNA) by reverse transcriptases. The cDNA molecules are subsequently amplified using PCR or in vitro transcription (IVT). Finally, the cDNA molecules are tagged with barcoding tags and other short sequences required by the sequencing technology.

We can investigate each cell and measure its individual contribution to the entire cell population—and its organism or environment—using single-cell technology [8]. This degree of precision is particularly useful when researching rare cells or investigating phenotypic changes within populations of the same cell type [20]. scRNA-seq has been effectively used to a wide range of biological studies [11, 12, 21]. Because of the immense variety of cell types in the mammalian brain and the difficulties in purifying and analyzing them using conventional techniques, one of the most beneficial applications was in neuroscience. Using scRNA-seq, mapping of the developing human brain has been created [22], providing a powerful resource for explaining how a normal human central nervous system develops and how differentiation and maturation of specific cell types is disrupted in neurodevelopmental diseases. Furthermore, snRNA-seq has been used to analyze human brain disease directly by profiling normal brain tissue and brain tissue from individuals with neurological disorders [5]. As a result, a snRNA-seq study of autism spectrum disorder (ASD) found that in this disease, a specific subtype of neurons in the human brain cortex changes substantially [23, 24]. Another study employed snRNA-seq to determine how various neuronal and glia cell types respond to and change in multiple sclerosis lesions [25].

Both techniques (scRNA-seq and microarray) require mRNA as starting material; thus, mRNA extraction is the first step in both. However, the starting material (mRNA) requirement for microarray analysis is higher—roughly 1 μg mRNA. RNA-seq, on the other hand, requires as little as 1 ng of total RNA or mRNA. When compared to RNA-seq, the microarray approach has a better throughput. Nonetheless, microarray is a low labor-intensive approach that relies mostly on instruments and microarray setup. On the other hand, the RNA-seq approach is a time-consuming method that requires personnel for sample preparation, sequencing, and data interpretation. As a result, the sensitivity and dynamic range of RNA-seq are greater than those of the microarray.

Surprisingly, both methods have a higher repeatability. Based on the probe employed in the array, the microarray can detect only splice variants, whereas the RNA-seq can detect splice variants as well as other changes such as SNPs. One of the most significant drawbacks of the microarray is the need for preexisting sequencing knowledge to construct the array. RNA-seq, on the other hand, requires no prior sequence information and can discover novel changes.

3.2 Methods in Neuroproteomics

Neuroproteomics is a sub-field of proteomics—a joint in science where the main object of interest is the complete set of proteins of an organ or an organism at a given time and under specific physiological conditions [26]. Changes in the environment, such as temperature, ionic strength, pH, oxidant levels, and other factors, can alter protein structure and/or function. A proteome is a broad concept that encompasses much more than just the identification of the proteins in the set [27]. Proteins may interact with a specific number of other proteins (or other molecules) in any given proteome, affecting how the protein operates as a component of the entire system. Neuroproteomics is devoted to answering these similar concerns about the nervous system's organs, tissues, and cells. The primary objectives in neuroproteomics are to identify all of the proteins of a given tissue, cell type, or organelle under specific circumstances at a specific time; to identify the post-translational modifications in all of the proteins at that time and under these conditions; to evaluate how this proteome alters as a function of time (age), changes in the environment, genetic factors, and disease; and to determine how these changes affect the organism as a whole [26, 27].

Mass spectrometry (MS) is the key technique, used in proteomics. This approach has a high sensitivity for detecting either digested peptides or intact proteins [28]. Targeting specific protein changes such as phosphorylation, oxidation, ubiquitination, and others is now achievable, applying this method of analysis [29]. Unlike other methods, such as x-ray diffraction, which requires the entire proteins in crystal form or solution, MS necessitates peptides or proteins to be analyzed as ions in the gas phase [26].

Mass spectrometers only have one function: they measure mass. The mass of a protein in proteomics provides information on its identity, chemical modifications, and structure. An analyzer, a source, and a detector are the three main components of a mass spectrometer. Mass spectrometers measure the masses of charged molecules, so the source must be capable of producing ions, the analyzer must be capable of separating these ions based on their mass (or, more precisely, mass-to-charge ratio), and the detector has to be sensitive enough to detect charged particles and amplify the response to produce a measurable signal.

About 40 years ago, two revolutionizing technologies were created and presented to the public, which later on in 2002 won the award for Nobel Prize in Chemistry. John Fenn and Koichi Tanaka developed electrospray and MALDI (matrix-assisted laser desorption/ionization) mass spectrometry [30, 31], methods now essential in proteomics analyses. They enable the vaporization and ionization (and consequently the analysis) of big, non-volatile macromolecules like proteins and peptides.

The mass-to-charge ratio is measured indirectly using time-of-flight (TOF) sensors by measuring the time it takes an ion to reach the detector in a field-free vacuum. At time = 0, "packets" of ions are released into the TOF analyzer, where they drift toward the detector and form a signal at time = final. The analyte's mass is proportional to its flight time (measured in nanoseconds); thus heavier ions travel more slowly, taking longer to reach the detector.

MALDI molecular imaging of brain tissues is a more recent use of MS to neuroproteomics research. Although this technique has made significant progress under the direction of Dr. Richard Caprioli, it is still in its infancy, and its applicability to neuroproteomics will be primarily for functional study of subsets of proteins and understanding the origin and course of neurological illnesses [26]. The ability to create three-dimensional molecular images of the brain, or brain parts, and then use these images to assess the dynamic evolution of sub-proteomes is a significant advantage of MALDI-based tissue imaging.

Scientists discovered that employing neuroproteomics to solve biological problems requires a variety of experimental techniques. Assays such as protein arrays, MS, 2D-DIGE, MS-based tissue imaging, scRNA-seq, surface plasmon resonance (SPR), RNA arrays, protein interaction network analysis, multidimensional liquid chromatography, and many more techniques are now routinely used. Understanding neurological issues, such as the protein networks affected by apoE genotypes in Alzheimer's disease patients, the structure and function of pre- and post-synaptic densities, the sets of proteins that are regulated by mental processes such as learning and memory formation, and the structure of synaptic protein complexes in animal models of epilepsy, is now not hard to be achieved applying knowledge from areas like neurotranscriptomics and neuroproteomics.

3.3 Methods in Epigenetics

Research interests in the area of epigenetics include gene expression, DNA patterning, interaction between distal chromosomal regions, and terminal differentiation for DNA regions. Chromosomal regions that register, signal, and preserve changes in our genes caused by cellular or systemic events are structurally modified. Epigenetics is a vast topic of study that has applications in genetic engineering, neuroscience, and personalized medicine.

Through the innovative use of systems engineering methodologies that integrate technologies and data analysis, the study of the epigenome at a system level and its regulation is a fast-growing subject. Examples include the use of bioinformatics for the analysis of epigenetic data, high-resolution imaging for the visualization of chromatin dynamics, genetic engineering for the modification of epigenomes, and bioengineering and systems engineering techniques for the integration of epigenomics and physiology. By discovering novel epigenetic markers to help in illness diagnosis and creating innovative treatment ways to interfere with pathogenic processes, it is used to solve biological issues [32, 33].

The main mechanism of genome plasticity is represented by the significant roles played by chromatin remodeling and genetic components in the regulation of gene expression. The enormous flexibility and complexity of the nervous system pose the biggest obstacle to epigenetic study in neuroscience. Numerous genes in the brain have epigenetic states, although this does not always result in changed gene expression. Anxious, depressive, or aggressive behaviors as well as devastating neurodegenerative diseases may instead result from subtle but long-lasting epigenetic modifications that are specific to certain cell types and brain regions. These modifications only occur in response to additional different conditions.

Apart from characteristics like hypotonia, hypogonadism, and hyperphagia that may lead to obesity, Prader–Willi syndrome (PWS) is a condition that exhibits symptoms similar to those of illnesses on the autistic spectrum disorder [34]. It is brought on by numerous genes on the paternal copy of 15q11-q13 losing their ability to function. This can happen through a variety of methods, such as imprinting mistakes that silence the male allele [35]. DNA methylation, covalent histone changes, and antisense RNA (ribonucleic acids) are some of the epigenetic processes that cause the paternal allele to be silenced in an abnormal way [34, 35].

Heritable gene silencing is brought on by the covalent modification of DNA known as methylation. By adding a methyl group to cytosine 5 on CpG dinucleotides, DNA methyltransferases (DNMTS) catalyze the process. This results in one of two silencing mechanisms. First off, methylation can directly obstruct a transcription factor's ability to bind to DNA recognition sites. Second, histone deacetylase (HDAC) or histone methyltransferase-containing corepressor complexes can be recruited by methyl-CpG binding domain proteins (MBPs), which can strengthen silencing (HMTs). MBP-associated HMTs provide a positive feedback loop between DNA methylation and methylated lysine 9 on histone 3, an additional epigenetic silencing marker [36–42].

The use of genome-editing technology enables an overview of the central nervous system on a molecular level. With the help of genome-editing tools, researchers can alter DNA, changing physical characteristics like eye color and the risk of contracting diseases. These tools cut the DNA at a precise location like scissors. The damaged DNA can then be repaired, expanded, or replaced. Engineered restriction enzymes called zinc finger nucleases (ZFNs) are made to target particular DNA sequences in the genome. The enzyme machinery can target a specific location in the genome and activate endogenous DNA repair mechanisms when a zinc finger DNA-binding domain is assembled with a DNA-cleavage domain. Each zinc finger nuclease (ZFN) subunit is made up of a peptide linker that joins the zinc finger domain to the nuclease, an N-terminal DNA-binding zinc finger domain called Cys2-His2 (C2H2), and a nonsequence-specific cleavage domain at the C-terminal that is catalyzed by the FokI nuclease. A DNA double-strand break (DSB) is induced by the restriction enzyme FokI as a catalytic dimer. So, two zinc finger subunits must attach to the gene target region in the opposite orientation for DNA cleavage to occur, which causes FokI dimerization. In living cells, the DNA double-strand break (DSB) brought on by ZFN at endogenous loci can predominantly be repaired by two processes [43–45], non-homologous end joining (NHEJ) and homology-directed repair (HDR), which is error-prone when donor DNA is present. This ZFN-mediated strategy easily achieves gene knockout in eukaryotic cells because tiny base-pair insertions or deletions can be directly produced by the NHEJ-driven DNA repair mechanism. By co-expressing ZFN and donor DNA, one or more transgenes can be inserted into the DSB site for HDR modifications. HDR mediated by ZFN is highly adaptable when donor DNA is homologous to the sequences flanking the DSB. This flexibility allows for the insertion of marker genes, the replacement of mutant genes with wild-type genes, or the insertion of several transgenes at the same or separate loci on chromosomes. HDR can be employed in eukaryotic cells/organisms where the gene sequence of the targeted locus is known, but the NHEJ pathway can be used in all eukaryotic cells without significant knowledge of the sequence of the targeted gene [43–45]. This technology and the more advanced approach in the face of CRSPR/Cas systems allow the researchers to tackle the nervous systems and obtain a more precise knowledge which can help in understanding the molecular origin of certain diseases.

In vitro cultivation of neurons is another applicable method, allowing the usage of induced pluripotent stem cells (iPSCs) produced from somatic cells (like fibroblasts) of healthy people or patients with neurological diseases. This method's ability to study genetic alterations in individuals with various human genetic origins is a significant benefit. Parkinson's, Alzheimer's, and Huntington's illnesses are only a few of the neurological disorders for which iPSC-based disease models have been developed. These models have shown to closely resemble the cellular and molecular characteristics of human diseases. The models may be used by genome-editing technologies to study the genetic relationship between risk variants and cellular pathways implicated in multigenic neurological illnesses in a high-throughput manner [46, 47].

Gene activation and repression can be mediated by long-distance chromosomal contacts, and both inter- and intrachromosomal associations have been found. Gene expression regulation can take place both locally and across significant genomic distances. In many cases, regulatory elements are located far upstream or downstream of the genes they regulate, and they can even affect the expression of genes on other chromosomes. When it comes to the place and degree of expression, as well as the timing of transcription during development and the cell cycle, the majority of genes show relatively distinctive traits. When sections of the same chromosome's genetic sequence are physically closer to one another than they are to intervening regions, a chromatin loop occurs. This straightforward definition does not address the level of proximity necessary to be functionally significant or the length of the intervening sequence. Numerous gene loci have been shown to contain looped structures that contrast major genetic components. Although they do not technically constitute loops in the traditional sense, physical interactions have also been discovered between components on different chromosomes, and these interactions may have the same purposes as loops.

Chromosome conformation capture (3C), also known as nuclear ligation assay, is now the most popular technique. This approach is predicated on the idea that protein complexes can chemically cross-link distant genomic regions that they have brought together. DNA fragments are created by subsequent restriction digestion and can be re-ligated to one another. The cross-linking and subsequent ligation of DNA fragments that are close to one another will occur more efficiently than those that are not. The quantities of distinct ligation products are determined using quantitative PCR (polymerase chain reaction) using primers spanning the ligated segments. Chromatin immunoprecipitation (ChIP) tests have also been used to infer the creation of chromatin loops. To determine whether a transcription factor is present in nearby genomic regions, the DNA ligation and PCR amplification steps of the 3C assay are combined with the ChIP assay. However, just because a transcription factor is found near the base of a loop does not mean that it is necessary for the creation of the loop [48–51]. The transcriptional status of linked genes may be affected by these interactions, which may make it easier for regulatory elements to communicate. The cytokine LCR and interferon, which are factors in the fate of TH1 cells, are associated intrachromosomally with naive helper cells. The connection disappears after differentiation, indicating that it controls the active gene expression. The X chromosome inactivation describes interactions between X chromosomes, polycomb response elements (PREs), olfactory receptor enhancers, and promoters that have also been seen. Inter-chromosomal interactions have brought attention to a potential role for nuclear architecture in the control of epigenetic processes. The phenomenon is likely worldwide since inter- and intrachromosomal association occurs in different types of genes and in different cell types [36–42]. X chromosome aneuploidy produces distinctive cognitive-behavioral and neuroanatomical traits, according to research that have been conducted so far. Through the addition of cohorts with a defective (TS) or supplemental (KS) complement of X chromosomes, we can explore the impact of X chromosome dosage on brain shape in early pubertal offspring [52, 53].

3.4 Flow Cytometry in Psychiatry and Neuroscience

Flow cytometry (FC) is a method using laser technology in which specific fluorescent-labeled antibodies identify the expression of intracellular or surface membrane antigens. It allows determination of absolute and relative number of cells; evaluation of infrequent events with high specificity, sensitivity, and speed of thousands of cells per second; as well as cell sorting for further analysis [54–57].

Innovative studies in neuroscience have made use of FC-irreplaceable tool for extraction of specific brain cells from diverse populations and assessment of their characteristics. Classification of activated neurons, antigen expression of neural cells, examination of neural stem cells, evaluation of the neurochemical environment and changes in morphology and cell density in different psychiatric and neurological disorders, as well as, after brain injuries, assessment of the therapy effect in neurodegenerative diseases are just a few of the possibilities of FC to be listed [58–61]. Blood and CSF (cerebrospinal fluid) are most commonly used samples. Because of its neighborhood to the central nervous system tissue, CSF is a good predictor of local disease. NKT, B lymphocytes, NK, plasmacytoid, and myeloid dendritic cells are present in very little amounts in normal CSF, which is mostly made up of CD4 T lymphocytes with a predominance of the central memory phenotype [62–64]. For the diagnosis of several infectious and inflammatory neurological disorders such as multiple sclerosis, paraneoplastic neurological syndrome, and Parkinson's disease, as well as for understanding the immunopathogenicity of these disorders, FC of CSF is crucial. For example, immune cell profiling of the brains of Parkinson's patients displays an increased number of classical and non-classical monocytes, and T lymphocytes and CD4+ lymphocytes tend to show increased absolute numbers. In case of leptomeningeal involvement in hematological malignancies, FC of CSF is an irreplaceable technique with prognostic and individualized therapeutic impact [65, 66]. Additional applications of FC include evaluation of the systemic immune activation in patients with depression; immune profiling of cell subsets in psychotic disorders; identification of potential drug targets for development of novel pharmaceutical treatments of neurological disorders and psychiatric illnesses by monitoring of the intracellular alterations in neural cells, such as calcium influx; the formation of cellular reactive oxygen species; and the activation of apoptosis [67, 68].

Multiparameter FC analysis is a potent tool, particularly for samples like CSF that contain few cells. However, a significant barrier to the practical deployment of this technology is the data reduction and analysis of the hundreds of detected cellular subphenotypes. This can be overcome by the use of automated data analysis softwares such as Kaluza, FlowJo, and their plugins [69].

3.5 ELISpot/FluoroSpot in Psychiatry and Neuroscience

The enzyme-linked immunospot (ELISpot) is one of the most widely used assays for the functional evaluation of single immune cells since it was first described in 1983.

ELISpot assay is a combination of a mixed lymphocyte reaction (MLR) assay and a conventional enzyme-linked immunosorbent assay (ELISA). While ELISA measures the total concentration of the molecules of interest produced from all cells in the sample, ELISpot assay evaluates the released cytokines from individual cells [70, 71].

In ELISpot method, peripheral blood mononuclear cells (PBMCs) are cultivated on the surface of a 96-well plate that has been treated with a particular capture antibody, either with or without external stimulation. The antibodies on the surface will bind to specific molecules released by the cells, such as cytokines. Then cells are removed after an appropriate incubation time, and the released substances are evaluated using a detection antibody. The addition of the substrate makes the product of stimulation visible as a "footprint" on the surface of the plate known as a spot. Every spot represents a singular cytokine-secreting cell [72, 73].

Advancement in the detection of double-stained spots was made with the invention of the FluoroSpot assay, which visualizes spots using fluorophores rather than enzyme and substrate combinations. However, automated spot-counting devices at the time still relied on the analysis of spot color. Clearly defining the difference between single and double (or even triple)-stained spots could only be made with the development of a new, two-level image analysis method that made the transition from single-function cell analysis to a multifunctional analysis [71–73].

The assay is widely used for the detection of specific cellular immune responses in infectious diseases such as active or latent tuberculosis (IFN-γ, IL-2) [74], COVID-19 (IFN-γ) [75], and organ rejection in transplantation and in clinical trials for the evaluation of biomarkers proving the potential of immunotherapeutic treatment [76].

In the pathogenesis of some neurological disorders such as multiple sclerosis, paraneoplastic neurological syndrome,

Parkinson's disease, and Alzheimer's disease, there is immune autoreactivity, which can be evaluated by the detection of cytokine production by ELISpot/FluoroSpot assay [77–80].

Multiple sclerosis (MS) is a central nervous system autoimmune disease caused by lymphocytes that cross the blood–brain barrier and lead to inflammation, demyelination of axons, and consequent neurological disorders. In addition to T-helper cells (CD4+) which are primarily responsible, there are some antigens like myelin basic protein (MBP), proteolipid protein (PLP), and myelin oligodendrocyte glycoprotein (MOG) that have been suggested as autoantigens in pathogenicity of MS. In several studies that investigate autoreactivity against MOG, PBMCs are stimulated by bead-bound MOG and evaluated for cytokine production in an INF-γ/IL-22/IL-17A FluoroSpot assay. The results showed significantly higher r values in the group of patients with MS compared to healthy controls [77, 78].

PBMCs from patients with Parkinson's disease (PD) were stimulated with an α-syn epitope pool and tested by triple-color IFN-γ, IL-5, and IL-10 FluoroSpot assay. In PD patients higher α-syn-specific reactivity compared to healthy controls is observed [80].

4 Conclusions

Bidirectional collaboration, personalized approach, and big data analysis (machine learning) are required for reliable biological evidence in psychiatric disorders. Based on reliable findings in patients, translational research must be conducted to understand the pathologies of mental, psychiatric, and stress-related disorders.

References

1. Takaesu Y, Watanabe K, Numata S, Iwata M, Kudo N, Oishi S et al (2019) Improvement of psychiatrists' clinical knowledge of the treatment guidelines for schizophrenia and major depressive disorders using the "Effectiveness of Guidelines for Dissemination and Education in Psychiatric Treatment (EGUIDE)" project: a nationwide dissemination, education, and evaluation study. Psychiatry Clin Neurosci [Internet] 73(10):642–648
2. Onitsuka T, Hirano Y, Nemoto K, Hashimoto N, Kushima I, Koshiyama D et al (2021) PCN FRONTIER REVIEW PCN trends in big data analyses by multicenter collaborative translational research in psychiatry. Available from: http://onlinelibrary.wiley.com/doi/10.1111/pcn.13311/full
3. Kolodziejczyk AA, Kim JK, Svensson V, Marioni JC, Teichmann SA (2015) The technology and biology of single-cell RNA sequencing. Mol Cell 58(4):610–620
4. Milward EA, Shahandeh A, Heidari M, Johnstone DM, Daneshi N, Hondermarck H (2016) Transcriptomics. Encycl Cell Biol 4: 160–165
5. Lowe R, Shirley N, Bleackley M, Dolan S, Shafee T (2017) Transcriptomics technologies. PLoS Comput Biol [Internet] 13(5): e1005457
6. Hartley SW, Coon SL, Savastano LE, Mullikin JC, Fu C, Klein DC (2015) Neurotranscriptomics: the effects of neonatal stimulus deprivation on the rat pineal transcriptome. PLoS One [Internet] 10(9):1–16
7. Cocolin L, Rantsiou K (2014) Molecular biology: transcriptomics. In: Encyclopedia of food microbiology, Second edn, Academic Press, Elsevier, Ltd., Amsterdam, pp 803–807
8. Liang KH (2013) Transcriptomics. In: Bioinformatics for biomedical science and clinical applications [Internet], pp 49–82

9. Taguchi YH (2018) Comparative transcriptomics analysis. In: Encyclopedia of bioinformatics and computational biology: ABC of bioinformatics, vol 1–3, pp 814–818
10. Sylwestrak EL, Rajasethupathy P, Wright MA, Jaffe A, Deisseroth K (2016. Feb 2 [cited 2022 Oct 6]) Multiplexed intact-tissue transcriptional analysis at cellular resolution. Cell [Internet] 164(4):792
11. Habib N, Li Y, Heidenreich M, Swiech L, Avraham-Davidi I, Trombetta JJ et al (2016) Div-Seq: single nucleus RNA-Seq reveals dynamics of rare adult newborn neurons. Science [Internet] 353(6302):925
12. Lake BB, Ai R, Kaeser GE, Salathia NS, Yung YC, Liu R et al (2016) Neuronal subtypes and diversity revealed by single-nucleus RNA sequencing of the human brain. Science [Internet] 352(6293):1586
13. Farrell RE (2010) Nucleic acid probe technology. In: RNA methodologies, Academic Press is an imprint of Elsevier, pp 261–282
14. Wages JM (2004) Nucleic acids – immunoassays. In: Encyclopedia of analytical science, Second edn, Elsevier Science Publishing Co Inc, pp 408–417
15. Luo Z, Geschwind DH (2001) Microarray applications in neuroscience. Neurobiol Dis 8(2):183–193
16. Haque A, Engel J, Teichmann SA, Lönnberg T (2017) A practical guide to single-cell RNA--sequencing for biomedical research and clinical applications. Genome Med [Internet] 9(1): 1–12
17. Shalek AK, Satija R, Adiconis X, Gertner RS, Gaublomme JT, Raychowdhury R et al (2013) Single-cell transcriptomics reveals bimodality in expression and splicing in immune cells. Nature 498:7453. [Internet]. 2013 May 19;498(7453):236–240
18. Tirosh I, Izar B, Prakadan SM, Wadsworth MH, Treacy D, Trombetta JJ et al (2016) Dissecting the multicellular ecosystem of metastatic melanoma by single-cell RNA-seq. Science [Internet] 352(6282):189
19. Wagner A, Regev A, Yosef N (2016) Revealing the vectors of cellular identity with single-cell genomics. Nat Biotechnol 34:11. [Internet]. 2016 Nov 8;34(11):1145–1160
20. Leitner DF, Mills JD, Pires G, Faustin A, Drummond E, Kanshin E et al (2021) Proteomics and transcriptomics of the hippocampus and cortex in SUDEP and high-risk SUDEP patients. Neurol Int 96(21):e2639
21. Kang HJ, Kawasawa YI, Cheng F, Zhu Y, Xu X, Li M et al (2011) Spatiotemporal transcriptome of the human brain. Nature 478(7370):483
22. Miller JA, Ding SL, Sunkin SM, Smith KA, Ng L, Szafer A et al (2014) Transcriptional landscape of the prenatal human brain. Nature 508(7495):199–206
23. Velmeshev D, Schirmer L, Jung D, Haeussler M, Perez Y, Mayer S et al (2019) Single-cell genomics identifies cell type-specific molecular changes in autism. Science 364(6441):685–689
24. Antunes ASLM, Martins-de-Souza D (2022) Single-cell RNA sequencing and its applications in the study of psychiatric disorders. Biol Psych Glob Open Sci 4:1–15
25. Healy LM, Stratton JA, Kuhlmann T, Antel J (2022) The role of glial cells in multiple sclerosis disease progression. Nat Rev Neurol 18:4. 2022 Feb 21;18(4):237–248
26. Alzate O (2010) Neuroproteomics. Neuroproteomics [Internet]. 2010 [cited 2022 Oct 10]. Available from: https://www.ncbi.nlm.nih.gov/books/NBK56018/
27. Anderson NL, Anderson NG (1998) Proteome and proteomics: new technologies, new concepts, and new words. Electrophoresis 19(11):1853–1861
28. Zetterberg H, Rüetschi U, Portelius E, Brinkmalm G, Andreasson U, Blennow K et al (2008) Clinical proteomics in neurodegenerative disorders. Acta Neurol Scand [Internet] 118(1):1–11
29. Becker M, Schindler J, Nothwang HG (2006) Neuroproteomics - the tasks lying ahead. Electrophoresis 27(13):2819–2829
30. Karas M, Ingendoh A, Bahr U, Hillenkamp F (1989) Ultraviolet-laser desorption/ionization mass spectrometry of femtomolar amounts of large proteins. Biomed Environ Mass Spectrom 18:841
31. Whitehouse CM, Dreyer RN, Yamashita M, Fenn JB (1985) Electrospray interface for liquid chromatographs and mass spectrometers. Anal Chem 57(3):675–679
32. Chen Z, Li S, Subramaniam S, Shyy JYJ, Chien S (2017) Epigenetic regulation: a new frontier for biomedical engineers. Annu Rev Biomed Eng 19:195–219
33. Sharma S, Chauhan A, Dobbal S, Kumar R (2022) Biology of plants coping stresses: epigenetic modifications and genetic engineering. S Afr J Bot 144:270–283
34. Karpova NN (ed) (2022) Epigenetic methods in neuroscience research. [cited 2022 Oct 10]. Available from: http://www.springer.com/series/7657

35. Driscoll DJ, Miller JL, Cassidy SB (1993) Prader-Willi syndrome. Neuropediatrics [Internet] [cited 2022 Oct 12];29:124–126
36. Kiefer JC (2007) Epigenetics in development. Dev Dyn 236(4):1144–1156
37. Berman MG, Kardan O, Kotabe HP, Nusbaum HC, London SE (2019) The promise of environmental neuroscience. Nat Hum Behav 3:5. [Internet]. 2019 Mar 18 [cited 2022 Oct 10];3 (5):414–417
38. Meloni M (2014) The social brain meets the reactive genome: neuroscience, epigenetics and the new social biology. Front Hum Neurosci 8 (May):309
39. Jiang Y, Langley B, Lubin FD, Renthal W, Wood MA, Yasui DH et al (2008) Mini-symposium epigenetics in the nervous system
40. Kwapis JL, Wood MA (2014) Epigenetic mechanisms in fear conditioning: implications for treating post-traumatic stress disorder. Trends Neurosci 37(12):706–720
41. Pickersgill M (2014) Neuroscience, epigenetics and the intergenerational transmission of social life: exploring expectations and engagements. Fam Relationsh Soc 3(3):481–484
42. Dernburg AF, Broman KW, Fung JC, Marshall WF, Philips J, Agard DA et al (1996) Perturbation of nuclear architecture by long-distance chromosome interactions. Cell 85(5):745–759
43. Geurts AM, Moreno C (2010. Oct 1 [cited 2022 Oct 19]) Zinc-finger nucleases: new strategies to target the rat genome. Clin Sci [Internet] 119(8):303–311
44. Jabalameli HR, Zahednasab H, Karimi-Moghaddam A, Jabalameli MR (2015) Zinc finger nuclease technology: advances and obstacles in modelling and treating genetic disorders. Gene 558(1):1–5
45. Chou ST, Leng Q, Mixson AJ (2022) Zinc finger nucleases: tailor-made for gene therapy. [cited 2022 Oct 19]; Available from: http://www.zincfingers.org/
46. Winden KD, Sundberg M, Yang C, Wafa SMA, Dwyer S, Chen PF et al (2019) Biallelic mutations in TSC2 Lead to abnormalities associated with cortical tubers in human iPSC-derived neurons. J Neurosci [Internet] [cited 2022 Oct 12] 39(47):9294–9305
47. Heidenreich M, Zhang F (2016) Applications of CRISPR-Cas systems in neuroscience, vol 17, Nature reviews neuroscience. Nature Publishing Group, London, New York, Berlin, Shanghai and Tokyo, pp 36–44
48. Miele A, Dekker J (2008. [cited 2022 Oct 19]) Long-range chromosomal interactions and gene regulation. Mol BioSyst [Internet] 4(11):1046
49. Kadauke S, Blobel GA (2009. [cited 2022 Oct 19]) Chromatin loops in gene regulation. Biochim Biophys Acta [Internet] 1789(1):17
50. Hahn S, Kim D (2013. [cited 2022 Oct 19]) Physical origin of the contact frequency in chromosome conformation capture data. Biophys J [Internet] 105(8):1786
51. Espeso-Gil S, Halene T, Bendl J, Kassim B, ben Hutta G, Iskhakova M et al (2020) A chromosomal connectome for psychiatric and metabolic risk variants in adult dopaminergic neurons. Genome Med [Internet] [cited 2022 Oct 19];12(1)
52. Hong DS, Hoeft F, Marzelli MJ, Lepage JF, Roeltgen D, Ross J et al (2014) Influence of the X-chromosome on neuroanatomy: evidence from turner and klinefelter syndromes. J Neurosci 34(10):3509–3516
53. Fowler KR, Hyppa RW, Cromie GA, Smith GR (2018) Physical basis for long-distance communication along meiotic chromosomes. Proc Natl Acad Sci U S A 115(40):E9333–E9342
54. Hausler M, Sellhaus B, Schweizer K, Ramaekers VT, Opladen T, Kleines M (2003) Flow cytometric cerebrospinal fluid analysis in children. Pathol Res Pract 199:667–675
55. Quijano S, Lopez A, Manuel Sancho J, Panizo C, Deben G, Castilla C, Antonio Garcia-Vela J, Salar A, Alonso-Vence N, Gonzalez-Barca E et al (2009) Identification of leptomeningeal disease in aggressive B-cell non-Hodgkin's lymphoma: improved sensitivity of flow cytometry. J Clin Oncol 27:1462–1469
56. Kraan J, Gratama JW, Haioun C, Orfao A, Plonquet A, Porwit A, Quijano S, Stetler-Stevenson M, Subira D, Wilson W (2008) Flow cytometric immunophenotyping of cerebrospinal fluid. CurrProtoc Cytom Chapter 6: Unit 6 25, 45
57. Picot J, Guerin CL, Le Van KC, Boulanger CM (2012) Flow cytometry: retrospective, fundamentals and recent instrumentation. Cytotechnology 64(2):109–130. https://doi.org/10.1007/s10616-011-9415-0
58. Behbahani H, Rickle A, Concha H, Ankarcrona M, Winblad B, Cowburn RF (2005) Flow cytometry as a method for studying effects of stressors on primary rat neurons. J Neurosci Res 82:432–444
59. McLarena FH, Svendsenb CN, Van der Meidec P, Jolya E (2001) Analysis of neural stem cells by flow cytometry: cellular differentiation modifies patterns of MHC expression. J Neuroimmunol 112:35–46
60. Muthuirulan P (2017) Flow cytometry: accelerating neuroscience research. Drug Target Rev, no. 3. Publisher's Version

61. Martin D, Xu J, Porretta C, Nichols CD (2017) Neurocytometry: flow cytometric sorting of specific neuronal populations from human and rodent brain. ACS Chem Neurosci 8(2):356–367. https://doi.org/10.1021/acschemneuro.6b00374
62. Svenningsson A, Hansson GK, Andersen O, Andersson R, Patarroyo M, Stemme S (1993) Adhesion molecule expression on cerebrospinal fluid T lymphocytes: evidence for common recruitment mechanisms in multiple sclerosis, aseptic meningitis, and normal controls. Ann Neurol 34:155–161
63. Svenningsson A, Andersen O, Edsbagge M, Stemme S (1995) Lymphocyte phenotype and subset distribution in normal cerebrospinal fluid. J Neuroimmunol 63:39–46
64. de Graaf MT, de Jongste AH, Kraan J, Boonstra JG, Sillevis Smitt PA, Gratama JW (2011) Flow cytometric characterization of cerebrospinal fluid cells. Cytometry B Clin Cytom 80(5):271–281. https://doi.org/10.1002/cyto.b.20603
65. Straus SE, Thorpe KE, Holroyd-Leduc J (2006) How do I perform a lumbar puncture and analyze the results to diagnose bacterial meningitis? JAMA 296:2012–2022
66. Pillny C, Nitsch L, Proske-Schmitz S, Sharma A, Wüllner U (2021) Abnormal subpopulations of monocytes in the cerebrospinal fluid of patients with Parkinson's disease. Parkinsonism Relat Disord 84:144. https://doi.org/10.1016/j.parkreldis.2021.02.017
67. Maes M, Lambrechts J, Bosmans E, Jacobs J, Suy E, Vandervorst C, Raus J (1992) Evidence for a systemic immune activation during depression: results of leukocyte enumeration by flow cytometry in conjunction with monoclonal antibody staining. Psychol Med 22(1):45–53. https://doi.org/10.1017/S0033291700032712
68. Räuber S, Heming M, Repple J, Ruland T, Kuelby R, Schulte-Mecklenbeck A et al (2021) Cerebrospinal fluid flow cytometry distinguishes psychosis spectrum disorders from differential diagnoses. Mol Psychiatry 268:455–466. https://doi.org/10.1038/s41380-021-01244-5
69. Cheung M, Campbell JJ, Whitby L, Thomas RJ, Braybrook J, Petzing J (2021) Current trends in flow cytometry automated data analysis software. Cytometry 99:1007–1021. https://doi.org/10.1002/cyto.a.24320
70. Cox JH, Ferrari G, Janetzki S (2006) Measurement of cytokine release at the single cell level using the ELISPOT assay. Methods [Internet]. Apr [cited 2022 Oct 21] 38(4):274–282
71. Janetzki S, Rueger M, Dillenbeck T (2014) Stepping up ELIspot: multi-level analysis in fluorospot assay. Cell 3(4):1102–1115
72. Czerkinsky CC, Nilsson LA, Nygren H, Ouchterlony O, Tarkowski A (1983) A Solid-Phase Enzyme-Linked Immunospot (ELISPOT) assay for enumeration of specific antibody-secreting cells. J Immunol Methods 65:109
73. Harriman WD, Collarini EJ, Sperinde GV, Strandh M, Fatholahi MM, Dutta A et al (2009) Antibody discovery via multiplexed single cell characterization. J Immunol Methods 341(1–2):135–145
74. della Bella C, Spinicci M, Grassi A, Bartalesi F, Benagiano M, Truthmann K et al (2018) Novel M. tuberculosis specific IL-2 ELISpot assay discriminates adult patients with active or latent tuberculosis. PLoS One 13(6):1102–1115
75. Tarke A, Sidney J, Methot N, Yu ED, Zhang Y, Dan JM et al (2021) Impact of SARS-CoV-2 variants on the total CD4+ and CD8+ T cell reactivity in infected or vaccinated individuals. Cell Rep Med 2(7):100355
76. Körber N, Behrends U, Hapfelmeier A, Protzer U, Bauer T (2016) Validation of an IFNγ/IL2 FluoroSpot assay for clinical trial monitoring. J Transl Med 14(1)
77. Bronge M, Ruhrmann S, Carvalho-Queiroz C, Nilsson OB, Kaiser A, Holmgren E et al (2019) Myelin oligodendrocyte glycoprotein revisited—sensitive detection of MOG-specific T-cells in multiple sclerosis. J Autoimmun 102(102):38–49
78. Bronge M, Asplund Högelin K, Thomas OG, Ruhrmann S, Carvalho-Queiroz C, Nilsson OB et al (2022) Identification of four novel T cell autoantigens and personal autoreactive profiles in multiple sclerosis [Internet]. Sci Adv 8. Available from: www.proteinatlas.org
79. Dhanwani R, Pham J, Premlal ALR, Frazier A, Kumar A, Pero ME et al (2020) T cell responses to neural autoantigens are similar in Alzheimer's disease patients and age-matched healthy controls. Front Neurosci 14:874
80. Kouli A, Williams-Gray CH (2020) Timing is everything: the T-cell response to α-Synuclein is maximal in early Parkinson's. Mov Disord 35(7):1137. https://doi.org/10.1002/mds.28122. Epub 2020 Jun 19

Chapter 3

Toward the Use of Research and Diagnostic Algorithmic Rules to Assess the Recurrence of Illness and Major Dysmood Disorder Features: The Diagnosis "Bipolar Disorder" Is Useless

Michael Maes

Abstract

The DSM/ICD mood disorder classifications are incorrect, and their dogmatic nature prevents both inductive and deductive remodeling of diagnostic criteria. Using machine learning, we (a) proposed new Research and Diagnostic Algorithmic Rules (RADAR) to score lifetime (LT), current suicidal ideation and attempts, recurrence of illness (ROI), and the phenome of mood disorders; (b) developed precision models of mood disorders based on ROI and the phenome, as well as immune, nitro-oxidative, and gut–autoimmune pathways; and (c) built a new endophenotype "major dysmood disorder (MDMD)," which is characterized by increased ROI, a severe phenome, and the above pathways. The current narrative review demonstrates that the diagnoses of major (MDD) and bipolar (BD) disorder and bipolar 1 (BP1) and BP2 are ineffective because (a) a common core underpins recurrence of depressive and manic episodes, indicating that they are manifestations of ROI; (b) patients with MDD, BP1, and BP2 can be classified as MDMD or non-MDMD, indicating that MDD/BD/BP1/BP2 do not reflect overall severity; (c) because ROI is a more important indicator in our precision models, the binary diagnoses of MDD/BD/BP1/BP2 are meaningless; and (d) there are no significant differences in the model parameters concerning ROI and the phenome between unipolar and BD as found using measurement invariance assessment of composite models (MICOM) and multi-group analysis (MGA). In conclusion, diagnoses such as MDD/BD/BP1/BP2 should be abandoned in favor of our RADAR scores and the MDMD case definitions. Future studies should investigate which pathways underlie MDD + BD combined, ROI, and the first (hypo)manic episode.

Key words Depression, Mood disorders, Psychiatry, Neuro-immune, Oxidative stress

1 Mood Disorder Concepts: The Ultimate Chaos

When diagnosing mood disorders, it appears that psychiatrists cannot communicate with one another and talk in different languages. These different languages include psychodynamic psychiatry, systemic therapy, psychoanalysis, cognitive behavioral therapy, folk psychology, self-system theory, and mind–brain dualism, along

Drozdstoy Stoyanov et al. (eds.), *Computational Neuroscience*, Neuromethods, vol. 199,
https://doi.org/10.1007/978-1-0716-3230-7_3,

with other non-falsifiable theories [1]. Other more falsifiable models include biological psychiatry (it's serotonin and other neurotransmitters), neurocognitive psychiatry (cognitive impairments in memory or attention are involved), molecular psychiatry (genes and intracellular networks explain mood disorders), and pan-omics and precision psychiatry (it's all about the genes, or pan-omics data will reveal the true nature of depression phenotypes or transdiagnostic pathway phenotypes) [1]. Nonetheless, the current approach to mood disorders is befuddled by a number of strange developments, including postpsychiatry (community development and engagements with communities are central, and boredom and depression are the defining moods of our time) and critical psychiatry, which accuses psychiatry of unethical and harmful strategies (psychiatric survivor networks question psychiatric knowledge base, treatment, practice, scientific methods, and the decontextualization).

If you believe that the self-proclaimed elite of psychiatric research (NIH–NIHM) brought some order to this chaos, you are mistaken. Thus, the NIHM's new Research Domain Criteria (RDoC) adds to the confusion by introducing new dimensions in the form of a matrix structure with columns and rows [2]. According to them, psychopathology consists of neural circuit disorders reflected by a matrix structure in which genes, molecules, cells, circuits, physiology, behavior, self-reports, and paradigms interact with memory, rewards, threats, and perception [2]. Nonetheless, as previously discussed, there is no evidence base for the RDoC matrix approach, which is developed top–down [3–5].

2 Diagnosis of Mood Disorders: The Ultimate Chaos

Diagnosing mood disorders using DSM or ICD criteria, which take into account diverse subtypes such as major depressive episodes (MDE), major depressive disorders (MDD), and bipolar disorders (BD), either type 1 (manic) or type 2 (hypomanic) episodes, is the gold standard in the West (and therefore worldwide) [6, 7]. Nevertheless, the DSM and ICD case definitions of MDD, MDE, and BD and BP1 and BP2 all have a number of significant flaws that need to be addressed. To begin with, the case definitions are frequently not reliable or replicable [8, 9] and sometimes show an intraclass kappa reliability of 0.28, indicating that there is low agreement among psychiatrists [1, 8, 9]. Studies have shown that only 42.9% of patients labeled with BD according to the DSM criteria actually match the diagnostic criteria [10], suggesting that BD is frequently overdiagnosed. Patients diagnosed with BD are frequently misdiagnosed as having MDD or other disorders at a rate that can reach up to 60% [11]. Another significant issue is that a diagnosis of BD is frequently overlooked in patients who present with a depressive index episode and an atypical history of manic or hypomanic symptoms [10]. The low threshold for distinguishing BD from other personality disorders, such as borderline personality disorder, polysubstance addiction, and attention deficit disorder, is another factor that contributes to the high likelihood of incorrect diagnoses [10].

The question of whether BD and MDD belong to a continuum (continuous theory) or whether they constitute distinct categories (discontinuous theory) is another topic that has been the subject of a great deal of controversy [12, 13]. It is believed that a number of findings support the continuity spectrum that exists between MDD and BD. These findings include the presence of mixed states (in which both manic and depressive symptoms co-occur), the fact that there is no real separation between MDD and MDE in BD, and the fact that many MDD patients may shift into BD [14]. Findings that support the categorical theory are that MDD is more common in females and BD has a more recurrent course than MDD and occurs more frequently in the relatives of people who have BD [12, 13]. The current situation with BP2 is an even more serious one. Based on proband studies, several studies have suggested that BP2 is a distinct group that ought to be differentiated from recurrent MDD and BP1 [15]. Nevertheless, some studies show that the reliability coefficient of BP2 is not more than that of chance [3, 4]. To complicate matters, some authors [16] offered a concept that they called the "bipolar spectrum." According to this theory, bipolarity can be found anywhere along a continuum, from mild to severe types of BD. Because of this, the increased prevalence of BD may be explained by the discovery of softer bipolar disorder phenotypes, such as rapid cyclers, cyclothymia, BP2, and BP3 [17]. The bipolar spectrum may also include MDD with hyperthymic traits, depressive mixed states with hypomanic symptoms including sexual arousal, ultrarapid-cycling forms, patients with lifelong temperamental dysregulation, and cyclic irritable–dysphoric, intermittently explosive, impulse-ridden clinical expression, lifelong temperamental dysregulation, and pseudounipolar or excited mixed depression [18]. Accordingly, a number of authors have loosened up the case definitions of BP2 by, for instance, utilizing updated hypomania checklists that include subsyndromal hypomania or subthreshold bipolarity [19, 20] which are both considered to be part of the mild bipolar spectrum. As a result, these authors employ this checklist, which demonstrates a sensitivity of 80% to find true bipolar patients and a specificity of 51% (computed versus MDD) to diagnose BD. These figures of merit are of course woefully inadequate [21], and therefore up to 79% of fibromyalgia patients suddenly belong to the bipolar spectrum. One wonders whether journals and their editors publishing these data have any self-respect. Overall, the dimensional approach to the idea of the mood disorder spectrum, as well as the overdiagnosis of BD with more inclusive diagnostic criteria, has resulted in a blurring of the lines between distinct diagnostic categories, which has resulted in a reduction in the diagnostic reliability of these mood disorders [22].

The extreme disorganization and crazy patterns that is typical of the current approach to BD are also typical of depression, as evidenced by the plethora of depression categories, subclasses, and labels that are now in use, including dysthymia, atypical depression, recurrent depressive disorder, melancholia, double depression, bipolar depression, mixed episodes, persistent depressive disorder, psychotic depression, treatment-resistant depression, and older labels that have disappeared (reactive depression, vital depression, situational depression, endogenomorph depression, endogenous depression, concealed depression, anxious depression, and hidden depression) [1].

Furthermore, no models distinguish MDD from common emotional distress responses such as loss, grief, and demoralization [1, 3, 4]. Folk psychologists and sociologists appear to believe that clinical depression is a boundary experience or that "psychiatry transformed normal sorrow into depressive disorder," thereby contributing to the medicalization of normal human behaviors such as sadness, grief, and demoralization [23–26]. This also indicates that journals like *Psychological Medicine* have degraded to the level of folk psychology. As a result, a serious medical illness, such as recurrent major depression, is frequently regarded as a boundary experience, and severe medical phenotypes are often lumped together with common emotional distress responses in the MMD/MDE group in psychiatric research [3, 4].

Given the aforementioned information, it should not come as a surprise that the DSM and ICD taxonomies lack reliability and validity and, as a result, are ineffective for the purposes of research [2–4, 27–29]. Because of this, it should come as no surprise that several scholars have concluded that all psychiatric diagnostic systems should be abolished [30].

3 Lack of a Correct Model Prevents Targeted Research

Recent reviews have underscored that there is no single model that adequately describes MDD, MDE, or BD [1, 3–5]. Not only is there not a model that adequately describes mood disorders, but there is also a wide variety of conceptual frameworks, classifications, and model conceptions—including one that doesn't even call itself a model—that are used to describe the same condition [1, 3, 4]. Without a wide agreement among psychiatrists, research on mood disorders is filled with too much noise and too many names and labels.

In reality, mood disorder diagnoses are post hoc, higher-order constructs based on clinical narratives of the condition [1, 3–

5]. MDD and MDE, for example, are based on the presence of a few symptoms over time, whereas BD is based on the presence of (hypo)mania episodes. These unreliable diagnoses are then frequently used as explanatory variables in clinical research, such as t-tests, ANOVAs, and GLM analyses, or as moderator variables in multiple regression analysis. As a result, using them as explanatory or moderator variables not only constitutes a serious conceptual error but also leads to a slew of errors and invalid conclusions because these post hoc diagnoses are incorrect [1, 3–5].

As previously stated, a correct model is (a) one that follows Popper's rules, which state that a model should be falsifiable, confirmable, parsimonious, progressive, truth value determinable, and changeable, and (b) one that has been validated and cross-validated and has sufficient replicability, convergence, and internal consistency [1, 3–5]. Needless to say, all "in the mind" theories do not meet any of these requirements and are thus merely folk psychology beliefs. But science is the last thing psychiatrists consider because they are unfamiliar with Popper's rules and, in any case, everything "is in the mind." Thus, falsifiability is not of interest to these physicians.

In fact, the clinical diagnosis should be used as a dependent variable in statistical tests and machine learning techniques only when a correct model could be constructed. As a result, most research (except genetic research) employs higher-order constructs as explanatory or moderator variables, whereas the proper approach is to employ a correct model as the dependent variable, in which genes, environmental factors, adverse outcome pathways (AOPs) (including biomarkers, pathways, intracellular signaling networks), neurocognitive disorders, and brain connectome circuits predict the outcome (or dependent variable) [1, 3–5]. One wonders what could be gleaned from the published data on mood disorders as a result of the current flawed approach. The best part is that everyone, including myself, is forced to use these incorrect models because the new generation of junior reviewers knows only how to employ t-tests and chi-square tests. As a result, most research on mood disorders is just pitiful.

4 Machine Learning Models

Given that machine learning allows for the creation and cross-validation of new correct models, including new endophenotype classes [1, 5], it is perplexing that psychiatric research continues to use invalid outcome measurements as an explanatory or moderator variable in research papers [1, 3–5]. Recently, we developed a new clinimetrics method based on unsupervised and supervised machine learning to construct new phenotype classes (clinical-based clusters of patients) or endophenotype classes (clinical- and biomarker-based clusters) and new pathway phenotypes (a combination of biomarkers and clinical domains), and this new method is now known as precision nomothetic psychiatry [1, 3–5, 31–33].

We used the label "nomothetic" to reflect our method of extracting rules or mathematical laws from a set of input (independent indicator) variables that explain the variability in the outcome variable (the mood disorder phenomenon) which allows one to build covering law models for mood disorders [1, 3, 4, 27, 28]. Most importantly, these precision nomothetic models of mood disorders are built around a theoretical framework comprising the relevant building blocks or components of mood disorders: the causome (genetic and environmental causes) versus the protectome (genetic and environmental protective factors), adverse outcome pathways (AOPs, including biomarkers, pathways, and intracellular signal molecules), and the phenome of mood disorders (symptoms and self-ratings). Precision models of mood disorders should include a recurrence of illness (ROI) index based on the recurrence of depressive episodes, hypomanic episodes, suicidal ideation, and suicide attempts [3, 4, 31, 34].

Recently, we constructed four different precision cross-validated machine learning models of mood disorders, namely, (1) a clinical precision model, (2) an immune precision model, (3) a nitro-oxidative stress model, and (4) a gut–autoimmune model [1, 3–5, 34–36]. These data-driven new models of mood disorders comprise a ROI index constructed using the recurrence of episodes and suicidal ideation and attempts and a new (endo)-phenotype cluster of patients labeled major dysphoric disorder (MDMD) versus simple DMD (SDMD) [1, 3–5, 34–36]. MDMD is a cluster of patients which is cross-validated by immune, nitro-oxidative, and gut–autoimmune pathways and shows increased ROI, more cognitive impairments in executive functions and verbal fluency, increased severity of depression and anxiety, more suicidal ideation, more disabilities, and lowered health-related quality of life (HR-QoL) [1, 3, 4, 31–33, 36]. For an overview of the biomarker pathway models 2, 3, and 4, we would refer the interested readers to our published papers [1, 3, 4, 31–36].

In the current paper, we will focus on the clinical precision model and, in particular, on the status of BD versus unipolar depression. Indeed, it remained unknown whether unipolar MDD and BD MDE are the same disease or two separate diseases or whether unipolar MDD and BD are two separate but not mutually exclusive disorders [37]. As such, "depression with and without mania may be considered the same condition," whereas "BD disorder" may be considered depression with or without (hypo)mania [37].

Figure 1 shows the theoretical framework linking the different building blocks of mood disorders [31]. Thus, the phenome of

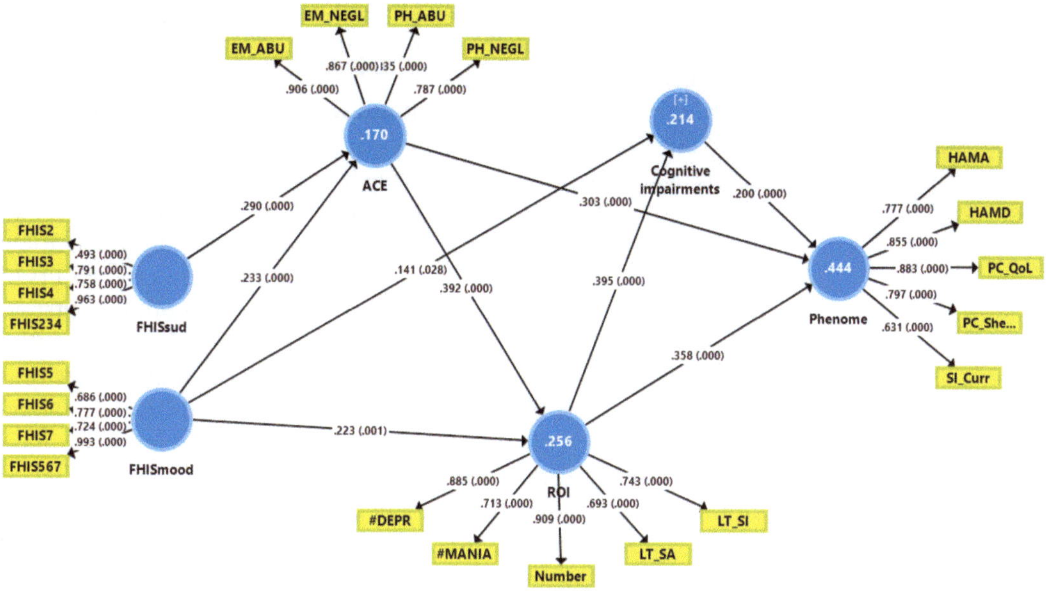

Fig. 1 Partial least squares model of mood disorders. This model shows the causal pathways from a family history of depression, bipolar disorder or suicide attempts (FHISmood), and substance use disorders (FHISsud) to adverse childhood experiences (ACEs; namely, physical and emotional abuse and neglect, PH_ABU, EM_ABU, PH_NEGL, EM_NEGL), to recurrence of illness (ROI), namely, the first factor extracted from a lifetime number of depressive (#dep) and (hypo)mania (#mania) episodes and suicidal ideation (LT_SI) and attempts (LT_SA), and to cognitive deficits in verbal fluency and executive functions (cognitive impairments). The output is the first factor extracted from the Hamilton depression (HAMD) and anxiety (HAMA) rating scale scores and the first factor extracted from the four domains of quality of life (PC_QoL), disability Sheehan (PC_She), and current suicidal ideation (SI_Curr) scores. We display only the significant paths with their path coefficients (with exact *p*-values) and the outer models loading (with *p*-values) of the outer models. Explained variances are shown in the blue circles. (Adapted from Maes et al. [31])

mood disorders is entered as a latent vector extracted from different rating scales including the Hamilton depression (HAMD) and anxiety (HAMA) rating scales [38, 39], the first principal component extracted from the four domains of the WHO QoL-BREF, namely, physical, psychological, social, and environmental domains [40, 41], and the first principal component extracted from the five domains of the Sheehan scale, namely, (a) work and school, (b) social life and recreation, (c) family life and household responsibilities, (d) days lost last week, and (e) days unproductive last week [42]. As such, this phenome latent score reflects an overall integrated index of the illness phenotype rather than an incorrect diagnosis such as BD or MDD. A number of input indicators predict the phenome, including neurocognitive dysfunctions (a composite score including executive impairment and fluency scores), a ROI index (a latent vector in a reflective model extracted from the number of depression and mania and lifetime number of suicidal attempts and ideation), and a latent vector extracted from adverse childhood scores of physical and emotional neglect and abuse (ACE). The latter is therefore an overall index of ACEs. The primary input variables are two latent vectors extracted from a family history of substance abuse disorders (SUDs) and depression, bipolar disorder, and suicidal attempts. As such, a clinical theoretical framework is built which links causome factors (family history and ACEs) with cognitive disorders and the phenome of mood disorders, while the effects of the causome on the phenome are at least partially mediated by ROI. Figure 1 shows the results of partial least squares (PLS)–SEM analysis (based on 5000 bootstrap samples), which was used to analyze and build the model. PLS showed that 44.4% of the variance in the phenome score was explained by cognitive impairments, ACE, and ROI. The model shows that 21.4% of the variance in neurocognitive impairments is explained by ROI and a family history of mood disorders (FHIS-mood), while 25.6% of the variance in ROI was explained by ACE and FHIS of depression, BD, or suicide. All outer and inner model quality fit metrics are more than adequate (including overall model fit as assessed using standardized root mean residual (SRMR) of 0.057 and convergence, replicability, and construct validity of all latent vectors).

Moreover, cluster analyses using all features showed three relevant clusters, namely, healthy controls (HCs), and two patient clusters, namely, the first labeled as major dysmood disorder (MDMD) and the second as simple dysmood disorder (SDMD) [1, 3, 4, 31–33]. This new MDMD data-driven mood disorder endophenotype class [31–33] is built by conducting a thorough examination of various phenome concepts such as depression, anxiety, current suicidal behaviors, HR-QoL scores, increased disabilities, ROI scores with and without immune, neuro-oxidative, and gut–autoimmune pathway biomarkers. Furthermore, this new endophenotype class is externally validated by increased ACE ratings and familial histories of mood disorders and SUDs [34–36]. In contrast, SDMD is not associated with these immune, nitro-oxidative stress, and gut–autoimmune pathways, indicating that MDMD, and not SDMD, is reified as a neuro-immune and neuro-oxidative nosological entity.

5 RADAR Scores and Plots

We also developed new Research and Diagnostic Algorithmic Rule (RADAR) scores based on the latent variable scores obtained from the key clinical features of mood disorders [3] including scores reflecting (1) family history of depression, BD, and suicide (FHIS-mood); (2) family history of substance abuse disorders (FHISsud); (3) adverse childhood experiences including physical and emotional abuse and neglect (ACE); (4) childhood sexual abuse (Sexabu);

(5) recurrent lifetime suicidal ideation (LT SI); (6) recurrent lifetime suicidal attempts (LT SA); (7) number of depression (#dep); (8) number of (hypo)manias (#mania); (9) ROI index; (10) current SI; (11) phenome score; (12) number of comorbid anxiety disorders, including generalized anxiety disorder (GAD), panic disorder (PD), post-traumatic stress disorder (PTSD), social phobia, special phobia, and obsessive compulsive disorder (OCD); (13) lifetime and current suicidal behaviors combined into overall severity index of suicidal behaviors; and (14) the lifetime trajectory of a mood disorder patient based on ACEs, ROI, and phenome scores [3].

Using these different RADAR scores, we made RADAR (or spider and star) graphs which show all 14 clinical feature scores of the disease. For example, Fig. 2 shows the RADAR scores of two patients with a typical MDMD profile and a typical SDMD profile. This RADAR graph illustrates the patient's location in relation to the mean values of normal controls set at a 0 score. There are 14 axes which represent the 14 feature RADAR scores expressed in z values on a continuous scale. As such, one can observe the distance between the 14 RADAR scores of the patients and the normal controls expressed in standard deviations. Moreover, the 14 axes are linked with each other, thereby creating grids which allow one to observe the differences between the mean RADAR scores of two study groups or the differences in the RADAR scores of two patients. For example, Fig. 2 shows that the patient with MDMD shows much higher RADAR scores of ACE, LS SI, LT SA, #dep, #mania, ROI, current SI, phenome, any anxiety disorder, all LT + current SB, and lifetime trajectory score than patients with SDMD. Moreover, since each patient is characterized by a specific profile of these 14 RADAR scores, an idiosyncratic fingerprint is constructed for each patient. Interestingly, the patient with MDMD in Fig. 2 shows high ACE scores, suggesting that these play a role in the development of the mood disorder in that patient, while the SMDM patient showed a higher family history load of depression, BD, and/or suicide.

6 Why the Diagnosis "Bipolar Disorder" Is Useless

6.1 Patients with BP1 and BP2 May Be Classified as SMDM or MDMD

Figure 2 depicts two individuals who were classified as BD (BP1) patients because they both exhibited at least one manic episode. Nonetheless, as described before, cluster analysis based on all aspects of the condition categorized these two BP1 patients as having MDMD versus SDMD, and their idiosyncratic fingerprint (RADAR) scores are entirely distinct. Consequently, the diagnosis of BP1 is not particularly instructive. In addition, Fig. 3 depicts the RADAR scores of a BP2 (female, 51 years old) patient and a BP1 (female, 31 years old) patient who were both classified as MDMD patients. So, diagnoses like BP1 and BP2 are not very useful.

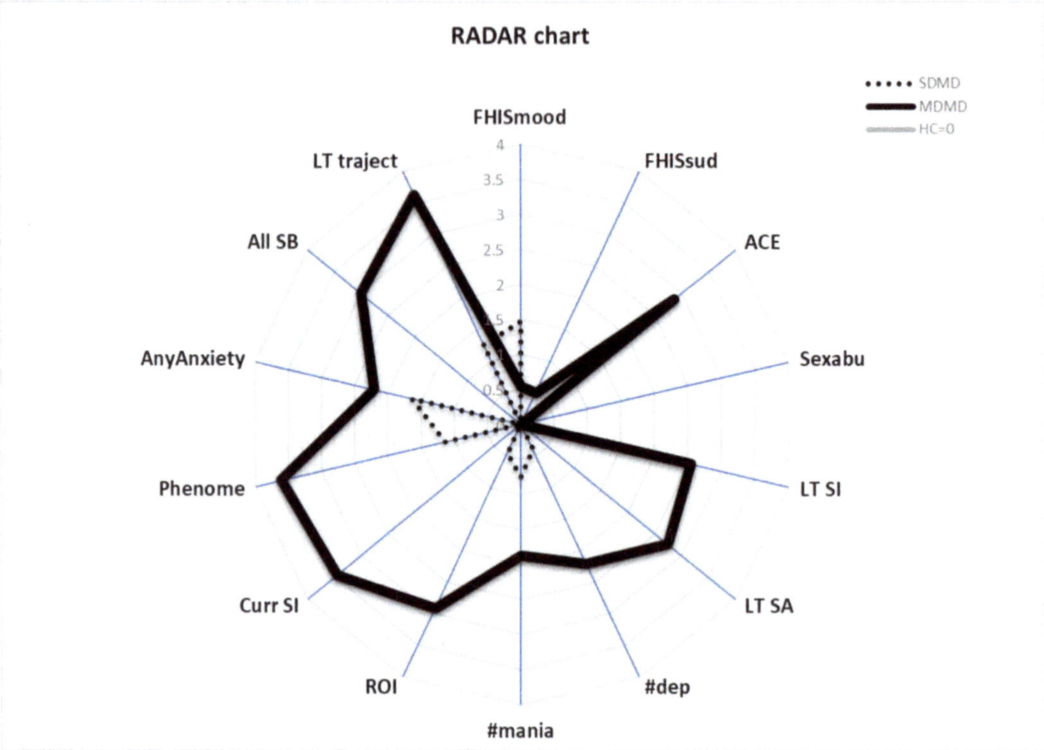

Fig. 2 RADAR graph displaying the Research and Diagnostic Algorithmic Rule (RADAR) scores of a patient with major dysmood disorder (MDMD) and another patient with simple dysmood disorder (SDMD). This graph depicts the subject's position relative to the common center point, which is defined as the mean value of all features of healthy controls (HCs) set to zero. All 14 feature scores are represented by standard deviations on 14 axes which are interconnected and divide the graph into grids representing the difference between the feature scores of both patients versus healthy controls. FHISmood/FHISsud: family history of depression, bipolar disorder or suicide, and substance use disorders, respectively. ACE adverse childhood experiences (physical and emotional neglect and abuse), Sexabu adverse childhood experiences, sexual abuse, LT SI lifetime suicidal ideation, LT SA lifetime suicidal attempts, #dep number of lifetime depressive episodes, #mania number of lifetime (hypo)mania episodes, ROI recurrence of illness, Curr SI current suicidal ideation, Phenome the phenome score as explained in Fig. 1, Any anxiety number of comorbid anxiety disorders, All SB all LT and current suicidal behaviors (SI and SA) combined, LT Traject lifetime trajectory score

RADAR scores, on the other hand, highlight the similarities and contrasts between the fingerprints of both patients.

It is clear to see that utilizing BP1 or BP2 as a diagnostic tool is utterly reductionist, whereas our RADAR ratings allow one to assess all the disorder's important components. In addition, as mentioned in the introduction, the DSM and ICD case definitions for mood disorders such as BP1 and BP2 are unreliable, and psychiatrists tend to differ with one another, with some categories being under- or overvalued [1, 3, 4]. To add insult to injury, DSM/ICD case definitions are generated using a top–down method based on a consensus among so-called professionals, and

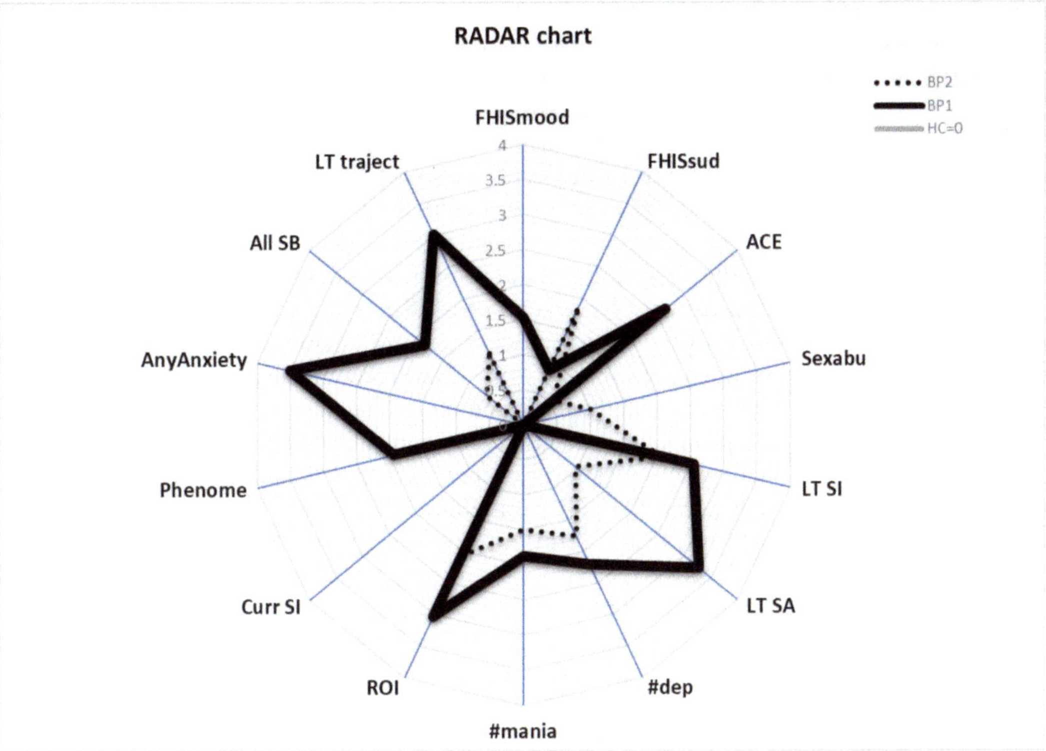

Fig. 3 RADAR graph displaying the Research and Diagnostic Algorithmic Rules (RADAR) scores of two patients with major dysmood disorder (MDMD), one with bipolar disorder type 1 (PB1) and another with BP2. See Fig. 2 for explanation about the RADAR graph

their dogmatic nature impedes falsification and, by extension, deductive and inductive learning [1, 3, 4]. Despite its evident shortcomings, the vast majority of psychiatric studies continue to employ the DSM/ICD as the primary explanatory (input) variable, which is another peculiarity of the current state of psychiatric research. Consequently, most of the psychiatric research uses a defective and highly controversial higher-order construct to establish misleading model assumptions (input and output variables are reversed) while employing improper statistical methods [1, 3–5]. As stated, the current state of research on mood disorders is pathetic. In contrast, our precision nomothetic models and RADAR scores can be checked and confirmed, and the model's truth value can be ascertained; furthermore, the model is progressive [1, 3, 4, 34, 43]. Our machine learning-based quantitative RADAR ratings of ACE, ROI, the phenome, and MDMD are therefore preferable to the incorrect DSM/ICD model assumptions and should be utilized as dependent variables in future research.

6.2 Depressive and Manic Episodes Are Manifestations of ROI

As previously indicated, a common core may be derived from the number of depressive and manic episodes, as well as the lifelong recurrence of suicide ideas and attempts. The discovery that the number of depressive and manic episodes is manifestations of the same latent construct shows that they are highly interrelated expressions of the same construct ROI which drives both manifestations. This demonstrates that recurrent depressive and (hypo)-manic episodes and suicidal behaviors are expressions of a shared pathophysiology. Earlier studies reported that increased levels of inflammatory markers, pro-inflammatory cytokines, and oxidative stress predicted the occurrence of episodes, suicide attempts, and hospitalizations [44–49]. More recent targeted research shows that ROI is strongly predicted by increased immune and growth factor responsiveness in unipolar depression [36], dysfunctional T regulatory activities and elevated T effector functions in bipolar disorder [50], and Q192R paraoxonase 1 (PON1) genotype X ACE interactions, resulting in impairments in the PON1–high-density lipoprotein (HDL) cholesterol complex in both MD and BD combined [34, 35]. Thus, the ROI and the recurrence of depressive, hypomanic, and suicidal behaviors are assumed to be the outcome of increasing sensitization in neuro-immune and neuro-oxidative stress pathways [33, 36, 44, 46].

6.3 The Diagnoses of MDD, MDE, BP1, and BP2 Are Irrelevant in Our Precision Models

Importantly, forced entry of BD, BP1, BP2, or MDD as categorical variables in two-step cluster analysis performed on MDD and BD combined showed that these diagnostic classes were not relevant after considering ROI and phenome scores. When phrased differently, the ROI concept is more predictive of the phenome, including present suicide behaviors, in the combined study group of MDD and BD patients than the BD, BP1, BP2, or MDD diagnosis. This is because ROI measures the number of (hypo)manic episodes, which indicates BD, but also contains information on the recurrence frequency of (hypo)manic and depressive episodes. As such, ROI and phenome RADAR scores contain more accurate information on mood disorders than the binary diagnoses of MDD, BD, BP1, and BP2. In addition, our results show that MDMD is in fact a more relevant diagnostic group that is more impactful than the classification into BD, BP1, BP2, and MDD and in fact crosses across these DSM/ICD diagnoses.

Figure 4 depicts the RADAR fingerprints of two separate patients, one classified as MDD (51-year-old female) by DSM criteria and the other as BP1 (50-years-old female). Nonetheless, both individuals were classified as MDMD using our RADAR technique. Both patients have similar numbers of depressive episodes, but one has substantially higher lifetime suicidal behaviors and a higher ROI score. Both patients exhibit significantly higher RADAR scores overall when compared to the SDMD patient's RADAR scores in Fig. 2. As a result, although our RADAR scores

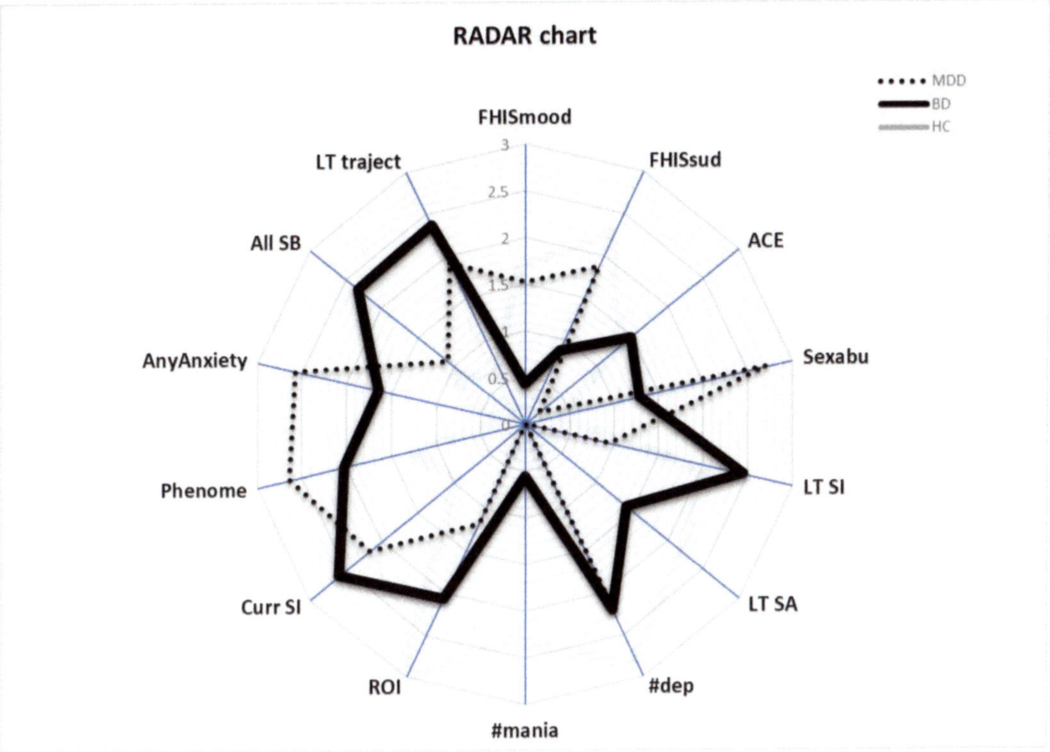

Fig. 4 RADAR graph displaying the Research and Diagnostic Algorithmic Rules (RADAR) scores of two patients with major dysmood disorder (MDMD), one with major depressive disorder (MDD) and another with bipolar disorder. See Fig. 2 for explanation about the RADAR graph

enable us to create a fingerprint of all MDMD traits (including ROI with the number of depressions and hypomanias), the DSM diagnosis is merely a binary indicator variable that indicates if MDD or BD is present. It is clear that by utilizing MDD/BD as diagnostic categories, one loses access to the RADAR graphs' most crucial data, such as a person's lifetime history of suicidal ideation and attempts, ROI, current phenome severity, and the lifetime trajectory severity.

Figure 5 depicts another RADAR graph containing the RADAR scores of two additional MDMD patients, one with MDD (a 31-year-old female) and one with BP2 (a 51-year-old female). Again, both patients had much higher RADAR ratings than the SDMD patient in Fig. 2. Surprisingly, while the BP2 patient has much higher hypomania episode and phenome scores, the overall lifetime + current suicidal behavior score is quite similar, while recurring lifetime suicidal attempts are higher in the MDD patient than in the BP2 patient. While BP2 is an incorrect and unreliable diagnostic category that only contains binary information (BP2 versus no BP2), the diagnosis MDMD indicates a more severe phenotype with a much more severe lifetime trajectory,

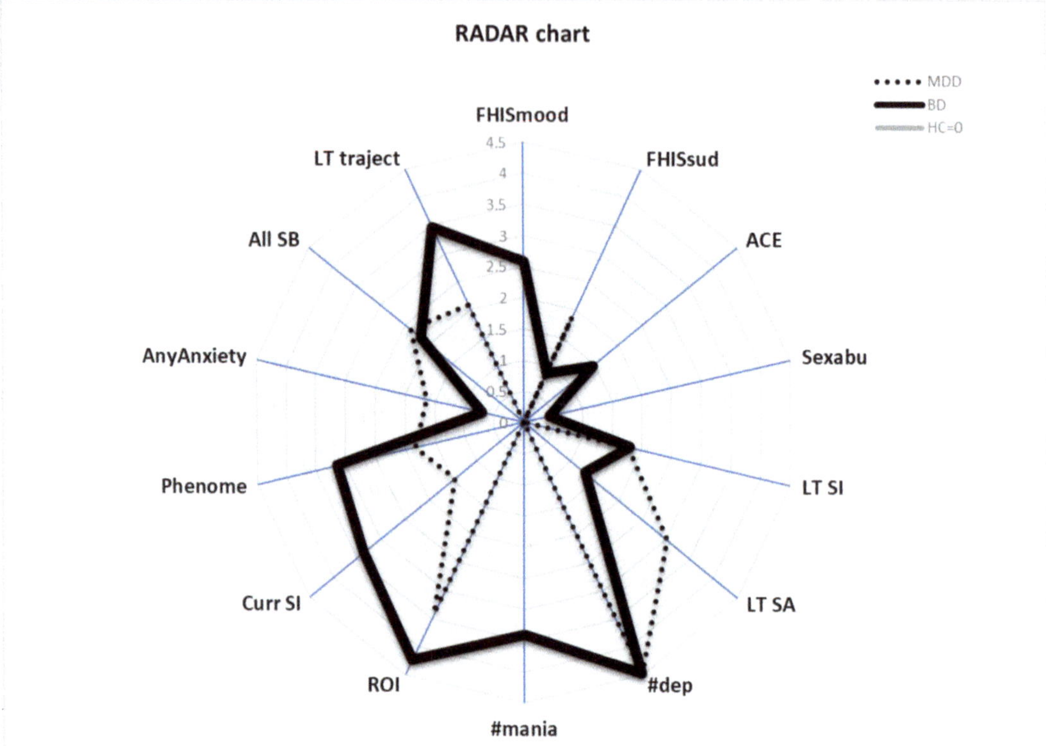

Fig. 5 RADAR graph displaying the Research and Diagnostic Algorithmic Rule (RADAR) scores in two patients with major dysmood disorder (MDMD), one with major depressive disorder (MDD) and another with bipolar disorder (BD). See Fig. 2 for explanation about the RADAR graph

whereas the RADAR fingerprint scores indicate all suicidal behaviors and ROI components, including the number of episodes either depressive or hypomanic. Using the DSM/ICD criteria for MDD/BD/BP1/BP2 (a) ignores the most important components of mood disorders that should be studied in psychiatric research (ROI, suicidal behaviors, severity of the phenome, and lifetime trajectory) and (b) ignores all of the most important information needed to make an effective treatment plan in clinical practice (adverse childhood experiences, ROI, suicidal behaviors).

6.4 No Model Differences Between Unipolar and Bipolar Disorders

It is clear from the above that the diagnosis of MDMD with accompanying RADAR scores has more merit than a binary diagnosis into MDD, BD, BP1, and BP2, which are additionally unreliable and incorrect. Nevertheless, another question is whether the PLS model characteristics (including the ROI–phenome associations) are different between MDD and BD. To examine possible differences between MDD and BD, we have carried out (a) measurement invariance assessment of composite models (MICOM) and (b) a multi-group analysis (MGA), both in PLS. MICOM in PLS assesses "configural invariance, compositional

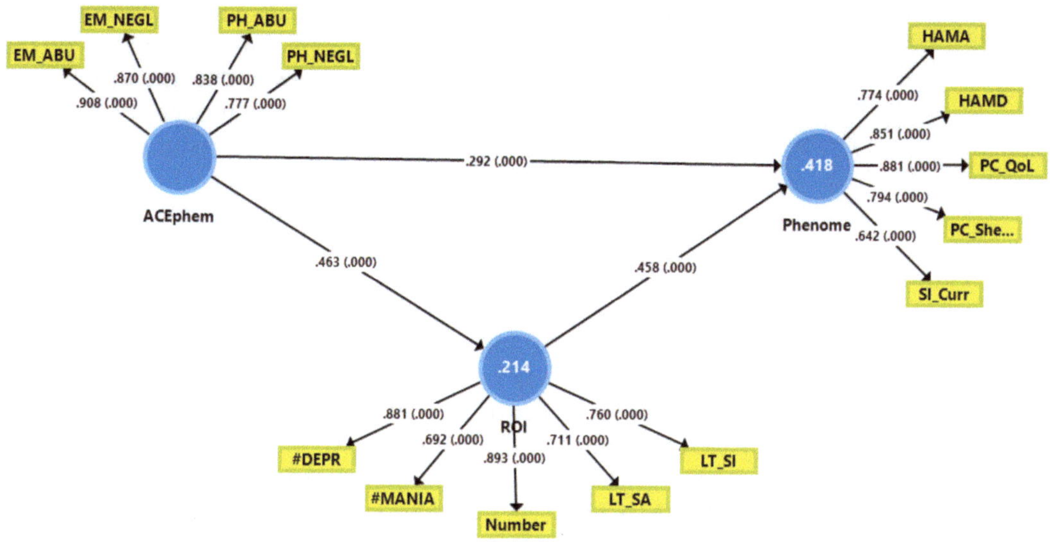

Fig. 6 Reduced partial least squares (PLS) model of mood disorders. This model shows the causal pathways from adverse childhood experiences (ACEs), namely, a factor extracted from physical and emotional abuse and neglect (PH_ABU, EM_ABU, PH_NEGL, EM_NEGL) to recurrence of illness (ROI), namely, a factor extracted from lifetime number of depressive (#dep) and (hypo)mania (#mania) episodes and suicidal ideation (LT_SI) and attempts (LT_SA). The final output is the phenome, namely, the first factor extracted from the Hamilton depression (HAMD) and anxiety (HAMA) rating scale scores, four domains of health-related quality of life (PC_QoL), five disability domain scores (Sheehan rating scale, PC_She), and current suicidal ideation (SI_Curr). We display only significant paths with their path coefficients and exact *p*-values and the outer model loadings with *p*-values. Explained variances are shown in the blue circles

invariance, and the equality of composite mean values and variances" [51–54]. The MGA-PLS was employed to examine whether predefined groups (MDD versus BD) differed significantly in the parameter estimates including path coefficients and outer model loadings [51–54].

Figure 6 shows a reduced model focusing on the associations between ACE, ROI, and the phenome, which are all three conceptualized as latent vectors extracted from its relevant manifestations. First, we extracted one latent vector from HAMA, HAMD, WHO HR-QoL, Sheehan disability data, and current suicidal ideation, and this construct shows adequate convergence and construct validity with average variance extracted (AVE) of 62.9%, Cronbach alpha = 0.849, and composite reliability = 0.893, while all indicators were significantly loaded on this factor (higher than 0.6). Also, ROI was validated as a construct extracted from the number of depressions, mania, total number of episodes, lifetime suicidal ideation, and attempts, with AVE = 0.628, composite reliability = 0.893, and Cronbach alpha = 0.849. The ACE latent vector was conceptualized as a construct extracted from four different ACE scores, namely, emotional abuse and neglect and physical abuse and neglect. AVE was 0.722, Cronbach alpha = 0.872, and composite reliability = 0.912. The construct cross-validated communalities were > 0.444 indicating sufficient replicability of the vectors. Confirmatory tetrad analysis showed that the three latent constructs were not mis-specified as reflective models. As such, we constructed a cross-validated model of mood disorders based on ACE, ROI, and the phenome. With a SRMR of 0.057, the model fit is more than adequate. After validating the outer and inner models, we may perform a complete PLS-SEM analysis, indicating that 41.8% of the variance in the phenome latent vector was explained by the regression on ROI and ACE and that part of the variance in ROI (21.4%) is explained by ACE. Analysis of specific and total indirect (or mediated) effects and total effects shows a specific indirect effect of ACE on the phenome which is mediated by ROI ($t = 5.02$, $p < 0.001$). Phrased differently, ROI is a partial mediator of the effects of ACE on the phenome of mood disorders.

The MICOM revealed that the permutation *p*-values for ACE ($p = 0.762$), phenome ($p = 0.489$), and ROI ($p = 0.085$) were all non-significant (at alpha = 0.05, two-tailed), confirming computational invariance. The mean original differences fell between the 2.5% and 97.5% limits, while the variance original differences of phenome and ROI did not, demonstrating partial invariance in composite equality. As a result, we performed MGA in PLS-SEM. There were no significant differences in the pathway coefficients from ACE to the phenome ($p = 0.326$), from ACE to ROI ($p = 0.126$), and from ROI to phenome ($p = 0.260$) using PLS-MGA, the parametric test, and the Welch–Satterthwait test (shown are the PLS-MGA *p*-values). Similarly, no significant differences in the specific indirect (mediated) and total effects were seen between MDD and BD. Furthermore, there was no significant difference between MDD and BD in the relationships between ACE and phenome ($p = 0.687$), ACE and ROI ($p = 0.064$), and ROI and phenome ($p = 0.559$) in the SmartPLS permutation output. These findings imply that there are no significant differences in the connections between the core components of mood disorders between MDD and BD, and hence both study groups should be integrated. Furthermore, we must emphasize that the approach we utilized to use BD and MDD groups is conceptually incorrect because (as explained above) BD and MDD are post hoc constructs that should be employed as dependent variables rather than as moderators in regression models. Nonetheless, the findings reveal that, even when BD and MDD are considered to be moderators (as most if not all psychiatrists do), there is no significant difference in the models between these diagnostic categories. It's worth noting that when we looked at the differences between males and women, those with metabolic syndrome against those without, and those with a family history of mood disorders or SUDs versus those without, we found full compositional invariance and composite equality.

7 Conclusions

We have reviewed that the DSM/ICD categories of mood disorders (MDD, BD, BP1, and BP2) are invalid post hoc higher-order constructs, while the dogmatic nature of these DSM/ICD classifications inhibits both inductive and deductive remodeling of the diagnostic criteria. Using supervised and unsupervised machine learning methods, we constructed new Research and Diagnostic Algorithmic Rules (RADAR) to score ACEs, ROI, lifetime (LT), current suicidal ideation and attempts, recurrence of illness (ROI), and the phenome of mood disorders. Based on these scores, we developed four new precision nomothetic models of mood disorders, namely, a first based on clinical indicators including ACE, ROI, and the phenome and three other models including immune-inflammatory, nitro-oxidative stress, and gut–autoimmune biomarkers. These four models revealed a new endophenotype class of mood disorders characterized by increased ACE, ROI, and a phenome scores and consequently was labeled as major dysmood disorder (MDMD), while the non-MDMD group was labeled simple dysmood disorder. MDMD was characterized by activated immune-inflammatory and growth hormone, nitro-oxidative stress, and gut–autoimmune pathways. ROI is linked to a higher family history of mood disorders and SUDs, as well as increased ACE and childhood sexual abuse, as well as ACE x PON1 gene interaction, immune-inflammatory and nitro-oxidative pathways, and specifically dysfunctions in the PON1–HDL cholesterol complex.

In this narrative review, we have provided evidence that the DSM/ICD diagnoses of MDD (or MDE), BD, BP1, and BP2 are not useful because (a) a common core underpins recurrence of depressive and manic episodes, indicating that they are manifestations of ROI; (b) patients with MDD, BP1, and BP2 may be either classified as MDMD or SDMD indicating that MDD/BD/BP1/BP2 do not reflect the overall severity of the clinical picture; (c) the binary diagnoses of MDD/BD/BP1/BP2 are irrelevant in our precision models because ROI and the phenome are more prominent indicators; and (d) there are no significant differences in model parameters concerning ROI and the phenome between unipolar MDD and BD as detected using MICOM and MGA-PLS. The diagnoses of MDD/MDE/BD/BP1/BP2 are other examples of the DSM/ICD and clinicians' obsession with transforming inaccurate man-made, top–down created, higher-order constructs based on descriptive narratives into incorrect gold standard diagnoses, which additionally are frequently mistakenly used as independent or moderator variables in statistical analyses. It follows that diagnoses such as MDD/BD/BP1/BP2 should be replaced with our RADAR scores and MDMD case criteria.

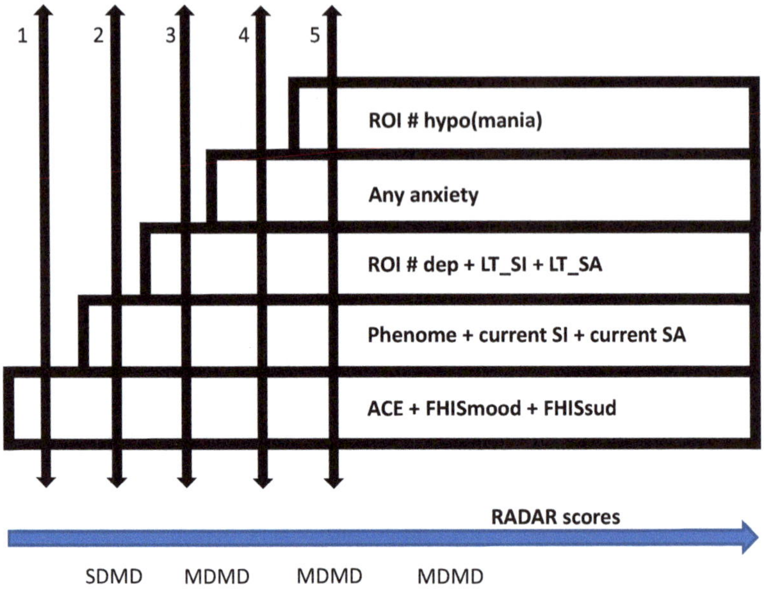

Fig. 7 Summarizing figure. RADAR: Research and Diagnostic Algorithmic Rules, SDMD/MDMD: simple and major dysmood disorder, ACE adverse childhood experiences (physical and emotional neglect and abuse), FHISmood/FHISsud: family history of depression, bipolar disorder or suicide, and substance use disorders, respectively, ROI: recurrence of illness, # dep number of depressive episodes, LT SI/SA: lifetime suicidal ideation and attempts, respectively, Any anxiety: any DSM-IV anxiety disorder, # (hypo)mania number of hypomania/mania episodes. (1) No mood disorder, (2) SDMD, (3) MDMD, (4) MDMD with an anxiety disorder, (5) MDMD with a bipolar phenotype

Figure 7 summarizes the main findings of the present study. Importantly, all mood disorder patients suffer from MDD or MDE, but only a fraction of them have a bipolar phenotype. Nonetheless, the DSM/ICD and the vast majority, if not all, of psychiatrists identify the latter individuals as BD and believe this group to be a distinct nosological entity from mood disorders without BD. This is perplexing, as the presence of depressive episodes is the predominant symptom of all individuals with mood disorders, while (hypo)-manic episodes superimpose on this depressive background. In actuality, it would be more accurate to identify people with mood disorders as depressive disorder patients, with or without a bipolar signature. The same is true for the DSM-IV anxiety disorders (OCD, PTSD, panic disorder, GAD, and phobias), which superimpose on a depressive background and are also more prevalent among BD patients. Therefore, it would be more accurate to identify people with mood disorders as depressive disorder patients with or without anxiety disorders and a bipolar signature. However, this would be more of a game than a scientific endeavor, as all of these diagnoses need to be replaced by RADAR scores and the diagnosis MDMD.

It follows that future studies should investigate which pathways underlie both phenotypes of mood disorders combined (MDD and BD), the ROI (including lifetime suicidal ideation and attempts), and phenome RADAR scores, as well as the onset of the first (hypo)manic episode.

Acknowledgments

We gratefully acknowledge the help of all psychiatry staff involved in the execution of this study.

Declarations

Ethics Approval and Consent to Participate

Approval for the study was obtained from the Institutional Review Board of the State University of Londrina, Brazil (protocol number: CAAE 34935814.2.0000.5231). All participants gave written informed consent prior to taking part in the study.

Availability of Data and Materials

The dataset (PLS models as HTM files) generated during and/or analyzed during the current study will be available from Prof. Dr. Michael Maes upon reasonable request and once the dataset has been fully exploited by the authors.

References

1. Stoyanov D, Maes MH (2021) How to construct neuroscience-informed psychiatric classification? Towards nomothetic networks psychiatry. World J Psychiatry 11(1):1–12. https://doi.org/10.5498/wjp.v11.i1.1. PMID: 33511042; PMCID: PMC7805251
2. Insel T, Cuthbert B, Garvey M, Heinssen R, Pine DS, Quinn K, Sanislow C, Wang P (2010) Research domain criteria (RDoC): toward a new classification framework for research on mental disorders. Am J Psychiatry 167(7): 748–751. https://doi.org/10.1176/appi.ajp.2010.09091379. PMID: 20595427
3. Maes M (2022) Research and Diagnostic Algorithmic Rules (RADAR) for mood disorders, recurrence of illness, suicidal behaviors, and the patient's lifetime trajectory. Acta Neuropsychiatr 35(2):104–117
4. Maes M (2022) Precision nomothetic medicine in depression research: a new depression model, and new endophenotype classes and pathway phenotypes, and a digital self. J Pers Med 12(3):403. https://doi.org/10.3390/jpm12030403. PMID: 35330403; PMCID: PMC8955533
5. Maes MH, Stoyanov D (2022) False dogmas in mood disorders research: towards a nomothetic network approach. World J Psychiatry 12(5):651–667. https://doi.org/10.5498/wjp.v12.i5.651. PMID: 35663296; PMCID: PMC9150032
6. American Psychiatric Association (2013) Diagnostic and statistical manual of mental disorders, Fifth edn. American Psychiatric Publishing, Arlington, pp 5–25. isbn:978-0-89042-555-8
7. World Health Organization (2004) ICD-10: international statistical classification of diseases and related health problems: tenth revision, 2nd edn
8. Regier DA, Kuhl EA, Kupfer DJ (2013) The DSM-5: classification and criteria changes. World Psychiatry 12(2):92–98. https://doi.org/10.1002/wps.20050. PMID: 23737408
9. Lieblich SM, Castle DJ, Pantelis C, Hopwood M, Young AH, Everall IP (2015) High heterogeneity and low reliability in the diagnosis of major depression will impair the

development of new drugs. Br J Psych Open 1(2):e5–e7. https://doi.org/10.1192/bjpo.bp.115.000786. PMID: 27703745
10. Ghouse AA, Sanches M, Zunta-Soares G, Swann AC, Soares JC (2013) Overdiagnosis of bipolar disorder: a critical analysis of the literature. Sci World J 2013:297087. https://doi.org/10.1155/2013/297087. PMID: 24348150; PMCID: PMC3856145
11. Lish JD, Dime-Meenan S, Whybrow PC, Price RA, Hirschfeld RM (1994) The National Depressive and Manic-depressive Association (DMDA) survey of bipolar members. J Affect Disord 31(4):281–294. https://doi.org/10.1016/0165-0327(94)90104-x. PMID: 7989643
12. Benazzi F (2005) The relationship of major depressive disorder to bipolar disorder: continuous or discontinuous? Curr Psychiatry Rep 7(6):462–470. https://doi.org/10.1007/s11920-005-0068-6. PMID: 16318825
13. Benazzi F (2007) Is there a continuity between bipolar and depressive disorders? Psychother Psychosom 76(2):70–76. https://doi.org/10.1159/000097965. PMID: 17230047
14. Akiskal HS (1996) The prevalent clinical spectrum of bipolar disorders: beyond DSM-IV. J Clin Psychopharmacol 16(2 Suppl 1):4S–14S. https://doi.org/10.1097/00004714-199604001-00002. PMID: 8707999
15. Simpson SG, McMahon FJ, McInnis MG, MacKinnon DF, Edwin D, Folstein SE, DePaulo JR (2002) Diagnostic reliability of bipolar II disorder. Arch Gen Psychiatry 59(8):736–740. https://doi.org/10.1001/archpsyc.59.8.736. PMID: 12150650
16. Berk M, Berk L, Moss K, Dodd S, Malhi GS (2006) Diagnosing bipolar disorder: how can we do it better? Med J Aust 184(9):459–462
17. Akiskal HS, Bourgeois ML, Angst J, Post R, Möller H, Hirschfeld R (2000) Re-evaluating the prevalence of and diagnostic composition within the broad clinical spectrum of bipolar disorders. J Affect Disord 59(Suppl 1):S5–S30. https://doi.org/10.1016/s0165-0327(00)00203-2. PMID: 11121824
18. Akiskal HS, Benazzi F, Perugi G, Rihmer Z (2005) Agitated "unipolar" depression re-conceptualized as a depressive mixed state: implications for the antidepressant-suicide controversy. J Affect Disord 85(3):245–258. https://doi.org/10.1016/j.jad.2004.12.004. PMID: 15780694
19. Angst J (2003) "Up and down". Bipolare Erkrankungen--unterschätzte Prävalenz und Komplexität "Up and down". Bipolar illnesses--underestimated prevalence and complexity. Krankenpfl J 41(10–12):224. German. PMID: 14746202
20. Angst J, Adolfsson R, Benazzi F, Gamma A, Hantouche E, Meyer TD, Skeppar P, Vieta E, Scott J (2005) The HCL-32: towards a self-assessment tool for hypomanic symptoms in outpatients. J Affect Disord 88(2):217–233. https://doi.org/10.1016/j.jad.2005.05.011. PMID: 16125784
21. Alciati A, Sarzi-Puttini P, Batticciotto A, Torta R, Gesuele F, Atzeni F, Angst J (2012) Overactive lifestyle in patients with fibromyalgia as a core feature of bipolar spectrum disorder. Clin Exp Rheumatol 30(6 Suppl 74):122–128. Epub 2012 Dec 14. PMID: 23261011
22. Phelps J, Angst J, Katzow J, Sadler J (2008) Validity and utility of bipolar spectrum models. Bipolar Disord 10(1 Pt 2):179–193. https://doi.org/10.1111/j.1399-5618.2007.00562.x. PMID: 18199236
23. van Os J, Kohne ACJ (2021) It is not enough to sing its praises: the very foundations of precision psychiatry may be scientifically unsound and require examination. Psychol Med 51(9):1415–1417. https://doi.org/10.1017/S0033291721000167. Epub 2021 Feb 9. PMID: 33557970; PMCID: PMC8311815
24. Köhne ACJ, van Os J (2021) Precision psychiatry: promise for the future or rehash of a fossilised foundation? Psychol Med 51(9):1409–1411. https://doi.org/10.1017/S0033291721000271. Epub 2021 Feb 22. PMID: 33612128
25. Summerfield D (2006) Depression: epidemic or pseudo-epidemic? J R Soc Med 99:161–162
26. Frances A (2013) Saving normal. Harper Collins, New York
27. Stoyanov D (2020) The reification of diagnosis in psychiatry. Neurotox Res 37:772–774. https://doi.org/10.1007/s12640-019-00139-2. PMID: 31811587
28. Hyman SE (2010) The diagnosis of mental disorders: the problem of reification. Annu Rev Clin Psychol 6:155–179. https://doi.org/10.1146/annurev.clinpsy.3.022806.091532. PMID: 17716032
29. Hyman SE (2011, 2011) Diagnosing the DSM: diagnostic classification needs fundamental reform. Cerebrum:6. PMID: 23447775. PMCID: PMC 3574782
30. Timimi A (2014) No more psychiatric labels: why formal psychiatric diagnostic systems should be abolished. Int J Clin Health Psychol 14(3):208–215. ISSN 1697-2600. https://doi.org/10.1016/j.ijchp.2014.03.004

31. Maes M, Moraes JB, Bonifacio KL, Barbosa DS, Vargas HO, Michelin AP, Nunes SOV (2021) Towards a new model and classification of mood disorders based on risk resilience, neuro-affective toxicity, staging, and phenome features using the nomothetic network psychiatry approach. Metab Brain Dis 36(3): 509–521. https://doi.org/10.1007/s11011-020-00656-6. Epub 2021 Jan 7. PMID: 33411213

32. Simeonova D, Stoyanov D, Leunis JC, Murdjeva M, Maes M (2021) Construction of a nitro-oxidative stress-driven, mechanistic model of mood disorders: a nomothetic network approach. Nitric Oxide 106:45–54. https://doi.org/10.1016/j.niox.2020.11.001. Epub 2020 Nov 10. PMID: 33186727

33. Maes M, Rachayon M, Jirakran K, Sodsai P, Klinchanhom S, Gałecki P, Sughondhabirom A, Basta-Kaim A (2022) The immune profile of major dysmood disorder: proof of concept and mechanism using the precision nomothetic psychiatry approach. Cell 11(7):1183. https://doi.org/10.3390/cells11071183. PMID: 35406747; PMCID: PMC8997660

34. Maes M, Congio A, Moraes JB, Bonifacio KL, Barbosa DS, Vargas HO, Morris G, Puri BK, Michelin AP, Nunes SOV (2018) Early life trauma predicts affective phenomenology and the effects are partly mediated by staging coupled with lowered lipid-associated antioxidant defences. Biomol Concepts 9(1):115–130. https://doi.org/10.1515/bmc-2018-0010. PMID: 30471214

35. Maes M, Moraes JB, Congio A, Bonifacio KL, Barbosa DS, Vargas HO, Michelin AP, Carvalho AF, Nunes SOV (2019) Development of a novel staging model for affective disorders using partial least squares bootstrapping: effects of lipid-associated antioxidant defenses and neuro-oxidative stress. Mol Neurobiol 56(9):6626–6644. https://doi.org/10.1007/s12035-019-1552-z. Epub 2019 Mar 25. PMID: 30911933

36. Maes M, Rachayon M, Jirakran K, Sodsai P, Klinchanhom S, Debnath M, Basta-Kaim A, Kubera M, Almulla AF, Sughondhabirom A (2022) Adverse childhood experiences predict the phenome of affective disorders and these effects are mediated by staging, neuroimmunotoxic and growth factor profiles. Cells 11(9): 1564. https://doi.org/10.3390/cells11091564. PMID: 35563878; PMCID: PMC9105661

37. Cuellar AK, Johnson SL, Winters R (2005) Distinctions between bipolar and unipolar depression. Clin Psychol Rev 25(3):307–339. https://doi.org/10.1016/j.cpr.2004.12.002. PMID: 15792852; PMCID: PMC2850601

38. Hamilton M (1959) The assessment of anxiety states by rating. Br J Med Psychol 32:50–55

39. Hamilton M (1960) A rating scale for depression. J Neurol Neurosurg Psychiatry 23:56–62

40. Skevington SM, Lotfy M, O'Connell KA, WHOQOL Group (2004) The World Health Organization's WHOQOL-BREF quality of life assessment: psychometric properties and results of the international field trial. A report from the WHOQOL group. Qual Life Res 13: 299–310

41. Fleck MP, Louzada S, Xavier M, Chachamovich E, Vieira G, Santos L, Pinzon V (2000) Application of the Portuguese version of the abbreviated instrument of quality life WHOQOL-BREF. Rev Saude Publica 34: 178–183

42. Sheehan DV, Harnett-Sheehan K, Raj BA (1996) The measurement of disability. Int Clin Psychopharmacol 11(Suppl 3):89–95

43. Popper KR (1962) Conjectures and refulation: the growth of scientific knowledge. Routledge, London, UK. ISBN: 9781135971441. https://doi.org/10.2307/2218271

44. Maes M, Ombelet W, De Jongh R, Kenis G, Bosmans E (2001) The inflammatory response following delivery is amplified in women who previously suffered from major depression, suggesting that major depression is accompanied by a sensitization of the inflammatory response system. J Affect Disord 63(1–3): 85–92. https://doi.org/10.1016/s0165-0327(00)00156-7. PMID: 11246084

45. Celik C, Erdem M, Cayci T, Ozdemir B, Ozgur Akgul E, Kurt YG, Yaman H, Isintas M, Ozgen F, Ozsahin A (2010) The association between serum levels of neopterin and number of depressive episodes of major depression. Prog Neuro-Psychopharmacol Biol Psychiatry 34(2):372–375. https://doi.org/10.1016/j.pnpbp.2010.01.002. Epub 2010 Jan 13. PMID: 20074610

46. Maes M, Mihaylova I, Kubera M, Ringel K (2012) Activation of cell-mediated immunity in depression: association with inflammation, melancholia, clinical staging and the fatigue and somatic symptom cluster of depression. Prog Neuro-Psychopharmacol Biol Psychiatry 36(1):169–175. https://doi.org/10.1016/j.pnpbp.2011.09.006. Epub 2011 Sep 16. PMID: 21945535

47. Hope S, Ueland T, Steen NE, Dieset I, Lorentzen S, Berg AO, Agartz I, Aukrust P, Andreassen OA (2012) Interleukin 1 receptor

antagonist and soluble tumor necrosis factor receptor 1 are associated with general severity and psychotic symptoms in schizophrenia and bipolar disorder. Schizophr Res 145(1–3): 36–42. https://doi.org/10.1016/j.schres.2012.12.023. Epub 2013 Feb 8. PMID: 23403415

48. Siwek M, Sowa-Kućma M, Styczeń K, Misztak P, Nowak RJ, Szewczyk B, Dudek D, Rybakowski JK, Nowak G, Maes M (2017) Associations of serum cytokine receptor levels with melancholia, staging of illness, depressive and manic phases, and severity of depression in bipolar disorder. Mol Neurobiol 54(8): 5883–5893. https://doi.org/10.1007/s12035-016-0124-8. Epub 2016 Sep 23. PMID: 27660275
49. Sowa-Kućma M, Styczeń K, Siwek M, Misztak P, Nowak RJ, Dudek D, Rybakowski JK, Nowak G, Maes M (2018) Are there differences in lipid peroxidation and immune biomarkers between major depression and bipolar disorder: effects of melancholia, atypical depression, severity of illness, episode number, suicidal ideation and prior suicide attempts. Prog Neuro-Psychopharmacol Biol Psychiatry 81:372–383. https://doi.org/10.1016/j.pnpbp.2017.08.024. Epub 2017 Sep 1. PMID: 28867391
50. Maes M, Nani JV, Noto C, Rizzo L, Hayashi MAF, Brietzke E (2021b) Impairments in peripheral blood T effector and T regulatory lymphocytes in bipolar disorder are associated with staging of illness and anti-cytomegalovirus IgG levels. Mol Neurobiol 58(1):229–242. https://doi.org/10.1007/s12035-020-02110-1. Epub 2020 Sep 11. PMID: 32914395
51. SmartPLS (2022) Measurement Invariance Assessment (MICOM). As accessed 11-10-2022. Measurement Invariance Assessment (MICOM) – SmartPLS
52. Henseler J, Ringle CM, Sarstedt M (2016) Testing measurement invariance of composites using partial least squares. Int Mark Rev 33(3): 405–431
53. Cheah J-H, Thurasamy R, Memon MA, Chuah F, Ting H (2020) Multigroup analysis using SmartPLS: step-by-step guidelines for business research. Asian J Bus Res 10(3):I–XIX. ISSN 2463-4522 e-ISSN 1778-8933. https://doi.org/10.14707/ajbr.200087
54. Sarsyedt M, Henseler J, Ringle CM (2011) Multi-group analysis in partial least squares (PLS) path modeling: alternative methods and empirical results. Adv Int Mark 22:195–218. ISSN: 1474-7979. https://doi.org/10.1108/S1474-7979(2011)0000022012

Part III

Neurophysiological Methods

Chapter 4

The Concept of Event-Related Oscillations: A Spotlight on Extended Applications

Vasil Kolev, Roumen Kirov, and Juliana Yordanova

Abstract

Event-related neuroelectric oscillations have provided important tools for exploring information processing in the brain. The concept of event-related oscillations (EROs) is linked to that of event-related potentials (ERPs). Both the ERPs and EROs are derived from electroencephalographic (EEG) recordings following the appearance of an event. There are, however, several essential advantages of the ERO approach. These refer to the ability (1) to analyze a variety of characteristics of neuroelectric responses reflecting their magnitude, frequency and phase; (2) to separate functionally specific but simultaneous mechanisms of information processing; and (3) to apply a physiological approach assuming a close relationship between the ongoing brain state and the mode of incoming information processing. Also, established methods, analytic tools, and parameters for assessment of EROs are outlined. The major focus of the chapter is on some less well recognized extended applications of the concept of EROs in neurocognitive research. Specifically, applications to (1) internal information processing, (2) event-related frequency tuning, (3) event-related spatial synchronization, and (4) detection of multi-second behavioral patterns are described.

Key words Event-related oscillations, Event-related potentials, EEG, Internal information processing, Frequency tuning, Synchronization, Multi-second oscillations, Time–frequency analysis

Abbreviations

ADHD	Attention-deficit/hyperactivity disorder
CSD	Current source density
CWT	Continuous wavelet transform
DMN	Default mode network
EEG	Electroencephalography
ERO	Event-related oscillation
ERP	Event-related potential
ERSS	Event-related spatial synchronization
FFT	Fast Fourier transform
Nc	Correct response negativity

Roumen Kirov (deceased)

Drozdstoy Stoyanov et al. (eds.), *Computational Neuroscience*, Neuromethods, vol. 199,
https://doi.org/10.1007/978-1-0716-3230-7_4,

Ne	Error response negativity
PLF	Phase-locking factor
PLI	Phase-lag index
PLV	Phase-locking value
RRP	Response-related potential
SW	Sleep slow wave
TF	Time frequency
TOTP	Total power

1 The Concept of Event-Related Oscillations

Shadows of the concept of event-related brain oscillations (EROs) can be noticed in scientific works published more than 40 years ago. The idea of EROs was inspired and elaborated by Erol Başar [1–3]. It was essentially grounded on the argument that the principles of general systems theory [4] could be applied to biological systems, including the neural system and the brain. In particular, the oscillation principle was emphasized, generalized, and proposed as a fundamental principle of all natural systems (physical, biological, physiological, social). These ideas have fostered the progress and applications of the ERO concept and methodology. It is notable that brain oscillations and synchronizations, oscillatory brain systems, and neural networks have become a major focus of research in neuroscience much later, with this hot topic getting an increasing relevance. Currently, event-related neuroelectric oscillations provide important tools for exploring information processing in the brain.

In standard practical applications, the EROs are closely linked to event-related potentials (ERPs). Both the ERPs and EROs are derived from electroencephalographic (EEG) recordings following the appearance of an event. There are, however, several essential advantages of the ERO approach: (1) analysis of a variety of characteristics of EEG responses including magnitude, time, frequency, and phase; (2) analysis of parallel processes in the brain; and (3) evaluation of the relationship between the ongoing brain state and the mode of incoming information processing. In this chapter, the conceptual background of EROs will be briefly presented. Also, established methods, analytic tools, and parameters for assessment of EROs will be outlined. The main advantages of the EROs in standard applications linked to refined assessment of external information processing captured by ERPs also will be exemplified. Our major aim, however, is to focus on some less-recognized extended applications of the concept of EROs in neurocognitive research. These extended applications are motivated by some of the general principles in the original theory of E. Başar and are also associated with relevant methodological developments. Specifically, applications to (1) internal information processing, (2) event-related frequency tuning, (3) event-related connectivity, and (4) detection of multi-second behavioral patterns will be described.

The amount of research of brain oscillations in the last decades is impressive. Research activities encompass many different aspects of oscillatory brain phenomena. They are performed with different techniques and refer to a variety of theories and contexts. It is not the goal of the chapter to review the vast amount of existing literature on brain oscillations. Because the ERO concept provides a unitary frame for brain electrical activity (spontaneous and event-related), we select here for demonstration some of our previous works as they are essentially based on the ERO concept, which would help highlight the major structuring standpoints and potential broad appliances of this theory.

1.1 Conceptual Framework

1.1.1 Event-Related Potentials

The electroencephalogram is a time-varying signal. It reflects the summated neuroelectric activity from various neural sources [5]. An EEG response that occurs in association with an eliciting event (sensory or cognitive stimulus) is defined as an event-related potential [6, 7]. The ERP may contain EEG activity not related to specific event processing, as well as electric activity from non-neural sources. To extract the EEG activity elicited by a particular event, a specific experimental protocol with repeated stimulus presentation and an averaging procedure are used. The averaging is applied with the assumption that the event triggers brain processes that are newly generated and reflected by stimulus-locked EEG responses. Within this assumption, the event-locked EEG is invariant, while the electric activity not related to the event is random. Hence, the averaging emphasizes the stimulus-related EEG signal that is time-locked to the moment of stimulus delivery and attenuates the random EEG components [5, 8]. The classical view, known as the additive model for ERPs, interprets brain responses to external stimuli as newly generated and therefore added to the ongoing (background) EEG activity.

The averaged ERP consists of consecutive positive and negative deflections called ERP components. Therefore, the averaged ERP is typically analyzed in the time domain. Time-domain ERP components are mainly characterized by their polarity, peak latency, distribution over the scalp, and specific sensitivity to experimental variables [6, 7, 9]. ERP components have been related to a variety of sensory and cognitive processes. They have been widely used in cognitive neuroscience as providing objective signatures of neurophysiologic mechanisms of information processing in different psychological and clinical settings [7, 9, 10].

1.1.2 Event-Related EEG Oscillations

ERP components are computed from single-trial EEG traces. Yet, oscillatory electric phenomena and frequency-specific rhythms characterize the EEG (rev. [7]). Based on the existence of EEG oscillations as well as on the principles of the systems theory, the ERPs have been proposed to originate from the spontaneous EEG rhythms rather than from newly generated time-locked EEG signals triggered by an event [3, 11–14]. It has been assumed that distributed oscillatory networks generate frequency-specific rhythms. Signals from such unitary frequency-specific systems have been linked to both the spontaneous and stimulus-related EEG activities. In spontaneous or rest states, the components of a distributed frequency-specific system may be uncoupled or coupled, thus producing resting-state rhythms. Upon exposition to external or internal stimulation, the activity of the resting-state network is reorganized (phase-reset) giving rise to enhanced and phase-locked EEG oscillations. Importantly, only task-relevant networks are responding, which results in synchronized oscillatory responses from specific (task-relevant) frequency bands. On these grounds, the specific class of oscillatory EEG phenomena termed event-related oscillations has been introduced. According to the ERO model, ERP components are regarded as reflecting a superposition of oscillatory EEG responses in various frequency ranges. Also, frequency-specific EROs are considered as originating from the reorganization of the ongoing activity of distributed networks. The most important indices of event-related reorganization of the ongoing EEG in a given frequency channel are:

1. Power or amplitude changes (increase or decrease) in the post-stimulus period relative to pre-stimulus EEG.
2. Phase-reordering and phase locking in relation to a stimulus, i.e., temporal phase locking.
3. Spatial synchronization in relation to a stimulus.

The EROs can be extracted from the ERP and analyzed using appropriate analytic tools and procedures (e.g., [15]).

1.2 Advantages

The important consequences of the ERO concept are that the EROs (1) allow to characterize EEG responses in several dimensions of the signal, which refines functional correlates; (2) allow to study brain processes that are run in parallel, in contrast to ERPs; and (3) provide a physiological approach to understand brain responses by emphasizing the dependence of the mode of processing on the background brain state.

1.2.1 A Full Characterization of Event-Related EEG Signals

An EEG signal can be described by (1) amplitude, (2) time, (3) frequency, and (4) phase relationships, thus promoting a complete signal description.

Typically, ERPs are analyzed in the time domain. This analysis has revealed the presence of both early (e.g., P1, N1) and late (e.g., P300, N400) ERP components with proven functional relevance (e.g., rev. [6, 7]). A classical time-domain representation of ERPs is

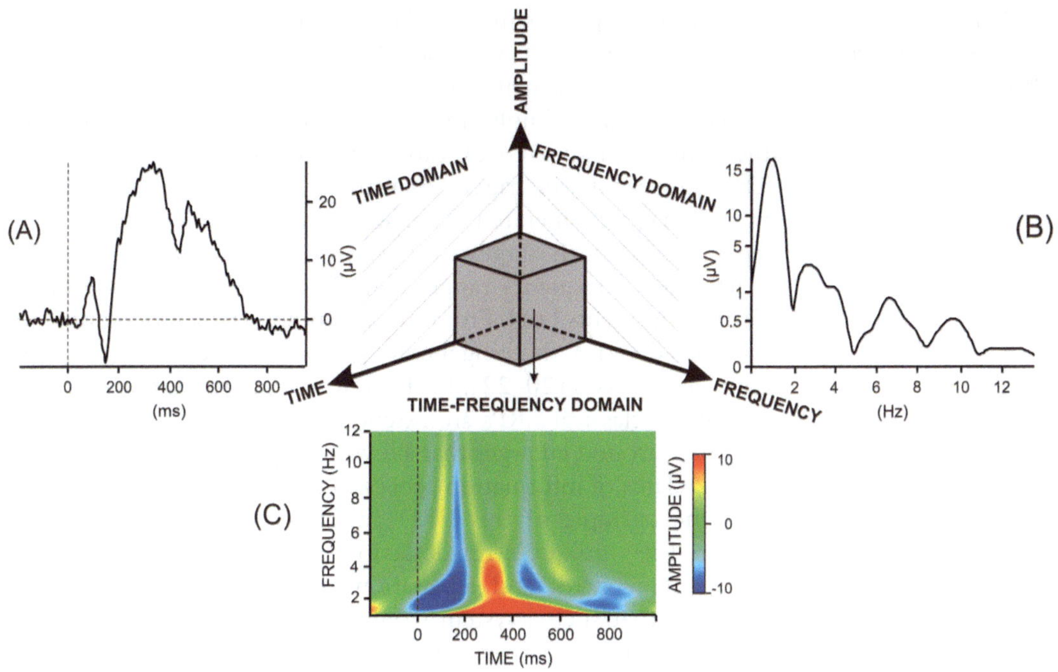

Fig. 1 Analysis dimensions of event-related oscillations. Averaged ERP presented in time, frequency, and time–frequency domains. (**a**) In the time-domain presentation, amplitude vs. time information is present, but no information exists about frequencies. (**b**) In the frequency-domain presentation, amplitude vs. frequency information is present, but no information exists about the timing of events. (**c**) In the time–frequency domain, events can be characterized by both time and frequency (with modifications from Yordanova and Kolev [24])

essential for analysis of the timing of the underlying neural events. As illustrated in Fig. 1a, the peak latencies of ERP components can be precisely determined. However, the frequency characteristics of these time-domain events remain obscure, and no information can be obtained about rhythmic or oscillatory events from various frequency bands present in the time-domain potential. Figure 1a shows that in the frequency domain, the same ERP is characterized by peaks from several frequency ranges—sub-delta (below 2 Hz), delta (2–4 Hz), theta (5–8 Hz), and alpha (around 10 Hz). The inability of time-domain ERPs to extract frequency characteristics of the signal seems to be a disadvantage because EEG activities from several frequency bands (theta, alpha, beta, gamma) have been associated consistently with sensory, cognitive, and motor processes (e.g., [16–20]). On the other hand, analysis only in the frequency domain does not reveal how frequency components vary over time and whether they are temporally linked to event processing. This can be achieved by time–frequency decomposition of ERPs, which provides information about time, frequency, and magnitude of the signal (Fig. 1c). The time–frequency decomposition further allows the evaluation of phase relationships of frequency-specific oscillations.

1.2.2 Evaluation of Parallel Processes in the Brain

Oscillatory responses from different frequency ranges can be generated simultaneously, with each frequency-specific response manifesting specific reactivity to task variables [13, 14]. Depending on the conditions, a stimulus may require different types of processing mechanisms supported by different frequency-specific networks. Thus, an event can trigger oscillations from different frequency ranges that are generated simultaneously and are temporally overlapped. It has been suggested that co-existent oscillatory potentials from various frequency ranges contribute to the formation of the ERP components [14]. From this viewpoint, the ERP components are considered as heterogeneous phenomena reflecting multiple parallel processes [20–22]. In this way, the interpretation and analysis of the classical ERPs are extended. In the ERO perspective, a new background emerges that allows to explore simultaneous neural correlates of information processing in the brain.

One example of this ERO advantage is a study which has explored if a unique error-detection neural system is involved in behavioral control and adaptation [20]. In that study, it was expected that such a system would generate error-specific signals after performance errors which would not appear after correct responses. Experimental data had shown, however, that correct and incorrect reactions elicited similar neuroelectric potentials in the time domain as reflected by the ERP component's error negativity (Ne) and correct response negativity (Nc). It was then hypothesized that such an error-specific signal might be frequency-specific and not obvious in the Ne as being overlapped by simultaneously generated oscillations. To test this hypothesis, correct and error negativities were elicited in a four-choice sensorimotor task performed with the fingers of the right and left hands in two modalities—auditory and visual. The time–frequency analysis of error negativity extracted two co-existent time–frequency components (Fig. 2). Independently of hand and modality, a unique error-specific sub-component from the delta (1.5–3.5 Hz) frequency band was found only after errors. It was associated with error detection at the level of overall performance monitoring. A second sub-component from the theta frequency band (4–8 Hz) was extracted after both correct and error responses (Fig. 2) in relation to motor response execution. As the theta sub-component also differed between error and correct reactions, it was proposed to indicate error detection at the level of movement monitoring. Time–frequency analysis thus helped implicate the existence of multiple functional systems operating in parallel at different levels of behavioral control.

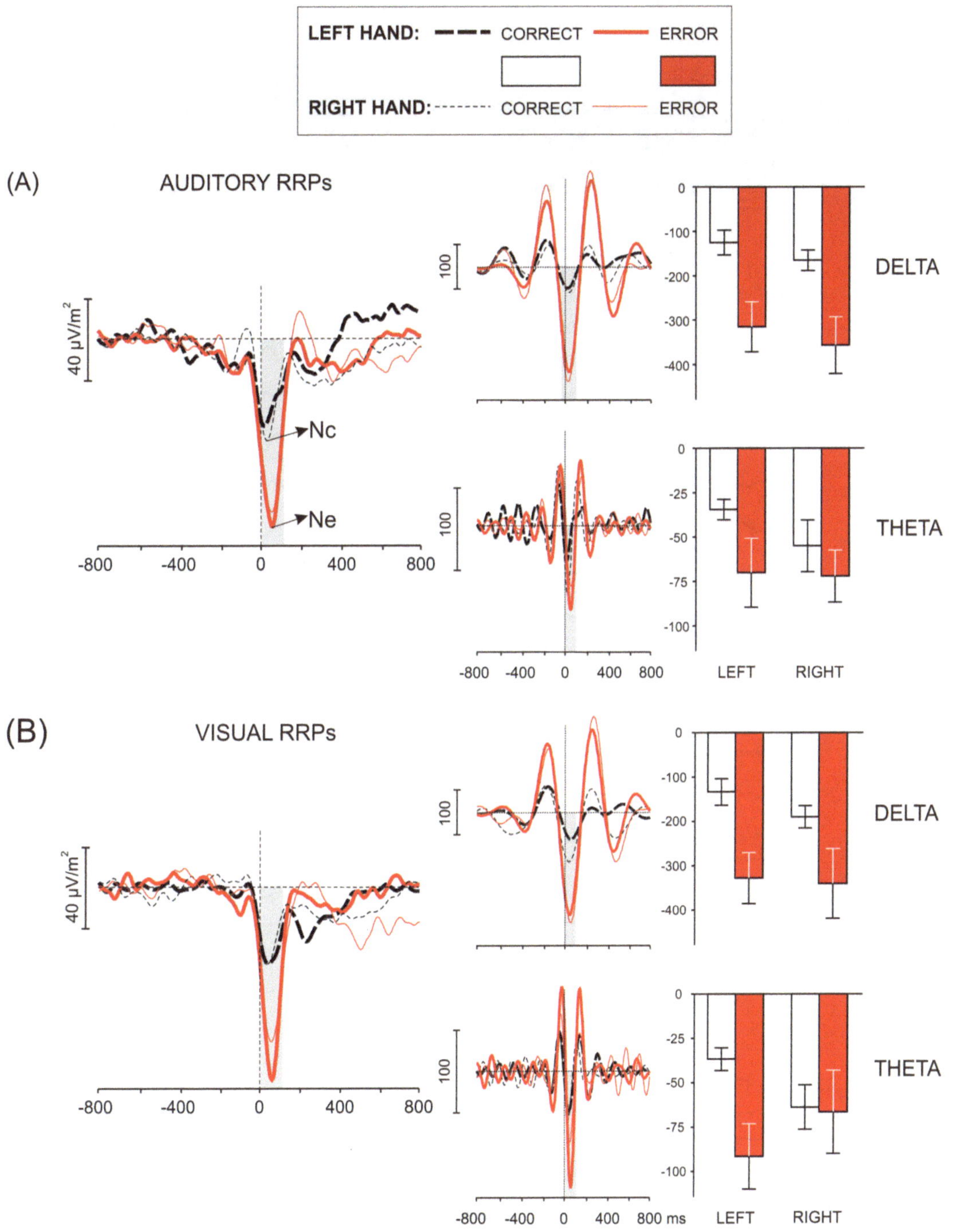

Fig. 2 Example for parallel processes in the brain. *Left panel:* Grand average time-domain response-related potentials at FCz elicited by correct (Nc) and error responses (Ne) with the left and right hand in the auditory (**a**) and visual (**b**) modality. Positivity is upward. *Middle panel:* Grand averaged time–frequency components of RRPs at FCz extracted from the delta (1.5–3.5 Hz) and theta (4–8 Hz) frequency ranges for correct and error responses produced by the left and right hands in the two modalities. *Right panel:* Corresponding to the curves of different TF components (presented at the middle panel), group means (±SE) are shown to illustrate statistical results (with modifications from Yordanova et al. [20])

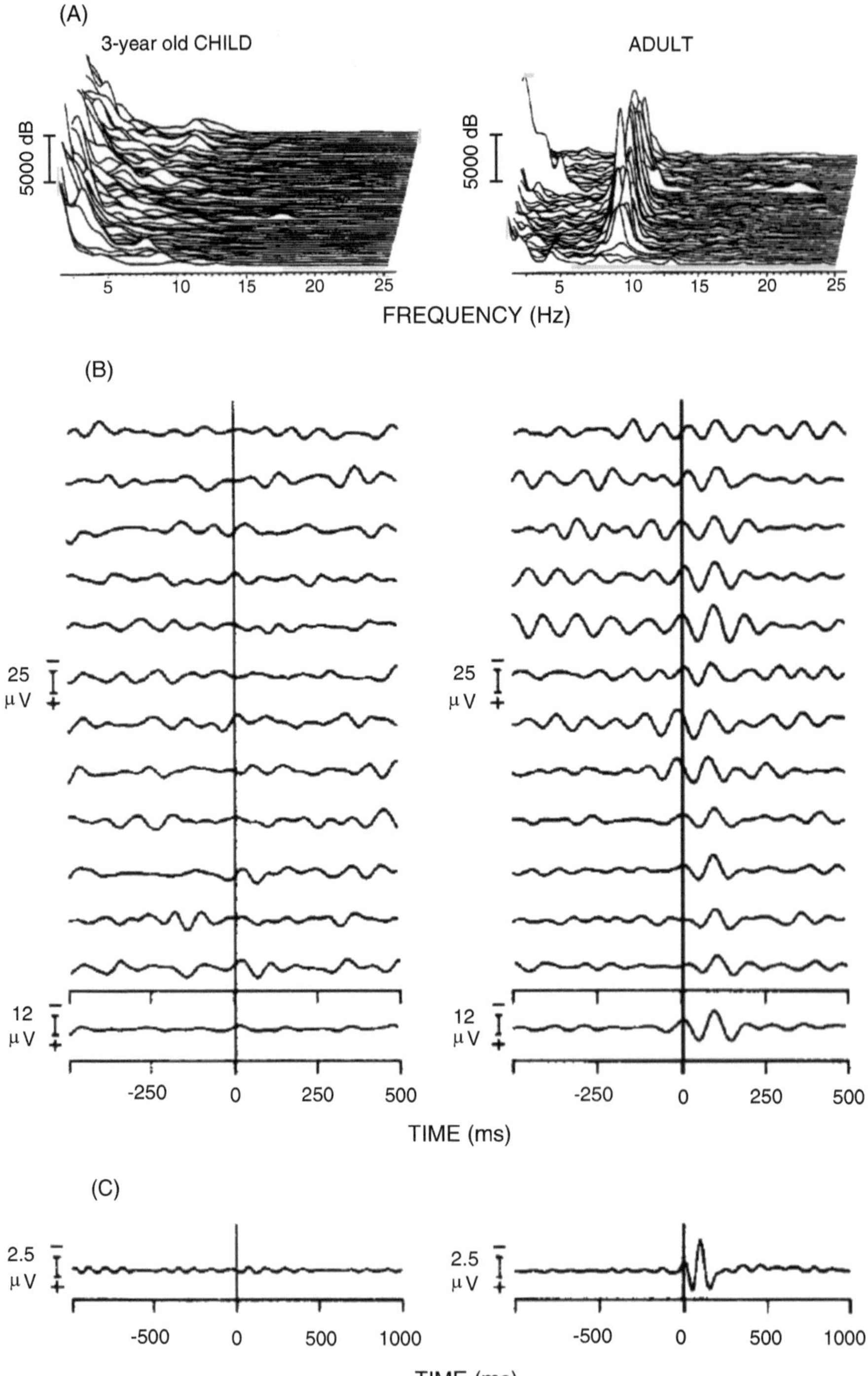

Fig. 3 Illustration of the susceptibility rule. (**a**) Instantaneous power spectra of EEG epochs recorded in one representative 3-year-old child and one representative adult subject. (**b**) Single-sweep EEG responses to auditory stimuli pass-band filtered in the range of 8–15 Hz for the same subjects. (**c**) The corresponding averaged potentials calculated from the presented single sweeps (8–15 Hz). Stimulus onset at 0 ms. Note the differences in the spectral characteristics in adults and children shown in (**a**), as well as the lack of alpha response in the 3-year-old child observed in (**c**) which corresponds to the lack of expressed alpha activity in its spontaneous EEG shown in (**a**) (with modifications from Basar-Eroglu et al. [28])

1.2.3 A Physiological Approach to the Event-Related EEG Activity

Within the ERO concept, the so-called susceptibility rule has been postulated [23]. According to this rule, a neural system has the ability to generate stimulus-triggered oscillatory responses in a defined frequency range only if this system has the ability to generate spontaneous oscillations in the same frequency range. The spontaneous oscillations reflect the neurobiological status of an oscillatory system (e.g., developmental, aging, pathology, state- or trait-related), whereas EROs reflect the competence for function-specific activation of this system [18, 19, 24–26]. This point of view is a physiologically oriented basis for the interpretations of interactions between internal brain states and active mechanisms of information processing in the brain.

The susceptibility rule is an important theoretical principle of the ERO concept. It was tested in our early studies [27, 28] where auditory and visual ERPs were recorded along with the spontaneous EEG in 3-year-old children and 20–22-year-old adults. The spectral characteristics of spontaneously recorded EEGs, and frequency ERP components were analyzed. As demonstrated in Fig. 3, the findings showed that (1) the main spectral components of the spontaneous EEG were in the alpha band for adults and in the slower theta and delta ranges for children (Fig. 3a); (2) frequency ERP components in 3-year-old children were "delta" and "theta responses" with delayed time courses in comparison to adult's ERPs; and (3) in response to external stimulations, adults generated phase-locked EEG alpha responses, in contrast to children (Fig. 3b, c). It was concluded that the evoked frequency components depended on the frequency content of the spontaneous EEG rhythms, providing evidence for the close associations between ongoing and event-related oscillations.

2 Methodology

Event-related oscillations contain information about timing, frequency, and time–frequency characteristics of EEG responses. Accordingly, relevant methodologies include methods which decompose the EEG responses in the time–frequency domain. According to the ERO concept, it is critical to evaluate the degree of phase alignment triggered by an event. In this regard, methods for evaluation of phase locking also are essential for analysis of EROs. In the last years, a remarkable progress occurred in the development and advancement of mathematical tools for analysis of brain oscillations and oscillatory neural networks. In the following, we briefly mention only some fundamental and best-established approaches and do not detail all analytic advances in this direction.

2.1 Analysis in the Frequency Domain

Frequency-domain analysis is performed by decomposing ERPs in the frequency domain by means of fast Fourier transform (FFT). FFT is chosen with respect to the timing of ERPs that is typically in the range of 1–2 s. FFT can be computed for both average and single-sweep ERPs. To focus the frequency-domain computation to the period of relevant information processing, it is important to choose precisely the analysis epoch and to compute FFT after applying a procedure reducing activity in the beginning and in the end of the analysis epoch and avoiding spurious edge effects (e.g., Hanning windowing). Frequency-domain analysis extracts the spectral components of event-related EEG responses and helps identify major frequency components tentatively associated with information processing. A detailed presentation of the mathematical procedures for frequency analysis of EEG can be found in a variety of relevant references (e.g., [3, 7, 12], etc.).

2.2 Analysis in the Time–Frequency Domain

Time–frequency (TF) analysis of EEG responses also can be performed for averaged or single-sweep waveforms. One established approach for TF decomposition of ERPs is the continuous wavelet transform (CWT, [29]) with Morlet wavelets as basis functions (e.g., [30]).

Complex Morlet wavelets $W(t,f)$ can be generated in the time domain for different frequencies, f, according to the equation:

$$W(t,f) = A\exp(-t^2/2\sigma_t^2)\exp(2i\pi f t),$$

where t is time, $A = (\sigma_t\sqrt{\pi})^{-1/2}$, σ_t is the wavelet duration, and $i = \sqrt{-1}$. For each specific analysis, wavelet family was characterized by a f_0/σ_f ratio, where f_0 is the central frequency and σ_f is the width of the Gaussian shape in the frequency domain. The choice of the ratio f_0/σ_f is oriented to expected presence of slower or faster phase-locked components in the ERPs. For faster frequency components (e.g., beta, gamma), the ratio f_0/σ_f should be greater than 5. In case of expected slower phase-locked components, although sub-optimal, the ratio f_0/σ_f can be smaller than 5, but this affects the shape of the Morlet wavelet and decreases its decay. For different f, time and frequency resolutions can be calculated as $2\sigma_t$ and $2\sigma_f$, respectively [30], which are related by the equation $\sigma_t = 1/(2\pi\sigma_f)$. It is to be noted that there are different ways of time–frequency decomposition based on both digital filtering of EEG responses (e.g., [12]) and various variants of the WT analysis (e.g., [31–33]) that have been successfully used for EROs analysis.

2.2.1 Phase-Locked Power

The phase-locked power of EROs in different frequency bands is computed as the power frequency-specific components of the averaged ERPs. It reflects the magnitude of only those TF components which are emphasized by the averaging and are present in the averaged ERP.

The time-varying energy $|E(t,f)|$ of the signal in a frequency band is the square norm of the result of the convolution (complex multiplication) of the complex wavelet $W(t, f)$ with the signal $S(t)$: $E(t,f) = |W(t,f) \times S(t)|^2$. Convolution of the signal by a family of wavelets provides a time–frequency representation of the signal.

2.2.2 Total Power

Total power (TOTP) comprises the phase-locked and non-phase-locked fractions of the ERPs. It is computed from TF decomposed single EEG trials related to an event and represents the total energy of EEG responses in specific frequency ranges. The usual procedure is to perform TF decomposition for each single trial. Afterward, for each trial, the time-varying power in relevant frequency bands (delta, theta, alpha, beta, gamma, etc.) is calculated by squaring the absolute value of the convolution of the signal with the complex wavelet.

When computing phase-locked or total power, the mean TF energy of the pre-stimulus (reference) period is typically considered as a baseline level and is subtracted from the post-stimulus energy for each frequency band. Relevant TF components can be extracted afterward according to the central frequency f_0 in the time domain and statistically analyzed.

2.2.3 Temporal Phase Locking

The temporal phase locking refers to the phase synchronization of EROs across trials after event appearance. To assess the temporal phase locking, a sequence of single EEG sweeps related to event occurrence is needed in order to analyze phase coherence across sweeps in the whole set of experimentally recorded EEG responses (between-sweep synchronization). A pronounced temporal phase synchrony is an index for phase reordering or phase alignment induced by the event. Hence, the temporal phase locking is an important marker for the reorganization of ongoing EEG oscillations providing support to the concept of EROs. The temporal phase locking is analyzed at single electrodes. Cortical regions with maximal temporal phase synchronization also can be identified.

The temporal phase locking has gained increased relevance after establishing that the ability of oscillatory systems to get synchronized in relation to event processing may reveal unique neural correlates of information processing mechanisms in normal and pathological conditions that are independent of signal power (e.g., [18, 19, 24, 34–36], etc.).

The methods for computing temporal phase locking have been refined after brain oscillations became the focus of neurocognitive research. Currently, the phase-locking factor (PLF) is the best-established quantifier. Here, we describe first a methodology proposed by our group for temporal phase synchrony analysis in earlier studies (the single-sweep wave identification method, SSWI). Second, we present the PLF. Although the PLF is currently widely being used, we demonstrate the SSWI because of the potential of this approach to be preferentially used for some extended applications of EROs. The employment of SSWI can be particularly beneficial for analysis of discrete time series produced by a variety psychophysiological, performance, or behavioral parameters.

Single-Sweep Wave Identification Method

For a quantitative evaluation of the phase locking of single sweeps, a simple method called the single-sweep wave identification (SSWI) method has been developed [37]. Figure 4a illustrates schematically the analysis procedure, which includes the following steps. First, all extrema (minima and maxima) are identified in the filtered (in this example 4–7 Hz) single EEG sweeps. To eliminate amplitude effects, maxima are replaced with +1, minima with −1, and modified single sweeps are stored. Critically, these modified single sweeps fix the detected time positions of minima and maxima along

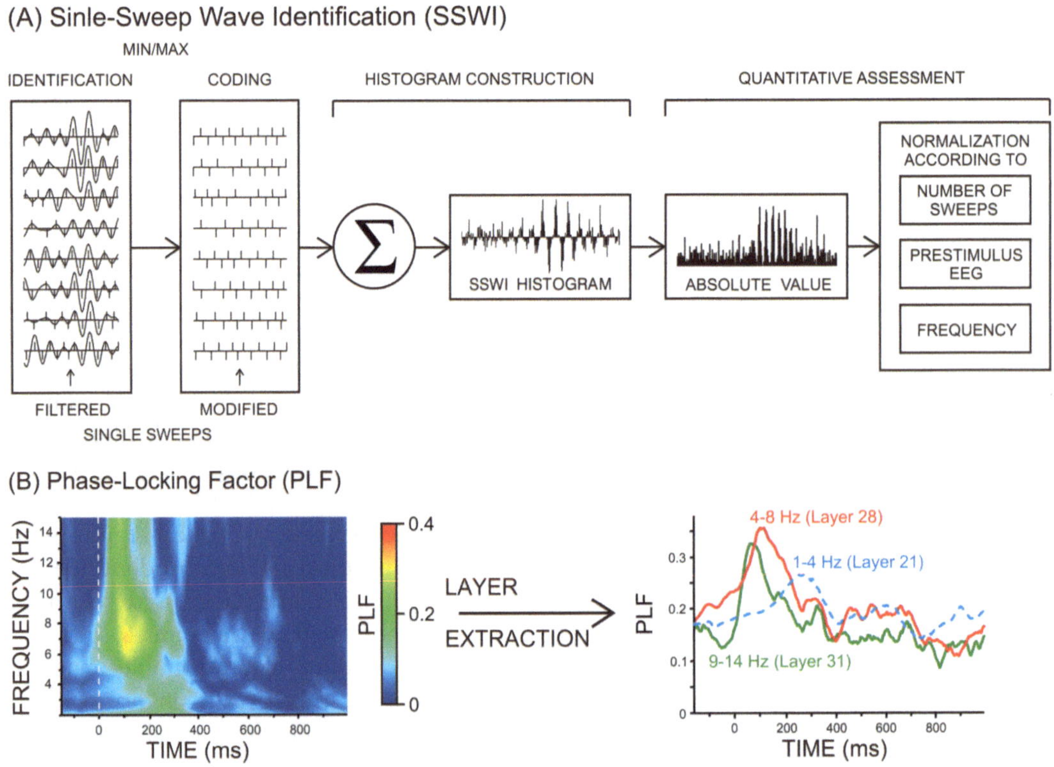

Fig. 4 Methods for evaluation of temporal phase locking. (**a**) A flowchart of the procedure for single-sweep wave identification (SSWI). Extremes in the filtered single sweeps are identified, and corresponding single sweeps are modified by using the values of +1 and − 1 only. Modified single sweeps are averaged so that a SSWI histogram is obtained. For measurements, the SSWI histogram is rectified (absolute values) and normalized according to the number of sweeps, pre-stimulus EEG, and frequency of the waves. (**b**) Phase-locking factor (PLF) is calculated after continuous wavelet transform (see text), a time–frequency plot of PLF is presented. Time courses of PLF for different frequency bands (layers) are shown on the right (with modifications from Yordanova and Kolev [24])

the time axis without the signals. Second, after summing the modified single sweeps (coded extrema) across trials, a histogram of the number of phase-locked single waves (SSWI histogram) is constructed.

For quantitative evaluation of phase locking, the SSWI histogram is rectified and normalized by dividing the bar values by the number of single sweeps included. To control for effects related to possible changes in frequency, the number of oscillations in each time window can be measured, and SSWI histograms can be normalized by the obtained value [19].

Phase-Locking Factor

The phase synchronization across trials is typically measured by means of the phase-locking factor (PLF, e.g., [30, 38]). The PLF also provides a measure of the between-sweep synchronization of oscillatory activity relatively independently of the signal's amplitude (Fig. 4b). The values of PLF yield a number between 0 and 1 reflecting the degree of between-sweep phase locking, where 1 indicates perfect phase alignment across trials, and values close to 0 reflect the highest phase variability. PLF is computed for different TF components after TF decomposition of single sweeps using the CWT (Fig. 4b).

The calculation procedure for PLF is as follows. The normalized complex time-varying energy P_i of each single trial i,

$$P_i(t,f) = \{W(t,f) \times S_i(t,f)\} / \mid W(t,f) \times S_i(f) \mid$$

is averaged across single trials, leading to a complex value describing the phase distribution in the TF region centered on t and f (*see* Subheading 2.2 for designations). The modulus of this complex value, ranging from 0 (non-phase-locked activity) to 1 (strictly phase-locked activity), is called phase-locking factor. To test whether an activity is significantly phase-locked to the stimulus, the Rayleigh statistical test of uniformity of angle can be used [30, 39].

2.2.4 Event-Related Spatial Synchronization: Spatial Phase Locking

The event-related spatial synchronization (ERSS) refers to the phase synchronization of EROs across trials and electrodes after event appearance. The ERSS is used to reflect the strength of phase coupling of oscillatory activity between distinct cortical regions. It is assumed that strongly synchronized regions are associated with a joint involvement of these regions in information processing and may represent components of a distributed oscillatory network with functional specificity. It is also interpreted as a marker of functional connectivity.

Phase-Locking Value

An acknowledged quantifier of ERSS is the phase-locking value (PLV). Following methodological recommendations [40], the PLV can be used because it is robust to time dynamics, time lag, frequency mismatches, frequency non-stationarities, and increased variance in phase stability. It is strongly recommended to compute PLV after spatially enhancing the EEG signals by calculation of, e.g., current source density (CSD, [41]).

PLV measures the extent to which oscillation phase angle differences between electrodes are consistent over trials at each time–frequency point (e.g., [42]). As a measure of spatial synchronization, PLV can be calculated for different TF scales at each time point t and trial j according to the equation:

$$\mathrm{PLV}_{k,l} = \left| \frac{1}{N} \sum e^{i(\rho_{j,k}(t,f_0) - \rho_{j,i}(t,f_0))} \right|,$$

where N is the number of single sweeps, k and l are indices for the pair of electrodes to be compared, and ρ is the instantaneous phase of the signal. $\mathrm{PLV}_{k,l}$ results in real values between one (constant phase difference) and zero (random phase difference).

Phase-Lag Index

PLV is not protected against volume conduction or against simultaneous independent activations at multiple regions during external stimulation. This is in contrast to the phase-lag index (PLI), another quantifier of phase-based spatial connectivity [43], which reflects the consistency of inter-regional phase relationships, but may be less sensitive to time dynamics, frequency non-stationarity, and increased variance, due to the elimination of simultaneous phase coupling. Therefore, to increase confidence in PLV results, PLI can be used for control analyses or as an independent parameter for functional connectivity.

The phase-lag index is calculated according to the eq. [43]:

$$\mathrm{PLI} = |\langle \mathrm{sign}[\Delta\rho(t_k)] \rangle|,$$

where PLI denotes the phase-lag index obtained from time series of phase differences $\Delta\rho(t_k)$, $k = 1 \ldots N$. PLI ranges between 0 (no coupling or coupling a with phase difference centered around 0 mod π) and 1 (perfect phase locking at $\Delta\rho$ different from 0 mod π) and is in fact an index of the asymmetry of the phase difference distribution [40].

3 Extended Applications

Since the 1990s, the concepts and methods of event-related oscillations have been intensively applied to study the neurophysiologic mechanisms of external information processing in the brain. This enhanced interest is demonstrated by the great variety of important publications found in the data bases, special issues, reviews (e.g., [14]), and books. As mentioned before, the major focus here is to present nontrivial extended applications of EROs. In the following, we are going to describe some of our previous studies, in which the advantages of the EROs were specifically applied to settings different from the traditional analyses of ERPs and external information processing.

3.1 Internal Information Processing

In the original formulations of Başar, oscillatory potentials associated with external or internal events are called *frequency EEG responses* or *event-related oscillations* [3, 12]. This definition introduces internal events that also should have the ability to trigger a reorganization of ongoing EEG, similar to external events (usually sensory stimuli). However, investigation of internal information processing by means of ERPs has remained less well studied because of one important constrain: While the onset of external events can be precisely fixed in order to define the trigger required for assessing temporal and spatial event-related synchronization as well as event-triggered power modulations of EEG responses, it is very difficult to define the onset of internal events. Indeed, the onset of mental events such as emerging thoughts or emotions is still highly problematic to be determined, and such events cannot be analyzed by means of EROs. There are, however, internal events that can be detected, and their electrical, mechanical, or performance markers can be used as appropriate triggers for EROs. In the following, we give examples with studies, in which we employed the EROs to investigate internal event processing.

3.1.1 Response-Related Potentials

Response-related potentials (RRPs) are a special class of ERPs. They are elicited by both voluntary movements and movements executed in response to sensory stimuli in sensorimotor tasks. The execution of a movement is an internal event as being internally initiated, guided, and generated. A movement can be slow or fast, can be initiated but later inhibited, and can be correct or incorrect. The characteristics of a movement strongly depend on brain states and functioning of cognitive control and performance monitoring systems [44, 45]. Thus, RRP components reflect internal information processing. By means of EROs, it is possible to study in a refined way the mechanisms supporting both the movement generation and cognitive brain systems regulating the actions.

Within this perspective, in several previous studies, we have explored RRPs [20, 46–48]. Our objective was to explain the sources of performance decline with aging. Analysis of ERP/RRP in the time domain has revealed one important correlate of the performance monitoring system in the brain, i.e., the error negativity ([49]; *see* Subheading 1.2.). The Ne is generated after error responses and is absent or significantly reduced after correct responses. As described in Subheading 1.2, by using the ERO concept and methodology, we have demonstrated that the Ne is composed of two main TF components from the delta and theta frequency ranges [20]. Although the two TF components are generated in parallel, only the delta component is associated with performance errors, thus representing a unique correlate of error processing in the brain.

Basing on these observations, we further aimed to explore if error processing might be specifically impaired with increasing age in humans [46, 47]. To determine age-related changes in the frequency content, temporal dynamics, and functional reactivity of error-related potentials, we studied two groups of subjects, young (mean age 22.5 years) and older (mean age 58.3 years) in a four-choice reaction time task. RRPs to correct and error responses were computed using a threshold level of a mechanogram as a trigger. In alignment with previous time-domain reports, we confirmed that error negativity was substantially reduced in older adults (Fig. 5a, b), but the origin of this reduction was not known. Therefore, by means of the wavelet transform, we

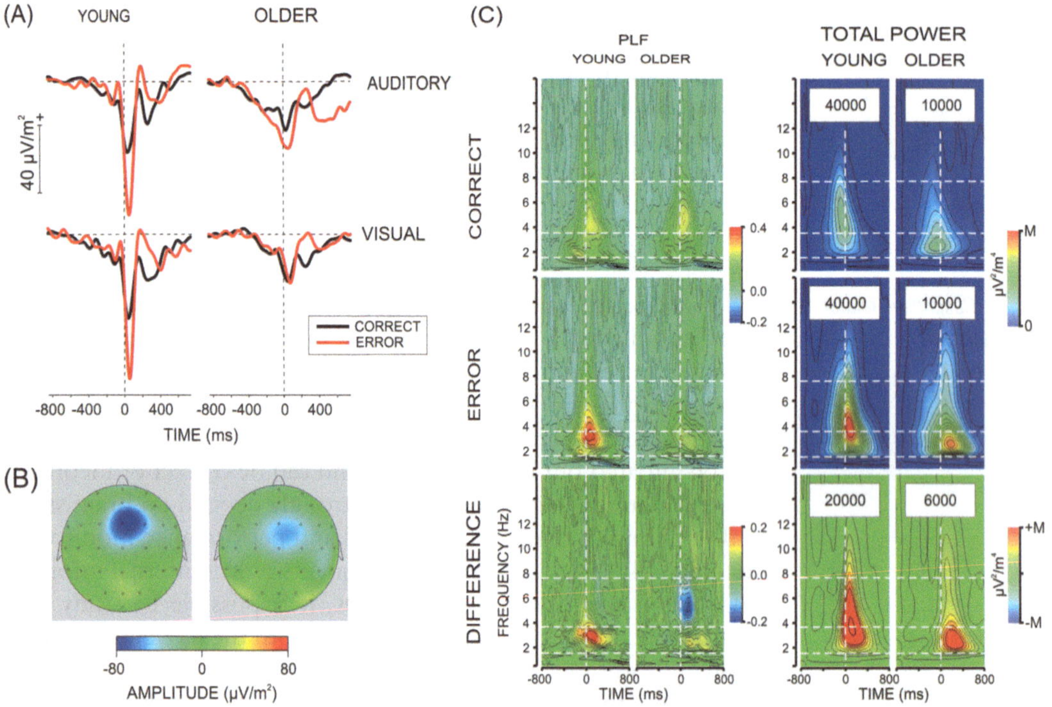

Fig. 5 Comparison between response-related negativities at FCz of young and older adults. (**a**) Grand average time-domain potentials from correct and error responses at FCz in two modalities (auditory and visual). Error negativity (Ne) and correct response negativity (Nc) peaking at about 80 ms after the response are demonstrated. Aging-related reduction of error negativity also is demonstrated. Response onset at 0 ms. Positivity is upward. (**b**) CSD maps of error-related scalp potentials for the same age groups. The two modalities and left–/right-hand responses are pooled together. (**c**) Phase-locking factor, PLF (left), and total power (right) for the two age groups. Time–frequency plots of PLF and total power at FCz for correct responses (upper panel), error responses (middle panel), and difference potentials (error minus correct) are presented. The two modalities and hands are pooled together. The scaling factor M is presented in the white boxes for the total power plots only. Response onset at 0 ms. Delta and theta frequency components are marked with horizontal lines (with modifications from Kolev et al. [46])

decomposed the Ne to identify time–frequency components in young and older subjects. The main observations are illustrated in Fig. 5c. Error-related potentials in both groups were characterized by phase-locked activity from the delta (1.5–3.5 Hz) frequency band. In contrast to young adults, no pronounced phase-locked theta (4–7 Hz) component was generated for either correct or error responses, nor did the total theta power manifest any sensitivity to error processing in older subjects. Thus, it was found that the frequency content of error-related potentials changed with aging. Error processing was not critically affected as implied by delta component observations. However, the theta TF component of the Ne that is assumed to play a major role in synchronizing and coordinating human actions [20, 50] was substantially attenuated implying a functional aging-related suppression of the underlying fronto-medial structures involved in movement regulation.

3.1.2 Coupling Between Slow Oscillations and Sleep Spindles

In another study [51], we applied the concept and methodology of EROs to study the temporal coupling between sleep events—slow oscillations and sleep spindles. These neuroelectric events are fundamental electrophysiological signatures of non-rapid eye movement (NREM) sleep [52]. It has been demonstrated that in humans, they may not be independent. Specifically, fast sleep spindles (13–16 Hz) may be coupled with the depolarizing up-state of cortical slow oscillations characterized by cortical activation, whereas slow sleep spindles (9–12 Hz) appear coordinated by the hyperpolarizing down-state associated with cortical inactivation [53].

Importantly, co-existent sleep spindles and slow waves have been viewed as a mechanism for offline information processing which supports memory consolidation during sleep. Accordingly, the neurodynamic characteristics and possible functions of temporally locked slow-wave and spindle activities have become a focus of increasing interest. In this regard, we aimed to refine the analysis of the coupling between slow oscillations and sleep spindles by employing the ERO methodology. We quantified the stability and timing of coupling between positive and negative phases of slow waves and sleep spindle activity during slow-wave sleep (SWS) in the following way (Fig. 6).

Slow-Wave Detection In a group of 53 subjects, sleep EEG was recorded from 28 electrodes. As demonstrated in Fig. 6a, to detect slow waves (SWs) in sleep EEG signals, sleep EEG was band-pass filtered in the frequency band of SWs (0.3–4 Hz). The goal was to determine the peaks of positive and negative half waves in order to use these peaks as triggers for the internal event (SW). A detailed description of the procedure is given in Fig. 6a [53]. As a result of the procedure, the peaks of negative and positive half SWs were

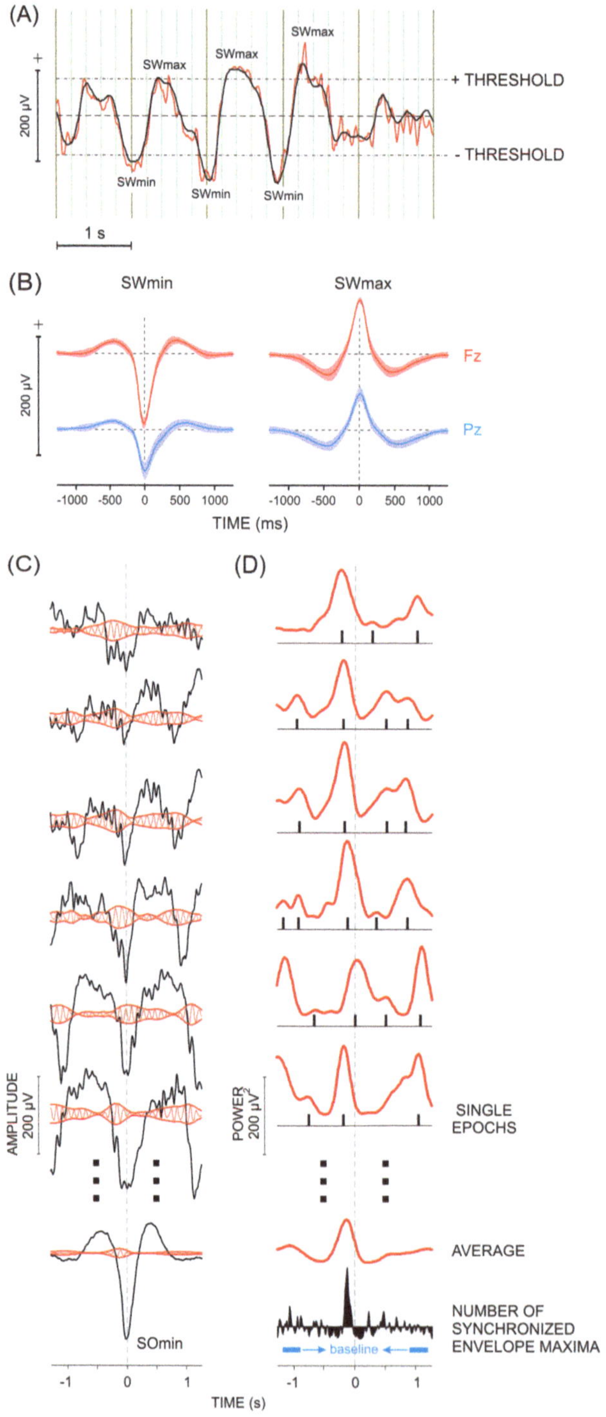

Fig. 6 Procedure for analysis of the coupling between slow-wave activity and sleep spindles. (**a**) Procedure for detection of SWs. Typical EEG recorded during slow-wave sleep at Fz. Sleep EEG is band-pass filtered within 0.3–4 Hz (frequency of SWs), and the resulting signal is level-triggered with a threshold of ±80 µV (Mölle et al. 53). Peaks exceeding the positive and negative thresholds are marked. SWmin/SWmax markers are thereafter transferred to the original EEG which is segmented around these markers. Thresholds and detected SW

marked as SWmin/SWmax. After detection, the markers were transferred to the original EEG and were used to segment the sleep EEG. Following the ERO approach, the EEG epochs were averaged for SWmin and SWmax triggers, with reference to a baseline (Fig. 6b), details given in the legend. Averages were then computed for each subject and electrode.

Temporal Coupling Between Spindle Activity and Slow Waves - Since spindles are discrete events of rhythmic oscillations from alpha (9–12 Hz) and sigma (13–16 Hz) frequency bands, single-trial SW-related EEG epochs were band-pass filtered in these bands to extract sleep spindle activity (Fig. 6c). As demonstrated and detailed in Fig. 6c, envelopes of filtered EEG epochs were calculated using complex demodulation to characterize the temporal dynamics of spindle activity. Afterward, the power of envelopes was computed and averaged with SWmin/SWmax triggers (Fig. 6d). In this way, any stable temporal grouping of modulated waves was extracted in the average of envelope power as a time-locked component. The SSWI method (*see* Subheading 2.2.3) was used to evaluate the congruency of envelope maxima across single epochs independently of power (Fig. 6d). As described in Subheading 2.2.3., the coding of the SSWI was applied to each single epoch (black bars below the curves on Fig. 6d, see figure caption for details). By averaging of these modified epochs, a histogram of the occurrence of envelope maxima was obtained. Two intervals both before and after the trigger again served as a baseline (Fig. 6d, bottom), and randomly occurring maxima were reduced to the baseline levels. The normalized number of synchronized envelopes and latency of the identified maxima in individual histograms were measured and analyzed to assess coupling variation and timing.

Fig. 6 (continued) extrema (SWmin and SWmax) are presented. (**b**) Grand average slow waves synchronized to their extrema at Fz and Pz and segmented to SWmin and SWmax (zero point) accordingly. SWmin/SWmax peaks are aligned to appear at time zero. These EEG segments are averaged separately for SWmin and SWmax epochs, with reference to the mean of two baselines of both before and after the trigger. (**c**) Sleep spindle analysis: Envelope amplitude quantification. Original EEG epochs (black curves) are superimposed with the band-pass filtered EEG (in this example 9–12 Hz, red curves). Envelopes of filtered EEG calculated by means of the complex demodulation method are presented in red. At the bottom, an average of 200 single epochs is presented. (**d**) Quantification of the phase synchronization of sleep spindle envelopes. The power of envelopes of filtered epochs is presented in red. After averaging, the total power of envelopes is obtained, and its maxima measured and further analyzed. To evaluate congruency of envelope maxima between single epochs independently of power, the SSWI method is used (*see* Subheading 2.2.3). All local envelope maxima in each single epoch are marked to produce modified single epochs containing only information about the position of the determined maxima (black bars below the red curves). Each bar in the modified single epoch has a value of 1. After averaging of the 200 modified epochs, a histogram of the number of synchronized envelope maxima is obtained (black curve at the bottom). For statistical analyses, the maximal values of histograms are obtained for each electrode and each subject. For normalization, a baseline activity is subtracted (designated in blue below the histogram) (with modifications from Yordanova et al. [51])

This approach helped reveal that spindle activity was temporally synchronized with internal events during NREM sleep—positive (up-state) and negative (down-state) slow half waves. It further demonstrated that the down state was associated with decoupling, whereas the up-state was associated with increased coupling of fast spindle activity over that hemisphere that was functionally activated before sleep [51].

3.1.3 Additional Internal Potentials: A Future Focus of Research

A variety of physiological signals can be used to explore internal information processing by means of the ERO methodology. Such examples are omitted stimuli, heart-beat potentials (brain potentials triggered by the e.g., ***R*** wave of the electrocardiogram), respiration-related very slow potentials, saccade-induced brain potentials, etc. How the brain processes such internal events is not sufficiently understood. However, as demonstrated above, approaches based on EROs can contribute to highlight these mechanisms.

3.2 Event-Related Frequency Tuning

As described in Subheading 1.1, according to the concept of EROs, the EEG can be regarded as reflecting the activity of neuronal ensembles producing oscillations in several frequency ranges, which are active in a very complex manner. Upon stimulation, functionally activated generators begin to act together in a coherent way. This transition is supposed to put the brain system from a disordered to an ordered state which brings the system to a resonance state and results in frequency stabilization, synchronization, and enhancement of the ongoing EEG activity [3].

In regard to the suggested transition of brain systems from a disordered to an ordered state, it was of major interest to verify such transitions and investigate their time evolution. To address this issue, in one of our previous studies [33], we quantified complex signal behavior in the ERP. In particular, the aim was (a) to show how external stimulus affects EEG frequency synchronization or frequency tuning to produce neuroelectric ordering and (b) to identify the temporal evolution of synchrony/desynchrony.

A new method for quantifying entropy in short-lasting EEG signals was developed to reflect the temporal evolution of order/disorder states in neuroelectric activity [33]. When applied to spectral EEG, low entropy values correspond to a narrow-band (monofrequency) activity characterizing highly ordered (regularized) bioelectric states, and high entropy values reflect a wide-band (multifrequency) activity [54]. To improve temporal resolution, the new method employed wavelet entropy instead of spectral entropy as a key approach [33]. Wavelet entropy was computed by means of orthogonal discrete wavelet transform of the EEG. This wavelet transform extracted overlapping frequency components with optimal time resolution. Three quantifiers were defined and analyzed: (a) relative wavelet energy, (b) total wavelet entropy (WE), and

(c) relative wavelet entropy (RWE). The relative wavelet energy was introduced to provide information about the relative energy of different frequency bands in relation to WE and RWE modulations. The WE was used to characterize the degree of order/disorder associated with a response to external stimulation, and the RWE was measured to reflect the degree of similarity between pre- and post-stimulus epochs [33].

The time window with minimal entropy value in the post-stimulus epoch was identified (t_m) to reflect the most ordered microstate in the ERP, and the time window with maximal RWE was identified (t_M) to reflect maximal dissimilarity between frequency distribution in the pre- and post-stimulus epochs. The latency (center of the respective time window) of t_m and t_M was measured (Fig. 7a).

This approach revealed a novel phenomenon in the event-related EEG activity [33, 55, 56]. As depicted in Fig. 7a, the time evolution of total WE demonstrated a minimum value of the WE at around 200 ms after external stimulus presentation (latency of t_{m} at 192 ms). Moreover, when this approach was applied to a developmental data set in 6–11-year-old children [57], it was found that a short-lasting ordering of the complex post-stimulus EEG signal occurred in each age group (Fig. 7b). The latency of this short-lasting ordered microstate decreased as a function of age and stemmed from a transient synchronization in the theta frequency channel (Fig. 7c). The developmental model indicated that, independently of the frequency content of the background EEG and ERPs in children at different ages, a highly ordered microstate in the ERP was always determined by theta frequency. Thus, transient dominance of synchronized theta oscillations might reflect an important functional mechanism sub-serving stimulus information processing. The precise functional relevance of this highly ordered microstate in the ERP supported by theta activity still remains to be explored.

3.3 Event-Related Spatial Synchronization

From the perspective of neuroelectric oscillatory events, EEG oscillations are produced by diffuse and selectively distributed brain systems operating in different frequency ranges [12, 13]. The generators within each oscillatory system may be active in a random manner, or they may be coupled and act coherently through local and long-distance synchronizing mechanisms [58, 59]. One important consequence of the understanding of frequency-specific oscillatory systems within the ERO concept is that the neural substrate of brain functions is spatially distributed. In support, neurophysiologic evidence has been provided that during motor response production, the activation of the motor cortical regions generating the movement emerges in close association with the activation of the medial frontal cortex implicated with performance monitoring

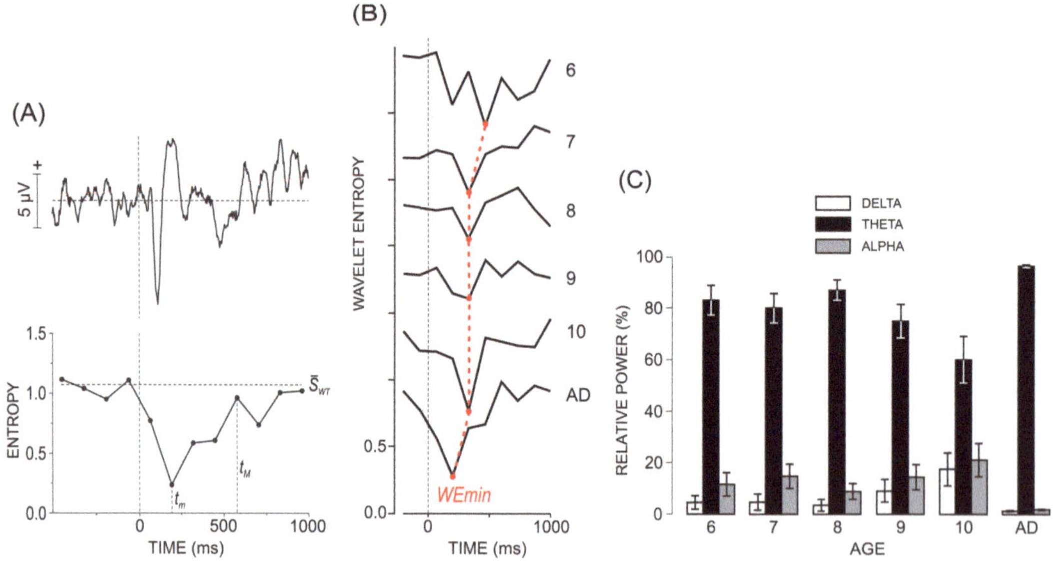

Fig. 7 Time evolution of wavelet entropy of event-related potentials. (**a**) ERP signal of a typical subject at Pz (*upper panel*). Time evolution of total WE (*bottom panel*). Time evolution of WE shows the latencies t_m (minimum value of the WE) at 192 ms and t_M (maximum value of the relative wavelet entropy) at 596 ms. The horizontal line represents the mean WE for the background EEG epoch ($\bar{S}_{WT} = 1.076$) used as a reference. (**b**) Time course of group mean wavelet entropy at Fz of six age groups (6, 6-year-olds; 7, 7-year-olds; 8, 8-year-olds; 9, 9-year-olds; 10, 10-year-olds; AD, adults). *WEmin* designates the minimum value of WE. Developmental reduction of the latency of *WEmin* is demonstrated. (**c**) Relative wavelet energies of three frequency bands: delta (0.1–4 Hz), theta (4–7 Hz), and alpha (8–15 Hz) during *WEmin* (beta and gamma energies are not shown as negligible). Designations of x-axes as in (**b**). The dominance of theta energy during WEmin is demonstrated (with modifications from (**a**) Rosso et al. 2001 [33] and (**b**) Yordanova et al. [56])

(e.g., [60]). Duprez et al. [50] have proposed that a brain system operating in the theta frequency range supports the synchronization and coordination of movements in different contexts.

Within the frequency-specific network perspective, we explored further the sources of aging-related decline with increasing age in humans (*see* Subheading 3.1.1) by analyzing the spatial synchronization of response-related theta oscillations. We tested the hypothesis that aging-related performance decline might originate from inefficient regulation of motor cortical regions due to impaired connections with the performance monitoring system.

The spatial event-related synchronization (*see* Subheading 2.2.4) of correct and error RRPs was computed in young and older adults. The analysis used PLV measures of pairs guided by the medial fronto-central electrode FCz (FCz-PLV). This analysis aimed at specifically assessing the connectivity between response generation (motor cortical) and response monitoring (medial fronto-central) regions. We further developed a mapping tool indicating the strength of connectivity between these cortical regions.

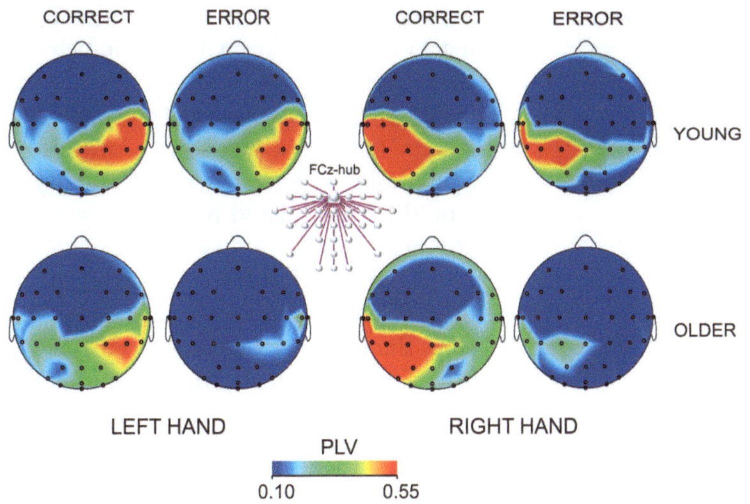

Fig. 8 Maps of the functional connectivity of FCz-guided connections with cortical regions in young and older adults. The computation of maps is based on PLV (spatial phase synchronization) analysis of all electrode pairs including the FCz electrode. Maps are presented for correct and error responses executed with the left or the right hand. The error-related reduction of the synchronization between the medial frontal FCz location and motor/sensorimotor cortical regions in older adults is demonstrated

Figure 8 demonstrates the maps of FCz-guided theta PLV of correct and error RRPs of young and older adults. It is shown that during both left- and right-hand responses, theta synchronization between FCz and movement-generating regions was significantly stronger for central and centro-parietal areas of the hemisphere contra-lateral to response. PLV analysis revealed that only during errors, the functional connectivity between performance monitoring and motor generation cortical regions was most substantially suppressed in older adults as compared to young subjects ($p < 0.01$). The application of the spatial dimension of the ERO concept and methods highlighted novel mechanisms of performance monitoring and cognitive control in the aging human brain.

3.4 Multi-Second Behavioral Patterns

In this section, we are going to demonstrate how the ERO concept and methodology can be applied not only to EEG signals but also to time series derived from other metrics (performance data). In this application, we aimed to explore the functioning of the default mode network (DMN) in healthy children and children with attention-deficit/hyperactivity disorder (ADHD). The DMN is a brain network that is active during passive brain states and is characterized by multi-second oscillations with frequency < 0.1 Hz. Goal-directed behavior is supported by the activation of attentional brain networks. The DMN is routinely attenuated during goal-directed task so that its activity does not interfere with attentional networks supporting an attentional focus and superior performance. It is well known that the behavior of children with ADHD is characterized by high variance, instability, and inaccuracy. We tested the hypothesis that an insufficient DMN by the attention system may be a key source of performance deficits in ADHD. Specifically, it tested if the variability of performance in ADHD would be modulated by the temporal characteristics of the DMN [61, 62].

In the study of Yordanova et al. [63], we analyzed the time dynamics of performance accuracy to search for multi-second periodic fluctuations of error occurrence. The aim was to study if attentional lapses in ADHD might originate from interferences from intrinsically oscillating DMN. Identifying periodic error fluctuations with a frequency < 0.1 Hz in patients with ADHD would provide a behavioral evidence for such interferences. Performance was monitored during a visual flanker task in 92 children (7 to 16 year olds), 47 with ADHD and 45 healthy controls who performed 10 blocks of the task.

The SSWI method (*see* Subheading 2.2.3) was used to detect periodic oscillations in error occurrence. To reveal cyclic dynamic patterns of error fluctuations, the timing of error occurrence was analyzed. The time distribution of error occurrence was quantified by coding the positions of error and correct trials (Fig. 9a). Then the coded trials were averaged across ten experimental blocks (Fig. 9b). If incorrect responses occur in a fully random manner, no peaks are expected to emerge in the average. In contrast, if errors occur with some consistent periodicity during a block, accumulation of incorrect responses at specific sequence positions would result in higher values in the average and accordingly in the formation of peaks in the histogram. The consecutive absolute values of individual averages were used as a time series signal (Fig. 9b).

This time series was analyzed in the frequency domain using FFT and in the time–frequency domain using WT (Fig. 9c). The presence of peaks in the fast Fourier transform would indicate that the time series of the analyzed signal contains rhythmic oscillations at specific frequencies. TF decomposition would reveal the time evolution of spectral components.

Major results based on group analyses are illustrated in Fig. 9d, e, f, g, h. They demonstrate that two prominent spectral peaks <0.1 Hz (at 0.05 Hz and 0.07–0.09 Hz) were significantly larger in the ADHD relative to the control group (Fig. 9e, g). The larger magnitude of error occurrence oscillations in the course of task block was also evident (Fig. 9d, f). The time–frequency analysis revealed that the "pathological" 0.05 Hz oscillation in ADHD persisted in the whole analysis epoch (the time of task performance in a block).

The application of EROs to performance data provided original data-based evidence that only in ADHD was there a "pathological"

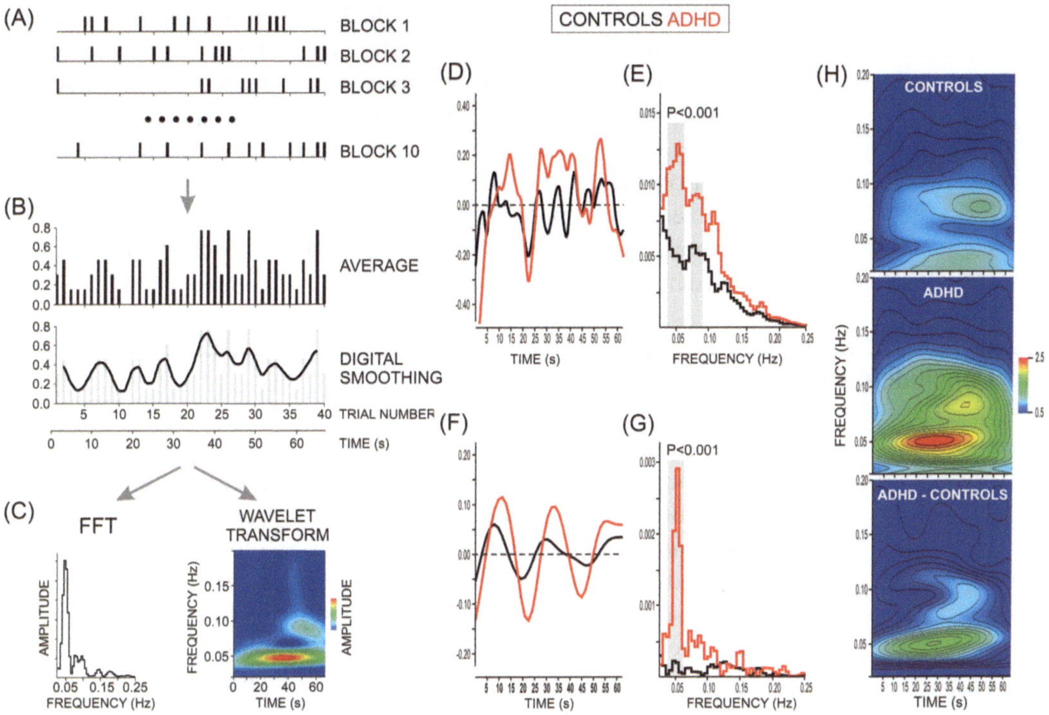

Fig. 9 Analysis of error dynamics. (**a**) Time distribution of error occurrence. Positions of errors are coded with +1, and positions of correct responses are coded with 0 in each of the ten blocks. (**b**) A point-by-point averaging is performed for each of the coded positions from the ten blocks. Temporal pattern of time distribution of error occurrence is obtained. Absolute values of the individual averages are measured. (**c**) Analysis of the time series of error occurrence in the frequency and time–frequency domains to explore periodic oscillations in error occurrence. Frequency representation (FFT) of a time series of error occurrence of one representative subject. Clear spectral peaks are seen at ~0.05 Hz and at ~0.07–0.09 Hz. Time–frequency (wavelet transform) decomposition plot of the same data. The ~0.05 Hz oscillations of error occurrence characterized the whole analysis epoch of the block. The faster periodic pattern of errors at ~0.07–0.09 Hz is evident only in the second half of the experimental blocks. (**d**) Comparison of group (ADHD and control) data: Group averages of the normalized time series of error occurrence. (**e**) FFT of the time series of error occurrence. Group FFTs are obtained by averaging individual FFTs in the frequency domain, which extracts the most stable across-subjects spectral components. Prominent spectral peaks <0.1 Hz (~0.05 Hz and ~0.07–0.09 Hz) are shaded. Significant between-group differences ($p < 0.001$) are marked (0.04, 0.05, and 0.06 Hz). (**f**) Comparison of group (ADHD and control) data: Superimposed ~0.05 Hz fluctuations of error occurrence. Larger oscillations are clearly observed in ADHD. (**g**) FFT of time-domain group averages emphasizing between-subjects consistency. The groups differ markedly for ~0.05 Hz oscillations. (**h**) Time–frequency plots of time series of error occurrence. The ~0.05 Hz error variations in ADHD persist in the whole analysis epoch. The ~0.07–0.09 Hz variations in the two groups occur mainly in the second half of the experimental block. The central between-group difference in the ~0.05 Hz fluctuation of error occurrence is further emphasized by the difference time–frequency TF plot shown at the bottom (ADHD minus controls) (with modifications from Yordanova et al. [63])

oscillation of error generation, which determined periodic drops of performance accuracy each 20–30 s. Thus, it was demonstrated that (i) attentive behavior of children is determined by multi-second regularities and (ii) a unique periodicity guides performance fluctuations in ADHD patients. These observations may re-conceptualize the understanding of attentive behavior beyond the executive top–down control and may reveal new origins of psychopathological behaviors in attention-deficit/hyperactivity disorder.

4 Concluding Remarks

The studies based on EROs presented in the chapter point to the relevance of the ERO concept for broad applications going beyond the standard goal to assess the processing of external information in the brain linked to ERPs. Important characteristics of any system in the body can be inferred from its reactions to interventions and responses to interactions with other systems functioning at different levels (molecular, physiological, psychological, or social). It is emphasized in the chapter that one critical task for such approaches is to enable the definition of the onset of the drive that would lead to a deviation from the spontaneous state. Another task is to parametrize with precision the temporal behavior of the system in discrete time series. In this perspective, the framework of ERO may turn especially useful for addressing newly emerging lines of research such as the interactions between systems at different levels of brain organization and functioning and interactions between the brain, mind, and body.

Acknowledgments

This study is supported by the National Research Fund by the Ministry of Education and Science, Sofia, Bulgaria (Project DN13-7/2017).

References

1. Başar E (1972) A study of the time and frequency characteristics of the potentials evoked in the acoustical cortex. Kybernetik 10:61–64. https://doi.org/10.1007/bf00292231
2. Başar E (1976) Biophysical and physiological systems analysis. Addison-Wesley Publishing Company Inc., Reading
3. Başar E (1980) EEG brain dynamics: relation between EEG and brain evoked potentials. Elsevier, Amsterdam
4. Wiener N (1961) Cybernetics. MIT Press, Wiley, New York
5. Regan D (1989) Human brain electrophysiology. Evoked potentials and evoked magnetic fields in science and medicine. Elsevier, Amsterdam
6. Luck SJ, Kappenman E (eds) (2012) The Oxford handbook of event-related potential components. Oxford University Press

7. Luck SJ (2014) An introduction to the event-related potential technique, 2nd edn. MIT Press
8. Ruchkin D (1988) Measurement of event-related potentials: signal extraction. In: Picton T (ed) Human event-related potentials. Handbook of EEG (revised series), vol 3. Elsevier, Amsterdam, pp 7–43
9. Picton TW, Bentin S, Berg P, Donchin E, Hillyard SA, Johnson R Jr, Miller GA, Ritter W, Ruchkin DS, Rugg MD, Taylor MJ (2000) Guidelines for using human event-related potentials to study cognition: recording standards and publication criteria. Psychophysiology 37:127–152. https://doi.org/10.1111/1469-8986.3720127
10. Polich J (1998) P300 clinical utility and control of variability. J Clin Neurophysiol 15:14–33. https://doi.org/10.1097/00004691-199801000-00004
11. Başar E, Bullock TH (eds) (1992) Induced rhythms in the brain. Birkhäuser, Boston
12. Başar E (1998) Brain function and oscillations, vol. 1. Brain oscillations. Principles and approaches. Springer, New York
13. Başar E (2004) Memory and brain dynamics. Oscillations integrating attention, perception, learning, and memory. CRC Press, Boca Raton
14. Başar E, Başar-Eroglu C, Karakaş S, Schürmann M (2001) Gamma, alpha, delta, and theta oscillations govern cognitive processes. Int J Psychophysiol 39:241–248. https://doi.org/10.1016/s0167-8760(00)00145-8
15. Cohen MX (2014) Analyzing neural time series data: theory and practice. The MIT Press, Cambridge, MA
16. Pfurtscheller G, Lopes da Silva FH (1999) Event-related EEG/MEG synchronization and desynchronization: basic principles. Clin Neurophysiol 110:1842–1857. https://doi.org/10.1016/s1388-2457(99)00141-8
17. Klimesch W (1999) EEG alpha and theta oscillations reflect cognitive and memory performance: a review and analysis. Brain Res Brain Res Rev 29:169–195. https://doi.org/10.1016/s0165-0173(98)00056-3
18. Yordanova J, Kolev V (1996) Developmental changes in the alpha response system. Electroencephalogr Clin Neurophysiol 99:527–538. https://doi.org/10.1016/s0013-4694(96)95562-5
19. Yordanova J, Kolev V (1998) A single-sweep analysis of the theta frequency band during an auditory oddball task. Psychophysiology 35:116–126. https://doi.org/10.1111/1469-8986.3510116
20. Yordanova J, Falkenstein M, Hohnsbein J, Kolev V (2004) Parallel systems of error processing in the brain. NeuroImage 22:590–602. https://doi.org/10.1016/j.neuroimage.2004.01.040
21. Kolev V, Demiralp T, Yordanova J, Ademoglu A, Isoglu-Alkaç Ü (1997) Time-frequency analysis reveals multiple functional components during oddball P300. Neuroreport 8:2061–2065. https://doi.org/10.1097/00001756-199705260-00050. ISSN 0959-4965
22. Yordanova J, Devrim M, Kolev V, Ademoglu A, Demiralp T (2000) Multiple time-frequency components account for the complex functional reactivity of P300. Neuroreport 11:1097–1103. https://doi.org/10.1097/00001756-200004070-00038
23. Başar E, Yordanova J, Kolev V, Başar-Eroglu C (1997) Is the alpha rhythm a control parameter for brain responses? Biol Cybern 76:471–480. https://doi.org/10.1007/s004220050360
24. Yordanova J, Kolev V (2007) Event-related brain oscillations in normal development. In: Schmidt LA, Segalowitz SJ (eds) Developmental psychophysiology: theory, systems, and methods. Cambridge University Press, New York, pp 15–68. https://doi.org/10.1017/CBO9780511499791
25. Yordanova J, Kolev V, Heinrich H, Woerner W, Banaschewski T, Rothenberger A (2002) Developmental event-related gamma oscillations: effects of auditory attention. Eur J Neurosci 16:2214–2224. https://doi.org/10.1046/j.1460-9568.2002.02286.x
26. Rothenberger A (2009) Brain oscillations forever - neurophysiology in future research of child psychiatric problems. J Child Psychol Psychiatry 50:79–86. https://doi.org/10.1111/j.1469-7610.2008.01994.x
27. Kolev V, Başar-Eroglu C, Aksu F, Başar E (1994) EEG rhythmicities evoked by visual stimuli in three-year-old children. Int J Neurosci 75:257–270. https://doi.org/10.3109/00207459408986308
28. Başar-Eroglu C, Kolev V, Ritter B, Aksu F, Başar E (1994) EEG, auditory evoked potentials and evoked rhythmicities in three-year-old children. Int J Neurosci 75:239–255. https://doi.org/10.3109/00207459408986307
29. Mallat S (1999) A wavelet tour of signal processing, 2nd edn. Academic Press, San Diego
30. Tallon-Baudry C, Bertrand O, Delpuech C, Permier J (1997) Oscillatory gamma-band (30-70 Hz) activity induced by a visual search task in humans. J Neurosci 17:722–734.

https://doi.org/10.1523/JNEUROSCI.17-02-00722.1997

31. Samar VJ, Swartz KP, Raghuveer MR (1995) Multiresolution analysis of event-related potentials by wavelet decomposition. Brain Cogn 27:398–438. https://doi.org/10.1006/brcg.1995.1028
32. Demiralp T, Yordanova J, Kolev V, Ademoglu A, Devrim M, Samar VJ (1999) Time-frequency analysis of single-sweep event-related potentials by means of fast wavelet transform. Brain Lang 66:129–145. https://doi.org/10.1006/brln.1998.2028
33. Rosso OA, Blanco S, Yordanova J, Kolev V, Figliola A, Schürmann M, Başar E (2001) Wavelet entropy: a new tool for analysis of short duration brain electrical signals. J Neurosci Methods 105:65–75. https://doi.org/10.1016/s0165-0270(00)00356-3
34. Uhlhaas PJ, Roux F, Singer W, Haenschel C, Sireteanu R, Rodriguez E (2009) The development of neural synchrony reflects late maturation and restructuring of functional networks in humans. Proc Natl Acad Sci U S A 106:9866–9871. https://doi.org/10.1073/pnas.0900390106
35. Uhlhaas PJ, Roux F, Rodriguez E, Rotarska-Jagiela A, Singer W (2010) Neural synchrony and the development of cortical networks. Trends Cogn Sci 14:72–80. https://doi.org/10.1016/j.tics.2009.12.002
36. Uhlhaas PJ, Singer W (2010) Abnormal neural oscillations and synchrony in schizophrenia. Nat Rev Neurosci 11:100–113. https://doi.org/10.1038/nrn2774
37. Kolev V, Yordanova J (1997) Analysis of phase-locking is informative for studying event-related EEG activity. Biol Cybern 76:229–235. https://doi.org/10.1007/s004220050335
38. Lachaux JP, Rodriguez E, Martinerie J, Varela FJ (1999) Measuring phase synchrony in brain signals. Hum Brain Mapp 8:194–208. https://doi.org/10.1002/(sici)1097-0193(1999)8:4%3C194::aid-hbm4%3E3.0.co;2-c
39. Jervis BW, Nichols MJ, Johnson TE, Allen E, Hudson NR (1983) A fundamental investigation of the composition of auditory evoked potentials. IEEE Trans Biomed Eng BME 30:43–49. https://doi.org/10.1109/tbme.1983.325165
40. Cohen MX (2015) Effects of time lag and frequency matching on phase-based connectivity. J Neurosci Methods 250:137–146. https://doi.org/10.1016/j.jneumeth.2014.09.005
41. Perrin F, Pernier J, Bertrand O, Echallier JF (1989) Spherical splines for scalp potential and current density mapping. Electroencephalogr Clin Neurophysiol 72:184–187. https://doi.org/10.1016/0013-4694(89)90180-6
42. Varela F, Lachaux JP, Rodriguez E, Martinerie J (2001) The brainweb: phase synchronization and large-scale integration. Nat Rev Neurosci 2:229–239. https://doi.org/10.1038/35067550
43. Stam CJ, Nolte G, Daffertshofer A (2007) Phase-lag index: assessment of functional connectivity from multi channel EEG and MEG with diminished bias from common sources. Hum Brain Mapp 28:1178–1193. https://doi.org/10.1002/hbm.20346
44. Ullsperger M, von Cramon DY (2001) Subprocesses of performance monitoring: a dissociation of error processing and response competition revealed by event-related fMRI and ERPs. NeuroImage 14:1387–1401. https://doi.org/10.1006/nimg.2001.0935
45. Cohen MX (2011) Error-related medial frontal theta activity predicts cingulate-related structural connectivity. Neuro Image 55:1373–1383. https://doi.org/10.1016/j.neuroimage.2010.12.072
46. Kolev V, Falkenstein M, Yordanova J (2005) Aging and error processing: time-frequency analysis of error-related potentials. J Psychophysiol 19:289–297. https://doi.org/10.1027/0269-8803.19.4.289
47. Kolev V, Beste C, Falkenstein M, Yordanova J (2009) Error-related oscillations: effects of aging on neural systems for behavioral monitoring. J Psychophysiol 23:216–223. https://doi.org/10.1027/0269-8803.23.4.216
48. Yordanova J, Falkenstein M, Kolev V (2020) Aging-related changes in motor response-related theta activity. Int J Psychophysiol 153:95–106. https://doi.org/10.1016/j.ijpsycho.2020.03.005
49. Falkenstein M, Hohnsbein J, Hoormann J, Blanke L (1990) Effects of errors in choice reaction tasks on the ERP under focused and divided attention. In: Brunia CHM, Gaillard AWK, Kok A (eds) Psychophysiological brain research 1990. Tilburg University Press, Tilburg, pp 192–195
50. Duprez J, Gulbinaite R, Cohen MX (2020) Midfrontal theta phase coordinates behaviorally relevant brain computations during cognitive control. NeuroImage 207:116340. https://doi.org/10.1016/j.neuroimage.2019.116340
51. Yordanova J, Kirov R, Verleger R, Kolev V (2017) Dynamic coupling between slow waves and sleep spindles during slow wave sleep in humans is modulated by functional pre-sleep activation. Sci Rep 7:14496. https://doi.org/10.1038/s41598-017-15195-x

52. Born J, Rasch B, Gais S (2006) Sleep to remember. Neuroscientist 12:410–424. https://doi.org/10.1177/1073858406292647
53. Mölle M, Marshall L, Gais S, Born J (2002) Grouping of spindle activity during slow oscillations in human non-rapid eye movement sleep. J Neurosci 22:10941–10947. https://doi.org/10.1523/jneurosci.22-24-10941.2002
54. Inouye T, Shinosaki K, Sakamoto H, Toi S, Ukai S, Iyama A, Katzuda Y, Hirano M (1991) Quantification of EEG irregularity by use of the entropy of power spectrum. Electroencephalogr Clin Neurophysiol 79:204–210. https://doi.org/10.1016/0013-4694(91)90138-t
55. Yordanova J, Kolev V, Rosso OA, Schürmann M, Sakowitz OW, Ozgören M, Başar E (2002) Wavelet entropy analysis of event-related potentials indicates modality-independent theta dominance. J Neurosci Methods 117:99–109. https://doi.org/10.1016/s0165-0270(02)00095-x
56. Yordanova J, Rosso OA, Kolev V (2003) A transient dominance of theta ERP component characterizes stimulus processing in an auditory oddball task. Clin Neurophysiol 114:529–540. https://doi.org/10.1016/S1388-2457(02)00415-7
57. Kolev V, Rosso OA, Yordanova J (2001) A transient dominance of theta ERP component characterizes passive auditory processing: evidence from a developmental study. Neuroreport 12:2791–2796. https://doi.org/10.1097/00001756-200109170-00008
58. von Stein A, Sarnthein J (2000) Different frequencies for different scales of cortical integration: from local gamma to long range alpha/theta synchronization. Int J Psychophysiol 38:301–313. https://doi.org/10.1016/s0167-8760(00)00172-0
59. Bressler SL, Menon V (2010) Large-scale brain networks in cognition: emerging methods and principles. Trends Cogn Sci 14:277–290. https://doi.org/10.1016/j.tics.2010.04.004
60. Urbano A, Babiloni C, Onorati P, Babiloni F (1998) Dynamic functional coupling of high resolution EEG potentials related to unilateral internally triggered one-digit movements. Electroencephalogr Clin Neurophysiol 106:477–487. https://doi.org/10.1016/s0013-4694(97)00150-8
61. Castellanos FX, Sonuga-Barke EJ, Scheres A, Di Martino A, Hyde C, Walters JR (2005) Varieties of attention-deficit/hyperactivity disorder-related intra-individual variability. Biol Psychiatry 57:1416–1423. https://doi.org/10.1016/j.biopsych.2004.12.005
62. Sonuga-Barke EJS, Castellanos FX (2007) Spontaneous attentional fluctuations in impaired states and pathological conditions: a neurobiological hypothesis. Neurosci Biobehav Rev 31:977–986. https://doi.org/10.1016/j.neubiorev.2007.02.005
63. Yordanova J, Albrecht B, Uebel H, Kirov R, Banaschewski T, Rothenberger A, Kolev V (2011) Independent oscillatory patterns determine performance fluctuations in children with attention-deficit/hyperactivity disorder. Brain 134:1740–1750. https://doi.org/10.1093/brain/awr107

Chapter 5

Quantitative EEG Analysis: Introduction and Basic Principles

Georgi Panov

Abstract

The use of mathematical models for electroencephalography (EEG) analysis has been going on for many years, and currently they are starting to find a place both in clinical practice and in parallel with other methods for evaluation of brain activity. The use and interpretation of these methods are not possible without knowledge of the basic mechanisms and processes that underlie the results obtained from their application. The advantages of the quantitative methods, the processes underlying their presentation, and the optimal parameters used, such as the number of electrodes and time intervals, have been evaluated. Attention has been paid to the advantages of the individual possibilities for EEG signal analysis—absolute and relative power as well as coherence. An analysis of these defined quantities provides an opportunity to enter into the intimate mechanisms underlying specific psychopathological phenomena.

Key words Electroencephalography (EEG), Quantitative EEG, Absolute power, Relative power, Coherence

1 Introduction

Electroencephalography is an interdisciplinary methodology based on the achievements of mathematics, physics, biophysics, neurophysiology, and clinical neurology and psychiatry. It provides guidelines for uncovering the intimate mechanisms underlying certain mental actions, attitudes, clinical phenomena, and neurophysiological processes.

The abduction of the electrical potentials coming from different parts of the brain and registered by the scalp by placing electrodes on its surface is done by technical devices that amplify the registered signal and present it as an electroencephalogram by analogy with the method of the electrocardiogram in cardiology. For decades, the standard electroencephalography (EEG) has become a routine examination in neurology and psychiatry. It is an indispensable tool when analyzing patients with epilepsy, and it is necessary to clarify the diagnosis of a specific syndrome or search

Drozdstoy Stoyanov et al. (eds.), *Computational Neuroscience*, Neuromethods, vol. 199,
https://doi.org/10.1007/978-1-0716-3230-7_5,

for the localization and topographical dynamics of the abducted epileptiform paroxysmal events.

Besides the search for these "bigger" neurophysiological phenomena such as the presence or absence of epileptiform paroxysmal episodes as well as focal slow-wave activity, the evaluation of more discrete local or diffuse changes in the EEG has always raised the question of subjectivity in the analysis of the electroencephalographic finding.

The factors associated with the precise interpretation of an EEG finding are related to the assessment of the age of the patient at the time of the examination, the time of the examination itself, the functional state of the patient and sleep, wakefulness, drowsiness, intake of certain medications with their effect on the EEG trace, the location of the study, etc. These variables that affect the characterization and interpretation of the study can and should be taken into account when conducting the examination. On the other hand, there remains another important aspect of EEG analysis, and that is the subjective judgment of the specialist conducting the study. What matters is his experience, his qualifications, and of course even his psychological state during the analysis itself.

The search for methods for more accurate objectification, as well as for the purpose of evaluating the visually hard-to-perceive changes in the EEG, has led to the use of mathematical models for the quantitative evaluation of the wave phenomena in the study. Quantitative EEG refers to the incorporation of topographic spatial mapping and mathematical analysis into EEG evaluation.

When using quantitative methods of analysis, it is necessary to understand what is actually behind these methods. The extraction of the information from the registered EEG is based on mathematical calculations and algorithms with the aim of configuring the "quantitative" EEG with the corresponding graphic and visual-cartographic representation of the distribution of potentials on the scalp. The use of such visualization (mapping) poses the question of the calculation or measurement of the values of the points standing between the located electrodes during EEG recording.

Various interpolation methods are used for this evaluation and calculation of the potential value of the points where no electrodes are placed. The most used method is with a three-point coordinate system/three-dimensional linear interpolation, where the values of the non-electrode locations are created and calculated based on the values of the three nearest electrodes. There are different approaches for data interpolation: linear interpolation, polynomial interpolation, as well as spline interpolation [1]. Most authors (more specifically the analysis systems used) use the triangular linear interpolation when evaluating the EEG for the purpose of building a brain cartographic image [2]. The main test of the reliability of an interpolation algorithm is to compare the calculated interpolated values with the actual measured ones. All current interpolation methods are based on purely mathematical interpolation theory, ignoring the neurophysiological basis and clinical reality of EEG signals, i.e., the possibility of artifactual activity in a given lead, and the presence of such would also lead to an inaccurate calculation in the interpolated values.

In this case, these extracerebral signals would have their impact on all calculated points in the analyzed section that depend on this calculation. In connection with this clinical reality, it is difficult and impossible to conduct the so-called “ideal” EEG recording with absolutely no artifacts; the author’s team looked for an option to further improve the efficiency of the interpolation. In those cases, when there is registration of “noise” and artifactual activity in one of the registered leads, they use a special methodology, which they call the standardized interpolation technique of the reference electrode. This method enables a more accurate construction of the cartographic picture of the distribution of potentials [3]. Using this technique, the authors found that with a 10% presence of “bad” channels, the reconstruction of the signal to avoid the artifacts was almost complete. With a greater presence of bad channels using their method, they were able to reduce both absolute and relative information analysis errors by more than 30% [3]. These studies and methods hold promise for entering more intensively into clinical work in order to isolate artifactual activity from the main information flows under analysis.

Another important issue when discussing quantitative EEG analysis and conducting interpolation methods is how many electrodes are needed to model a reliable cartographic visualization when conducting the analysis. There are numerous studies that generally indicate that the more electrodes there are, the more accurate the interpolation should be. Is this really the case?

A team of authors performed an analysis in which they considered to use two cases: a subject-specific case, in which the optimal number and positions of electrodes are determined for each subject individually, and a study subject-universal case, in which the electrodes are put in the same positions constantly (according to the 10–20 system). A 64-channel EEG study was conducted on 90 subjects, and the authors were surprised to note that the correlation between the actual and the reconstructed envelopes first increased with a decrease in the number of electrodes, with an optimum at about 20 electrodes; in general the data showed that the number can be reduced to 32 channels without affecting the quality of the analysis [4]. They found that it is not necessary to do a study with a large number of electrodes. Moreover, the larger number of electrodes would also raise the question of the artifactual activity underlying as a potential possibility in each of them at a certain moment of the study. This explanation shows that as the number of electrodes increases, so does the probability of noise activity occurred in a larger part of them. This point of view also leads us to the conclusion of searching for an optimal number of electrodes when conducting cartographic studies, as well as searching for a way to build the "ideal" connection between the electrode and the scalp.

Spectral analysis of brain electrical activity provides the decomposition of the EEG signal into its constituent frequency components: delta (0–4 Hz), theta (4–8 Hz), alpha (8–13 Hz), and beta (>13 Hz). Some authors question the analysis of gamma activity as well, assuming beta for the range 13–30 and gamma in the range 30–80 Hz [5]. Such a decomposition makes it possible to trace the distribution and power of individual frequency ranges in individual brain regions under certain physiological and pathological conditions both before and during the administration of a given treatment.

Different researchers use different frequency ranges for the standard alpha, beta, theta, and delta frequency spectra. Acceptance and use of these ranges are an essential step toward standardization of EEG diagnostic studies. The following frequency ranges are commonly used in the literature for future research: delta (0.5–3.5), theta1 (4–5.5 Hz), theta2 (6–7.5), alpha1 [8–10], and alpha2 (10.5–12.5). There is no unanimous consensus on the limits of beta ranges. Kubicki et al. [6] used the following ranges: beta1 (12–18 Hz), beta2 (18–21 Hz), and beta3 (21–30 Hz). Galderisi et al. [7] proposed the following ranges: beta1 (12.7–15 Hz), beta2 (15.2–26 Hz), and beta3 (26.2–35 Hz). Galderisi's recommended bands [7] cover the widest frequency range and are recommended by some researchers. There is no consensus in the literature regarding the frequency bands used, and, in general, publications with different ranges can be found. The main problem related to the use of different parameters in individual publications is the impossibility of comparing the results as well as their translation to clinical practice. Another problem that needs to be discussed is what new information this additional subdivision of the main signals can give us and what this extended analysis of the additional decay of the main frequency ranges can bring us as a correlation in clinical practice, apart from the additional clutter of information and the use of redundant attempts for classification modeling of the obtained raw EEG information.

Another important issue in the analysis of raw EEG data is how long-time frames of the recording should be analyzed. There are different views on the duration of the time frame under analysis and evaluation. Researchers accept that a duration of about 20 s, but not more than 40 s for an EEG epoch, is sufficient to assess EEG activity [8]. It is appropriate to use several epochs in order to average the results. On the other hand, at longer epochs, it is more likely that signals of other than brain origin will be included in the analysis, which will "contaminate" the obtained data and distort their analysis. When comparing patients, it is necessary to perform the EEG under the same technical conditions, such as situation, temperature, and lighting, in order to compare the same physiological state. This is also the reason why some authors recommend shorter epochs—up to 8 s in order to exclude artifacts from the analysis field [9]. Other authors consider that duration is not essential. They analyzed the duration of study epochs (those of 4, 8, 10, and 20 s) and found no difference in the distribution of individual spectral powers across studies [10]. These data show that the epochs of the EEG recording for which it is most important to be free of artifact manifestations should be selected individually according to the immediate situation.

When evaluating the power of the EEG activity, two aspects are analyzed—absolute and relative power. Absolute power reflects the value of the measured activity for a given frequency range in absolute value, i.e. regardless of the activity recorded in the other frequency ranges. In addition to the absolute power for each individual frequency range, the evaluation of the entire global power is also possible. It appears as the result of the summation of the power in the delta, theta, alpha, and beta spectra. With relative power, the percentage for each individual frequency spectrum of the total power that occupies the individual frequency range is evaluated. Relative power is measured in percentages. This means that if 100% is the total power, then each frequency range occupies a certain percentage of it.

The main question that logically follows is the analysis of which of the two powers, absolute versus relative, is more informative and indicative in the assessment of certain pathological conditions. The author's team analyzes patients with brain tumors and cerebrovascular accidents. They found that when using relative power, tumors and vascular pathology yielded 83 and 70% confirmation of the finding, respectively. Using absolute power gives 50% and 31%, respectively, in these pathological conditions. The authors concluded that relative power is a more indicative method than absolute power [11].

There are also opposite findings. When analyzing the frequency spectra in hypoxic ischemic encephalopathy in childhood, authors find that the use of absolute power is more indicative [12].

In general, it can be said that it is appropriate to use both types of power of the EEG activity when conducting quantitative analyses. Absolute power is more sensitive to the presence of artifact activity, which automatically affects the absolute value of the given frequency range. Relative power seems to be more "stable" to the presence of artifacts (which is no excuse for the presence of such) due to their reflection in individual frequency spectra, and for this reason, it seems to "blur" when calculated as a percentage for the individual spectra relative to each other.

Essential when evaluating EEG activity is coherence analysis. Coherence is an indicator of the functional connectivity between individual brain regions depending on neuronal oscillations.

Synchronous oscillation (phase synchronization) between neuronal populations in defined brain regions provides information about the integrity of the brain connections between them in a certain physiological state. Coherence estimation is a mathematical model that can be used to evaluate these relationships. It indicates whether two or more regions assessed by the measured activity from the electrodes placed above them have the same or similar oscillations.

In assessing coherence, it is important to track how this uniformity (phase synchronization) changes over time. On the other hand, it is important to evaluate the changes in coherence between individual regions in different pathological conditions involving the central nervous system (CNS) [13].

The evaluation of EEG quantitative indicators appears important for the analysis of the relationship between the clinical characteristics of certain nosological constructs with their corresponding neurophysiological correlates. On the one hand, they enable detailed tracking of discrete changes in the individual frequency spectra, both in the dynamics of the development of the registered clinical phenomena and the effectiveness of the applied treatment and its impact on the values of the main frequency bands. On the other hand, these data make it possible to analyze the functional connectivity between individual regions in the CNS.

The routine use of quantitative EEG in clinical practice enables an additional look at the overall development of pathological processes in the given patient, as well as the identification of certain "patterns" characteristic of certain diseases and disorders.

References

1. Steffensen JF (2006) Interpolation, 2nd edn, Mineola, N.Y. 978-0-486-15483-1, OCLC, 86777089
2. Casaglia DC, Pantaleo GG (1993) Brain mapping: a contribution to linear interpolation. Brain Topogr 5:283–288. https://doi.org/10.1007/BF01128995
3. Dong L, Zhao L, Zhang Y, Yu X, Li F, Li J, Lai Y, Liu T, Yao D (2021) Reference electrode standardization interpolation technique (RESIT): a novel interpolation method for scalp EEG. Brain Topogr 34(4):403–414. https://doi.org/10.1007/s10548-021-00844-2. Epub 2021 May 5. PMID: 33950323; PMCID: PMC8195908
4. Montoya-Martínez J, Vanthornhout J, Bertrand A, Francart T (2021) Effect of number and placement of EEG electrodes on measurement of neural tracking of speech. PLoS One 16(2):e0246769. https://doi.org/10.1371/journal.pone.0246769. PMID: 33571299
5. Crone NE, Korzeniewska A, Franaszczuk PJ (2011) Cortical gamma responses: searching high and low. Int J Psychophysiol 79(1):9–15. https://doi.org/10.1016/j.ijpsycho.2010.10.013
6. Kubicki S, Herrmann WM, Fichte K, Freund G (1979) Reflections on the topics: EEG frequency bands and regulation of vigilance. Pharmakopsychiatr Neuropsychopharmakol 12(2): 237–245
7. Galderisi S, Mucci A, Mignone ML, Maj M, Kemali D (1991) CEEG mapping in drug-free schizophrenics. Differences from healthy subjects and changes induced by haloperidol treatment. Schizophr Res 6(1):15–23
8. Gudmundsson S et al (2007) Reliability of quantitative EEG features. Clin Neurophysiol 118:2162–2171
9. Möcks J, Gasser T (1984) How to select epochs of the EEG at rest for quantitative analysis. Electroencephalogr Clin Neurophysiol

58(1):89–92. https://doi.org/10.1016/0013-4694(84)90205-0. PMID: 6203708

10. Yan MM, Xu XH, Huang ZL et al (2011) Selection of optimal epoch duration in assessment of rodent sleep-wake profiles. Sleep Biol Rhythms 2011(9):46–55. https://doi.org/10.1111/j.1479-8425.2010.00484.x
11. Dan P, Mircea O, Mihaela M (2021) EEG relative power versus absolute power mapping-advantages, disadvantages. 2000. 21–33
12. Govindan RB, Massaro A, Vezina G, Tsuchida T, Cristante C, du Plessis A (2017) Does relative or absolute EEG power have prognostic value in HIE setting? Clin Neurophysiol. 128(1):14–15. https://doi.org/10.1016/j.clinph.2016.10.094. Epub 2016 Nov 5. PMID: 27866114; PMCID: PMC7319225
13. Bowyer SM (2016) Coherence a measure of the brain networks: past and present. Neuropsychiatr Electrophysiol 2:1. https://doi.org/10.1186/s40810-015-0015-7

Chapter 6

QEEG and ERP Biomarkers of Psychotic and Mood Disorders and Their Treatment Response

Sevdalina Kandilarova and Igor Riečanský

Abstract

Electroencephalography (EEG) is widely accessible and can be relatively easily applied in clinical settings. This chapter is focused on selected quantitative electroencephalographic (QEEG) and event-related potential (ERP) measures as promising diagnostic and prognostic markers of psychotic and mood disorders. We first review the evidence linking mismatch negativity (MMN) and P300 ERPs to psychosis. While not specific only for psychotic disorders, MMN and P300 represent promising markers of the onset of psychosis in high-risk individuals as well as of the outcome in psychotic patients. The second part of this chapter covers the application of QEEG techniques in the search for diagnostic and prognostic biomarkers of depression. In particular, research suggests that increased power of alpha EEG oscillations is associated with depression and response to treatment with antidepressants. Furthermore, early decrease in theta cordance predicts good treatment response to both pharmacological and non-pharmacological treatments of depression. Research including large samples of patients is needed to confirm clinical utility and cost-effectiveness of these biomarkers.

Key words QEEG, ERP, Biomarker, Psychiatric disorder, Treatment response, Psychosis, Depression

1 The Perspective of Clinical Utility of Mismatch Negativity and P300 Event-Related Potentials in Psychotic Disorders

Event-related potentials (ERPs) are fluctuations of electroencephalogram (EEG) that are time-locked to external or internal events (such as presentation of a sensory stimuli or preparation of movements). Usually, ERPs become evident after averaging several epochs of the EEG time-locked to the same event since the event-related EEG fluctuations are too small to be detected in a single EEG segment. ERPs have temporal resolution in the millisecond range and reflect rapid dynamics of neural activity related to brain information processing. ERPs can be obtained relatively easily and at low costs, which is important from the perspective of their potential clinical applications in psychiatry. Stimulation paradigm importantly determines whether data can be obtained easily in

Drozdstoy Stoyanov et al. (eds.), *Computational Neuroscience*, Neuromethods, vol. 199,
https://doi.org/10.1007/978-1-0716-3230-7_6,

clinical settings. Patients with psychotic disorder suffer from a very severe disturbance of mental state and are often not able to complete complex tasks, which require focusing and maintaining attention over longer time period. Here we shortly summarize the findings on mismatch negativity (MMN) and P300 wave, two ERPs which are elicited using a simple stimulation protocol, the so-called oddball paradigm. In this paradigm, a series of stimuli are presented involving frequent standard and infrequent deviant stimuli. The load imposed on the participant to cooperate and complete this task is minimal so that the ERPs can be recorded also in psychotic patients, yielding valuable data potentially useful to assess risk, treatment options, and prognosis of patients with psychotic disorders.

1.1 MMN

Mismatch negativity (MMN) is a negative ERP deflection elicited by presentation of rare stimuli in the oddball paradigm. MMN occurs 100–250 ms after the onset of the deviant stimulus and is maximal at frontocentral sites. To evoke MMN, auditory stimuli are usually used with deviants differing in duration or frequency from the standard tones. MMN does not require to direct attention or respond to the stimuli and has been attributed to automatic sensory processing, adaptation, and memory [1]. MMN has been linked to glutamatergic neurotransmission since glutamate NMDA (N-methyl D-aspartate) receptor antagonists, such as ketamine, diminish MMN amplitude [2]. Given the requirement for the participant to undergo MMN assessment is minimal, MMN can be obtained also in patients with severely altered mental state, unable to perform complex cognitive tasks. Furthermore, MMN can be recorded in experimental animals, enabling direct translation of findings from animal models [3, 4].

MMN amplitude is diminished in several neuropsychiatric disorders, in particular schizophrenia [5, 6] and bipolar disorder [7–9], but MMN deficits have also been documented in schizotypal personality disorder [10], major depressive disorder [11], attention-deficit/hyperactivity disorder [12], autism spectrum disorder [13, 14], alcoholism [15], as well as several neurological conditions [16]. Furthermore, the amplitude of MMN decreases with increasing age [17, 18], which is accelerated in cognitively impaired and demented individuals [19, 20].

The progress in the research of MMN has been mostly driven by the studies focusing on psychotic disorders. It is well established that patients with schizophrenia have diminished auditory MMN, in particular when the deviant stimuli differ in duration from the standards [5, 6]. It has been suggested that the MNN deficit in schizophrenia reflects impaired predictive processing associated with dysfunction of signaling via NMDA receptors [21, 22]. However, contrary to this account, MMN deficits show no clear relationship with severity of psychotic symptoms and rather correlate with patients' functional impairment [18, 23, 24]. The MMN reduction is larger in chronic compared with first-episode schizophrenia patients, but there is no clear evidence for further amplitude decrease with illness progression [5, 6].

In some (but not all) studies, first-degree relatives of schizophrenia patients displayed diminished MMN amplitudes compared to general population, indicating a link between MMN and genetic risk of schizophrenia [25]. A recent study, however, has suggested that reduced MMN in relatives of schizophrenia patients is related to sub-threshold schizophrenia-like symptoms rather than increased generic vulnerability for the disorder [26]. In a study of twins concordant and discordant for schizophrenia, genetic correlation of MMN amplitude with schizophrenia has been estimated at about 40%, indicating that a significant proportion of genetic factors contributing to schizophrenia also affect the MMN amplitude [3].

Importantly, it has been shown that MMN is reduced in individuals with clinical high risk for psychotic disorder. These individuals show subtle changes in perception, thinking, mood, interests, social behavior, etc., which do not reach the intensity of psychosis, and have been together termed as psychosis risk syndrome, attenuated syndrome, or prodromal syndrome [27]. Conversion rates to full-blown psychosis for these individuals have been estimated at about 20% after 1 year and 36% after 3 years [28]. A number of studies have now confirmed that MMN amplitude is smaller in future converters to psychosis compared with non-converters [29–36]. This finding is of considerable interest since it indicates that the assessment of MMN may be useful in predicting whether an individual at high risk will later develop full psychosis [37, 38]. Potentially, MMN could be used to stratify the high-risk individuals in order to provide them with the appropriate health care.

A few studies investigated whether MMN amplitude predicts long-term outcome in patients with schizophrenia or affective psychosis. Kaur et al. [39] reported that greater deficits in baseline MMN, in a mixed sample of patients with schizophrenia- and affective-spectrum psychosis, were associated with increased levels of work disability 12–24 months later. In contrast, in a study by Higgins et al. [40], which included patients with first episode of psychosis, MMN was not predictive of functional outcomes at a 12-month follow-up. On the other hand, Nakajima et al. [41] revealed that first-episode patients who did not achieve remission of psychosis (during an average interval of 3 years) had lower baseline MMN amplitude when compared with remitters. In addition, MMN amplitude at baseline correlated with future cognitive functioning of the patients. In a study by Kim and co-workers [42], reduced MMN amplitude in first-episode patients at baseline was a significant predictor of resistance to treatment. It has also been reported that in the course of 6 months, schizophrenia patients with lower baseline MMN showed lower remission rates and more severe symptoms, and required higher dose of antipsychotic medication [43]. These studies thus suggest that MMN amplitude may be a useful marker for predicting clinical outcome and future functioning of psychotic patients.

1.2 P300

P300 is a positive ERP which occurs 200–500 ms following the onset of a deviant or novel stimulus in the oddball task. In contrast to MMN, P300 is elicited if the individual is asked to focus the attention and respond to the rare stimulus (e.g., to count the occurrence of the target stimulus). As in the case of MMN, auditory oddball paradigm has been used most often in clinical studies to elicit P300. P300 consists of two subcomponents, smaller earlier P3a and larger later P3b. P3a is maximal over frontocentral sites and has been attributed to automatic attention orienting to the infrequent stimulus. This wave is elicited also if no response to rare stimuli is required. P3b is maximal at centroparietal electrodes and has been related to controlled processing, conscious target detection, and working memory updating [44–46]. Nevertheless, P3a and P3b are not always distinguished in clinical studies, which usually assess the maximum amplitude in the time window of P300, thus corresponding to P3b. It should be also noted that complex cognitive tasks involving target detection, evaluation, processing, decision-making, and response yield positive potentials in the time range of P300 and later. In research studies employing such tasks, the positive wave is often termed P300, but one should be aware that it may reflect neural processes beyond those engaged by a simple oddball task. Nieuwenhuis et al. [47] suggested that P300 reflects the activity of the locus coeruleus—noradrenergic system regulating (potentiating) arousal and cognitive brain processing. However, recent studies combining the measurements of P300 and of the pupillary response, which is under noradrenergic control, have not confirmed a strong relationship between concurrently recorded P300 and pupillary responses, indicating that relationship between P300 and the activation of locus coeruleus is less straightforward than had been originally suggested by Nieuwenhuis et al. [48, 49].

Decreased amplitude and increased latency of P300 are well-established findings in schizophrenia [50–52] but also occur in several other mental disorders including mood disorders [53, 54], substance use disorders [55], attention-deficit/hyperactivity disorder [56], autism spectrum disorder [57], obsessive–compulsive disorder [58], post-traumatic stress disorder [59], borderline personality disorder [60], Alzheimer's disease [61, 62], or Parkinson's disease [63]. In bipolar disorder, the reduction of P300 seems to persist also over euthymic periods, while in unipolar depression, the reduction is present only during the disorder's episodes, and the

P300 amplitude normalizes with remission of symptoms [64]. Also in schizophrenia the deficit in auditory P300 seems to be present as a stable trait marker over time despite changing severity of the psychopathological symptoms in patients [65, 66]. On the other hand, several studies reported that higher symptom severity and longer illness duration are associated with stronger P300 deficits [67–70]. Furthermore, deficits in P300 are related to several aspects of cognitive dysfunction in patients, in particular deficits in attention and memory [71–73]. P300 is diminished (and P300 latency is prolonged) already in first-episode patients with schizophrenia [52], and smaller P300 is associated with poorer real-life functioning at later stages of the illness across the psychosis (schizophrenia-bipolar) spectrum [74].

There is firm evidence that P300 is reduced and prolonged also in unaffected relatives of schizophrenia and bipolar patients [25, 75, 76], indicating a link between the deficits of P300 and genetic risk of psychosis. Indeed, twin studies have revealed that heritability of P300 amplitude and latency is high and both markers show substantial genetic correlation with schizophrenia [77, 78]. However, identification of significant associations of P300 characteristics at the level of single genes has not yet been successful and warrants further investigations [79].

Robust evidence also exists for diminished P300 in individuals at clinical high risk for psychosis [35, 36, 38], and the P300 deficit is stronger in future converters to psychosis and is associated with time to psychosis onset [30, 59, 80–82]. Nieman et al. [83] have estimated that a 1-μV reduction of P300 amplitude in clinical high-risk individuals is related to an increased risk of psychosis onset in next 3 years by 27%. Interestingly, P300 is also reduced in individuals with family history of substance abuse [55], and a recent study (which adopted a visual oddball paradigm) revealed that reduced P300 in alcohol-naive adolescents predicted drinking at older age [84]. This indicates that P300 deficits may share underlying neuropathological processes across several disorders which opens wider horizons for identifying the pathophysiological pathways to understand the etiopathogenesis of mental illness.

In summary, the deficits of the ERPs recorded using the simple oddball paradigm, MMN and P300, are not specific for psychotic disorders. Nevertheless, research suggests that MMN and P300 represent promising markers of the onset of psychosis in high-risk individuals as well as of outcome in psychotic patients. This indicates that assessment of MMN and P300 could usefully be applied in clinical psychiatry. Moreover, these ERPs are valuable for the study of pathophysiological mechanisms of mental disorders, warranting further investigations of these biomarkers in both basic and clinical research.

2 Quantitative EEG Biomarkers of Depression and Antidepressant Treatment Response

Quantitative electroencephalography (QEEG) refers to the process of extraction of various parameters from digitized EEG used for subsequent spectral and/or topographic analysis. Among the most common QEEG measures are absolute and relative power of the different frequency bands—delta (1–3 Hz), theta (4–7 Hz), alpha (8–12 Hz), beta (13–30 Hz), and gamma (30–100 Hz). The QEEG method has been used to explore possible diagnostic markers of depression since the late decades of the previous century. In a series of publications in 1988, Lieber et al. reported their observations on the diagnostic and differential diagnostic potential of QEEG pointing at measures of interhemispheric coherence, beta activity, and slow-wave excess as potent discriminators. Patients with depression were discriminated from healthy controls (HC) with an accuracy of 86% and from other psychiatric conditions (alcohol use disorder, primary dementia) with an overall accuracy of 76% [85]. In terms of differential diagnosis between unipolar (UD) and bipolar depressions (BD) (which is crucial for accurate treatment), Lieber and Newbury found that patients with UD showed excessive alpha activity and deficient beta activity, while those with BD demonstrated reduced alpha activity and increased beta activity [86].

Detailed review articles from this period pointed at replicated findings of elevated alpha and beta in unmedicated depressed patients as compared to control individuals with less evidence for delta and theta changes [87]. However conflicting reports were published as well, which confronted the validity of this potential biomarker. Another often reported finding was the alpha asymmetry (lower right and higher left alpha power) in major depression which, however, was found to demonstrate low heritability, moderate stability over time, and influence by differences in cranial and brain parenchymal asymmetries in bone thickness and differences in EEG montages [88], thus rendering it not suitable for clinical use.

Further endeavors focused on the use of the more advanced QEEG method called LORETA (low-resolution electromagnetic tomography analysis) which allows for better spatial localization and visualization of the electrical signal in a three-dimensional space [89]. By means of LORETA, decrease in absolute power in all frequency bands was found in depressed females as compared to HC [90]. In addition, relative delta/theta and beta powers were increased, while alpha activity was decreased. Moreover, the alpha asymmetry index demonstrated right frontal hyper- and left frontal hypoactivation which correlated with the Hamilton depression scale [90]. Lubar et al. reported that chronic depression compared with control group exhibited a left-to-right alpha 2 current density dominance in the left postcentral gyrus [91]. Moreover, the between-groups comparison of spectral power revealed decreased activity in the right middle temporal gyrus in patients (significant in the delta band only). Interestingly, LORETA analyses revealed greater theta and alpha current density (i.e., reduced brain activity) in depressed versus healthy adolescents, particularly in the left dorsolateral prefrontal cortex [92].

The low response and remission rates following treatment with antidepressant have pushed clinically oriented researchers to explore the potential of QEEG in response prediction as well. Several studies found that responders had greater alpha power compared to non-responders although this difference was more prominent at different sites—occipital or frontal [93, 94]. More recent studies reported a tendency for responders to fluoxetine to demonstrate higher alpha power although they had significantly more pronounced alfa asymmetry [95]. Using current source density technique for EEG analysis, a study by Tenke et al. reported higher alpha power in responders to selective serotonin reuptake inhibitors (SSRIs) and serotonin and noradrenaline reuptake inhibitors (SNRIs) but found no difference in asymmetry [96]. Our research team found converging evidence for response to treatment being linked to higher total alpha power as well as mean alpha power in the frontal and parietal left sites in a naturalistic sample of depressed patients undergoing treatment with various antidepressants [97].

Another potential marker for response was slow EEG power at baseline in different treatment modalities. Earlier reports linked, for example, response to imipramine with significantly less overall theta power prior to treatment and significantly greater anterior theta increments following the first 2 weeks [98]. At baseline, frontal relative theta power was significantly lower in treatment responders than in non-responders to SNRIs [99]. Non-response to repeated transcranial magnetic stimulation (rTMS) was reported to be linked to frontocentral theta power [100]. Response to fluoxetine, on the other hand, was not found to relate to theta power [101]. Thus, the significance of theta band measures for treatment response prediction remained controversial.

In 1994 Leuchter et al. proposed a complex measure called cordance that combined absolute and relative powers [102]. Theta cordance was found to be associated with response to fluoxetine where non-responders demonstrated more often a discordant state as opposed to responders [103]. Moreover, placebo responders showed a significant increase in prefrontal cordance starting early in treatment that was not seen neither in placebo non-responders nor in medication responders or non-responders [104]. In another trial with fluoxetine, venlafaxine, or placebo, medication responders were characterized by significant decreases in prefrontal cordance at day 2 and day 7 of treatment start [105]. Interestingly, response to electroconvulsive treatment was more likely to occur in depressed patients with higher pretreatment central theta cordance [106].

Further naturalistic study by the same group confirmed that early decreases in prefrontal cordance differentiated responders from non-responders in treatment resistant cases managed with SSRIs [107]. Moreover, independent replication was achieved by Bares et al. in a sample of patients with treatment-resistant depression using various antidepressants [108]. Response to venlafaxine and bupropion in treatment-resistant depression was found to be highly correlated with early changes in theta cordance [109, 110].

Later, the usefulness of theta cordance as a treatment prediction marker was confirmed in a study on bipolar depression [111]. In accordance with earlier reports, change in theta cordance (reduction during the first week) was found to predict favorable outcome to treatment with venlafaxine or with rTMS [112]. Notably, treatment-induced changes of EEG cordance emerge as the most replicated and powerful predictor of treatment response across modalities (various antidepressants, TMS) and diagnostic entities (unipolar, bipolar depression).

A recent study by the same group explored the predictive potential of alpha 1, alpha 2, and theta power, along with prefrontal theta cordance (PFTC), frontal and occipital alpha 1, alpha 2 asymmetry (FAA1/2, OAA1/2) at baseline, and their changes at week 1 in predicting response to SSRIs/SNRIs in a larger sample or 103 patients [113]. Compared to non-responders, responders to antidepressants showed a decrease in PFTC, while unfavorable results of the treatment were associated with an increase of OAA1/OAA2 at week 1. The study failed to replicate some previous reports on the association of theta and alpha power with treatment outcome.

In conclusion, decades of application of QEEG techniques in the search of diagnostic and prognostic biomarkers of depression yielded a couple of promising results. First, increased alpha power was repeatedly found as a characteristic of depression as compared to healthy controls with less evidence for the reported alpha asymmetry. Second, greater alpha was consistently associated with better response to treatment with antidepressants. Last but not least, early decrease in theta cordance predicted good treatment response to various pharmacological and instrumental treatments of depression.

Acknowledgements

This work was supported by the Scientific Grant Agency of the Ministry of Education, Science, Research, and Sport of the Slovak Republic (project no. VEGA 2/0118/21).

References

1. Näätänen R, Paavilainen P, Rinne T, Alho K (2007) The mismatch negativity (MMN) in basic research of central auditory processing: a review. Clin Neurophysiol 118:2544–2590. https://doi.org/10.1016/J.CLINPH.2007.04.026
2. Rosburg T, Kreitschmann-Andermahr I (2016) The effects of ketamine on the mismatch negativity (MMN) in humans - a meta-analysis. Clin Neurophysiol 127:1387–1394. https://doi.org/10.1016/J.CLINPH.2015.10.062
3. Todd J, Harms L, Schal LU, Michie PT (2013) Mismatch negativity: translating the potential. Front Psych 4:171. https://doi.org/10.3389/FPSYT.2013.00171/XML/NLM
4. Tada M, Kirihara K, Mizutani S et al (2019) Mismatch negativity (MMN) as a tool for translational investigations into early psychosis: a review. Int J Psychophysiol 145:5–14. https://doi.org/10.1016/J.IJPSYCHO.2019.02.009
5. Umbricht D, Krljesb S (2005) Mismatch negativity in schizophrenia: a meta-analysis. Schizophr Res 76:1–23. https://doi.org/10.1016/J.SCHRES.2004.12.002
6. Erickson MA, Ruffle A, Gold JM (2016) A meta-analysis of mismatch negativity in schizophrenia: from clinical risk to disease specificity and progression. Biol Psychiatry 79:980–987. https://doi.org/10.1016/J.BIOPSYCH.2015.08.025
7. Chitty KM, Lagopoulos J, Lee RSC et al (2013) A systematic review and meta-analysis of proton magnetic resonance spectroscopy and mismatch negativity in bipolar disorder. Eur Neuropsychopharmacol 23:1348–1363. https://doi.org/10.1016/J.EURONEURO.2013.07.007
8. Donaldson KR, Novak KD, Foti D et al (2020) Associations of mismatch negativity with psychotic symptoms and functioning transdiagnostically across psychotic disorders. J Abnorm Psychol 129:570–580. https://doi.org/10.1037/ABN0000506
9. Raggi A, Lanza G, Ferri R (2022) Auditory mismatch negativity in bipolar disorder: a focused review. Rev Neurosci 33:17–30. https://doi.org/10.1515/REVNEURO-2021-0010/MACHINEREADABLECITATION/RIS
10. Niznikiewicz MA, Spencer KM, Dickey C et al (2009) Abnormal pitch mismatch negativity in individuals with schizotypal personality disorder. Schizophr Res 110:188–193. https://doi.org/10.1016/J.SCHRES.2008.10.017
11. Tseng YJ, Nouchi R, Cheng CH (2021) Mismatch negativity in patients with major depressive disorder: a meta-analysis. Clin Neurophysiol 132:2654–2665. https://doi.org/10.1016/J.CLINPH.2021.06.019
12. Cheng CH, Chan PYS, Hsieh YW, Chen KF (2016) A meta-analysis of mismatch negativity in children with attention deficit-hyperactivity disorders. Neurosci Lett 612:132–137. https://doi.org/10.1016/J.NEULET.2015.11.033
13. Schwartz S, Shinn-Cunningham B, Tager-Flusberg H (2018) Meta-analysis and systematic review of the literature characterizing auditory mismatch negativity in individuals with autism. Neurosci Biobehav Rev 87:106–117. https://doi.org/10.1016/J.NEUBIOREV.2018.01.008
14. Chen TC, Hsieh MH, Lin YT et al (2020) Mismatch negativity to different deviant changes in autism spectrum disorders: a meta-analysis. Clin Neurophysiol 131:766–777. https://doi.org/10.1016/J.CLINPH.2019.10.031
15. Rangaswamy M, Porjesz B (2014) Understanding alcohol use disorders with neuroelectrophysiology. Handb Clin Neurol 125:383. https://doi.org/10.1016/B978-0-444-62619-6.00023-9
16. Näätänen R, Sussman ES, Salisbury D, Shafer VL (2014) Mismatch negativity (MMN) as an index of cognitive dysfunction. Brain Topogr 27:451–466. https://doi.org/10.1007/S10548-014-0374-6/FIGURES/2
17. Cheng CH, Hsu WY, Lin YY (2013) Effects of physiological aging on mismatch negativity: a meta-analysis. Int J Psychophysiol 90:165–171. https://doi.org/10.1016/J.IJPSYCHO.2013.06.026

18. Erickson MA, Albrecht M, Ruffle A et al (2017) No association between symptom severity and MMN impairment in schizophrenia: a meta-analytic approach. Schizophr Res Cogn 9:13. https://doi.org/10.1016/J.SCOG.2017.05.002
19. Näätänen R, Kujala T, Kreegipuu K et al (2011) The mismatch negativity: an index of cognitive decline in neuropsychiatric and neurological diseases and in ageing. Brain 134:3435–3453. https://doi.org/10.1093/BRAIN/AWR064
20. Horvath A, Szucs A, Csukly G et al (2018) EEG and ERP biomarkers of Alzheimer's disease: a critical review. Front Biosci Landmark 23:183–220. https://doi.org/10.2741/4587/4587.PDF
21. Wacongne C (2016) A predictive coding account of MMN reduction in schizophrenia. Biol Psychol 116:68–74. https://doi.org/10.1016/J.BIOPSYCHO.2015.10.011
22. Randeniya R, Oestreich LKL, Garrido MI (2018) Sensory prediction errors in the continuum of psychosis. Schizophr Res 191:109–122. https://doi.org/10.1016/J.SCHRES.2017.04.019
23. Hamilton HK, Perez VB, Ford JM et al (2018) Mismatch negativity but not P300 is associated with functional disability in schizophrenia. Schizophr Bull 44:492–504. https://doi.org/10.1093/SCHBUL/SBX104
24. Li X, Deng W, Xue R et al (2021) Auditory event-related potentials, neurocognition, and global functioning in drug naïve first-episode schizophrenia and bipolar disorder. Psychol Med 53:785. https://doi.org/10.1017/S0033291721002130
25. Earls HA, Curran T, Mittal V (2016) A meta-analytic review of auditory event-related potential components as Endophenotypes for schizophrenia: perspectives from first-degree relatives. Schizophr Bull 42:1504–1516. https://doi.org/10.1093/SCHBUL/SBW047
26. Donaldson KR, Larsen EM, Jonas K et al (2021) Mismatch negativity amplitude in first-degree relatives of individuals with psychotic disorders: links with cognition and schizotypy. Schizophr Res 238:161–169. https://doi.org/10.1016/J.SCHRES.2021.10.006
27. Fusar-Poli P, Borgwardt S, Bechdolf A et al (2013) The psychosis high-risk state: a comprehensive state-of-the-art review. JAMA Psychiatry 70:107–120. https://doi.org/10.1001/JAMAPSYCHIATRY.2013.269
28. Fusar-Poli P, Bonoldi I, Yung AR et al (2012) Predicting psychosis: meta-analysis of transition outcomes in individuals at high clinical risk. Arch Gen Psychiatry 69:220–229. https://doi.org/10.1001/ARCHGENPSYCHIATRY.2011.1472
29. Bodatsch M, Ruhrmann S, Wagner M et al (2011) Prediction of psychosis by mismatch negativity. Biol Psychiatry 69:959–966. https://doi.org/10.1016/J.BIOPSYCH.2010.09.057
30. Bodatsch M, Brockhaus-Dumke A, Klosterkötter J, Ruhrmann S (2015) Forecasting psychosis by event-related potentials-systematic review and specific meta-analysis. Biol Psychiatry 77:951–958. https://doi.org/10.1016/J.BIOPSYCH.2014.09.025
31. Perez VB, Woods SW, Roach BJ et al (2014) Automatic auditory processing deficits in schizophrenia and clinical high-risk patients: forecasting psychosis risk with mismatch negativity. Biol Psychiatry 75:459–469. https://doi.org/10.1016/J.BIOPSYCH.2013.07.038
32. Hamilton HK, Roach BJ, Bachman PM et al (2022) Mismatch negativity in response to auditory deviance and risk for future psychosis in youth at clinical high risk for psychosis. JAMA Psychiatry 79:780. https://doi.org/10.1001/JAMAPSYCHIATRY.2022.1417
33. Shaikh M, Valmaggia L, Broome MR et al (2012) Reduced mismatch negativity predates the onset of psychosis. Schizophr Res 134:42–48. https://doi.org/10.1016/J.SCHRES.2011.09.022
34. Tateno T, Higuchi Y, Nakajima S et al (2021) Features of duration mismatch negativity around the onset of overt psychotic disorders: a longitudinal study. Cereb Cortex 31:2416–2424. https://doi.org/10.1093/CERCOR/BHAA364
35. Perrottelli A, Giordano GM, Brando F et al (2021) EEG-Based Measures in At-Risk Mental State and Early Stages of Schizophrenia: A Systematic. Rev Front Psychiatr 12:653642. https://doi.org/10.3389/fpsyt.2021.653642
36. Lepock JR, Mizrahi R, Korostil M et al (2018) Event-related potentials in the clinical high-risk (CHR) state for psychosis: a systematic review. Clin EEG Neurosci 49:215–225. https://doi.org/10.1177/1550059418755212
37. Näätänen R, Shiga T, Asano S, Yabe H (2015) Mismatch negativity (MMN) deficiency: a break-through biomarker in predicting psychosis onset. Int J Psychophysiol 95:338–

344. https://doi.org/10.1016/J.IJPSYCHO.2014.12.012
38. Wang B, Zartaloudi E, Linden JF, Bramon E (2022) Neurophysiology in psychosis: the quest for disease biomarkers. Transl Psychiatry 12:100. https://doi.org/10.1038/S41398-022-01860-X
39. Kaur M, Lagopoulos J, Lee RSC et al (2013) Longitudinal associations between mismatch negativity and disability in early schizophrenia- and affective-spectrum disorders. Prog Neuro Psychopharmacol Biol Psychiatr 46:161–169. https://doi.org/10.1016/J.PNPBP.2013.07.002
40. Higgins A, Lewandowski KE, Liukasemsarn S, Hall MH (2021) Longitudinal relationships between mismatch negativity, cognitive performance, and real-world functioning in early psychosis. Schizophr Res 228:385–393. https://doi.org/10.1016/J.SCHRES.2021.01.009
41. Nakajima S, Higuchi Y, Tateno T et al (2021) Duration mismatch negativity predicts remission in first-episode schizophrenia patients. Front Psych 12:777378. https://doi.org/10.3389/FPSYT.2021.777378
42. Kim M, Kim T, Hwang WJ et al (2021) Forecasting prognostic trajectories with mismatch negativity in early psychosis. Psychol Med 53:1–11. https://doi.org/10.1017/S0033291721003068
43. Kim JS, Kwon YJ, Lee HY et al (2020) Mismatch negativity indices as a prognostic factor for remission in schizophrenia. Clin Psychopharmacol Neurosci 18:127–135. https://doi.org/10.9758/CPN.2020.18.1.127
44. Kok A (2001) On the utility of P3 amplitude as a measure of processing capacity. Psychophysiology 38:557–577. https://doi.org/10.1017/S0048577201990559
45. Polich J (2007) Updating P300: an integrative theory of P3a and P3b. Clin Neurophysiol 118:2128–2148. https://doi.org/10.1016/J.CLINPH.2007.04.019
46. Verleger R (2020) Effects of relevance and response frequency on P3b amplitudes: review of findings and comparison of hypotheses about the process reflected by P3b. Psychophysiology 57:e13542. https://doi.org/10.1111/PSYP.13542
47. Nieuwenhuis S, Aston-Jones G, Cohen JD (2005) Decision making, the P3, and the locus coeruleus-norepinephrine system. Psychol Bull 131:510–532. https://doi.org/10.1037/0033-2909.131.4.510
48. Kamp SM, Donchin E (2015) ERP and pupil responses to deviance in an oddball paradigm. Psychophysiology 52:460–471. https://doi.org/10.1111/PSYP.12378
49. LoTemplio S, Silcox J, Federmeier KD, Payne BR (2021) Inter- and intra-individual coupling between pupillary, electrophysiological, and behavioral responses in a visual oddball task. Psychophysiology 58:e13758. https://doi.org/10.1111/PSYP.13758
50. Bramon E, Rabe-Hesketh S, Sham P et al (2004) Meta-analysis of the P300 and P50 waveforms in schizophrenia. Schizophr Res 70:315–329. https://doi.org/10.1016/j.schres.2004.01.004
51. Jeon YW, Polich J (2003) Meta-analysis of P300 and schizophrenia: patients, paradigms, and practical implications. Psychophysiology 40:684–701. https://doi.org/10.1111/1469-8986.00070
52. Qiu YQ, Tang YX, Chan RCK et al (2014) P300 aberration in first-episode schizophrenia patients: a meta-analysis. PLoS One 9:e97794. https://doi.org/10.1371/JOURNAL.PONE.0097794
53. Morsel AM, Morrens M, Dhar M, Sabbe B (2018) Systematic review of cognitive event related potentials in euthymic bipolar disorder. Clin Neurophysiol 129:1854–1865. https://doi.org/10.1016/J.CLINPH.2018.05.025
54. Kangas ES, Vuoriainen E, Lindeman S, Astikainen P (2022) Auditory event-related potentials in separating patients with depressive disorders and non-depressed controls: a narrative review. Int J Psychophysiol 179:119–142. https://doi.org/10.1016/J.IJPSYCHO.2022.07.003
55. Euser AS, Arends LR, Evans BE et al (2012) The P300 event-related brain potential as a neurobiological endophenotype for substance use disorders: a meta-analytic investigation. Neurosci Biobehav Rev 36:572–603. https://doi.org/10.1016/J.NEUBIOREV.2011.09.002
56. Szuromi B, Czobor P, Komlósi S, Bitter I (2011) P300 deficits in adults with attention deficit hyperactivity disorder: a meta-analysis. Psychol Med 41:1529–1538. https://doi.org/10.1017/S0033291710001996
57. Cui T, Wang PP, Liu S, Zhang X (2017) P300 amplitude and latency in autism spectrum disorder: a meta-analysis. Eur Child Adolesc Psychiatry 26:177–190. https://doi.org/10.1007/S00787-016-0880-Z
58. Raggi A, Lanza G, Ferri R (2021) A review on P300 in obsessive-compulsive disorder. Front Psych 12:2119. https://doi.org/10.3389/FPSYT.2021.751215

59. Hamilton HK, Boos AK, Mathalon DH (2020) Electroencephalography and event-related potential biomarkers in individuals at clinical high risk for psychosis. Biol Psychiatry 88:294–303. https://doi.org/10.1016/J.BIOPSYCH.2020.04.002
60. Kutcher SP, Blackwood DHR, Clair DS et al (1987) Auditory P300 in borderline personality disorder and schizophrenia. Arch Gen Psychiatry 44:645–650. https://doi.org/10.1001/ARCHPSYC.1987.01800190065010
61. Hedges D, Janis R, Mickelson S et al (2016) P300 amplitude in Alzheimer's disease. Clin EEG Neurosci 47:48–55. https://doi.org/10.1177/1550059414550567
62. Tarawneh HY, Mulders WHAM, Sohrabi HR et al (2021) Investigating auditory electrophysiological measures of participants with mild cognitive impairment and Alzheimer's disease: a systematic review and meta-analysis of event-related potential studies. J Alzheimers Dis 84:419–448. https://doi.org/10.3233/JAD-210556
63. Xu H, Gu L, Zhang S et al (2022) (2022) N200 and P300 component changes in Parkinson's disease: a meta-analysis. Neurol Sci 1:1–12. https://doi.org/10.1007/S10072-022-06348-6
64. Barreiros AR, Breukelaar IA, Chen W et al (2020) Neurophysiological markers of attention distinguish bipolar disorder and unipolar depression. J Affect Disord 274:411–419. https://doi.org/10.1016/J.JAD.2020.05.048
65. Mathalon DH, Ford JM, Pfefferbaum A (2000) Trait and state aspects of p300 amplitude reduction in schizophrenia: a retrospective longitudinal study. Biol Psychiatry 47:434–449. https://doi.org/10.1016/S0006-3223(99)00277-2
66. Devrim-Üçok M, Keskin-Ergen Y, Üçok A (2016) Lack of progressive reduction in P3 amplitude after the first–episode of schizophrenia: a 6-year follow-up study. Psychiatry Res 243:303–311. https://doi.org/10.1016/J.PSYCHRES.2016.02.065
67. Ford JM, Mathalon DH, Marsh L et al (1999) P300 amplitude is related to clinical state in severely and moderately ill patients with schizophrenia. Biol Psychiatry 46:94–101. https://doi.org/10.1016/S0006-3223(98)00290-X
68. O'Donnell BF, Faux SF, McCarley RW et al (1995) Increased rate of P300 latency prolongation with age in schizophrenia. Electrophysiological evidence for a neurodegenerative process. Arch Gen Psychiatry 52:544–549. https://doi.org/10.1001/ARCHPSYC.1995.03950190026004
69. Mathalon DH, Ford JM, Rosenbloom M, Pfefferbaum A (2000) P300 reduction and prolongation with illness duration in schizophrenia. Biol Psychiatry 47:413–427. https://doi.org/10.1016/S0006-3223(99)00151-1
70. Wang J, Hirayasu Y, Hiramatsu KI et al (2003) Increased rate of P300 latency prolongation with age in drug-naive and first episode schizophrenia. Clin Neurophysiol 114:2029–2035. https://doi.org/10.1016/S1388-2457(03)00207-4
71. Kruiper C, Fagerlund B, Nielsen MO et al (2019) Associations between P3a and P3b amplitudes and cognition in antipsychotic-naïve first-episode schizophrenia patients. Psychol Med 49:868–875. https://doi.org/10.1017/S0033291718001575
72. Giordano GM, Perrottelli A, Mucci A et al (2021) Investigating the relationships of P3b with negative symptoms and Neurocognition in subjects with chronic schizophrenia. Brain Sci 11:1632. https://doi.org/10.3390/BRAINSCI11121632
73. Perrottelli A, Giordano GM, Brando F et al (2022) Unveiling the associations between EEG indices and cognitive deficits in schizophrenia-Spectrum disorders: a systematic review. Diagnostics 12:2193. https://doi.org/10.3390/DIAGNOSTICS12092193
74. Monaghan CK, Brickman S, Huynh P et al (2019) A longitudinal study of event related potentials and correlations with psychosocial functioning and clinical features in first episode psychosis patients. Int J Psychophysiol 145:48–56. https://doi.org/10.1016/J.IJPSYCHO.2019.05.007
75. Bramon E, McDonald C, Croft RJ et al (2005) Is the P300 wave an endophenotype for schizophrenia? A meta-analysis and a family study. NeuroImage 27:960–968. https://doi.org/10.1016/J.NEUROIMAGE.2005.05.022
76. Bestelmeyer PEG, Phillips LH, Crombie C et al (2009) The P300 as a possible endophenotype for schizophrenia and bipolar disorder: evidence from twin and patient studies. Psychiatry Res 169:212–219. https://doi.org/10.1016/J.PSYCHRES.2008.06.035
77. Hall MH, Schulze K, Rijsdijk F et al (2006) Heritability and reliability of P300, P50 and duration mismatch negativity. Behav Genet 36:845–857. https://doi.org/10.1007/S10519-006-9091-6

78. Hall MH, Rijsdijk F, Picchioni M et al (2007) Substantial shared genetic influences on schizophrenia and event-related potentials. Am J Psychiatry 164:804–812. https://doi.org/10.1176/AJP.2007.164.5.804/ASSET/IMAGES/LARGE/R721T4.JPEG
79. Hederih J, Nuninga JO, van Eijk K et al (2021) Genetic underpinnings of schizophrenia-related electroencephalographical intermediate phenotypes: a systematic review and meta-analysis. Prog Neuro Psychopharmacol Biol Psychiatr 104:110001. https://doi.org/10.1016/J.PNPBP.2020.110001
80. Van Tricht MJ, Nieman DH, Koelman JHTM et al (2010) Reduced parietal P300 amplitude is associated with an increased risk for a first psychotic episode. Biol Psychiatry 68:642–648. https://doi.org/10.1016/J.BIOPSYCH.2010.04.022
81. Hamilton HK, Roach BJ, Bachman PM et al (2019) Association between P300 responses to auditory oddball stimuli and clinical outcomes in the psychosis risk syndrome. JAMA Psychiatry 76:1187–1197. https://doi.org/10.1001/JAMAPSYCHIATRY.2019.2135
82. Tang Y, Wang J, Zhang T et al (2020) P300 as an index of transition to psychosis and of remission: data from a clinical high risk for psychosis study and review of literature. Schizophr Res 226:74–83. https://doi.org/10.1016/J.SCHRES.2019.02.014
83. Nieman DH, Ruhrmann S, Dragt S et al (2014) Psychosis prediction: stratification of risk estimation with information-processing and premorbid functioning variables. Schizophr Bull 40:1482–1490. https://doi.org/10.1093/SCHBUL/SBT145
84. Harper J, Malone SM, Iacono WG (2021) Parietal P3 and midfrontal theta prospectively predict the development of adolescent alcohol use. Psychol Med 51:416–425. https://doi.org/10.1017/S0033291719003258
85. Lieber AL, Prichep LS (1988) Diagnosis and subtyping of depressive disorders by quantitative electroencephalography: I. discriminant analysis of selected variables in untreated depressives. Hillside J Clin Psychiatry 10(1):71–83
86. Lieber AL, Newbury ND (1988) Diagnosis and subtyping of depressive disorders by quantitative electroencephalography: III. Discriminating unipolar from bipolar depression. Hillside J Clin Psychiatry 10(2):165–172
87. Pollock VE, Schneider LS (1990) Quantitative, waking EEG research on depression. Biol Psychiatry 27(7):757–780
88. Olbrich S, Arns M (2013) EEG biomarkers in major depressive disorder: discriminative power and prediction of treatment response. Int Rev Psychiatry 25(5):604–618. https://doi.org/10.3109/09540261.2013.816269
89. Pascual-Marqui RD, Michel CM, Lehmann D (1994) Low resolution electromagnetic tomography: a new method for localizing electrical activity in the brain. Int J Psychophysiol 18(1):49–65
90. Saletu B, Anderer P, Saletu-Zyhlarz GM (2010) EEG topography and tomography (LORETA) in diagnosis and pharmacotherapy of depression. Clin EEG Neurosci 41(4):203–210. https://doi.org/10.1177/155005941004100407
91. Lubar JF, Congedo M, Askew JH (2003) Low-resolution electromagnetic tomography (LORETA) of cerebral activity in chronic depressive disorder. Int J Psychophysiol 49(3):175–185. https://doi.org/10.1016/s0167-8760(03)00115-6
92. Auerbach RP, Stewart JG, Stanton CH, Mueller EM, Pizzagalli DA (2015) Emotion-processing biases and resting EEG activity in depressed adolescents. Depress Anxiety 32(9):693–701. https://doi.org/10.1002/da.22381
93. Ulrich G, Renfordt E, Frick K (1986) The topographical distribution of alpha-activity in the resting EEG of endogenous-depressive inpatients with and without clinical-response to pharmacotherapy. Pharmacopsychiatry 19:272–273
94. Suffin SC, Emory WH (1995) Neurometric subgroups in attentional and affective disorders and their association with pharmacotherapeutic outcome. Clin Electroencephalogr 26(2):76–83
95. Bruder GE, Sedoruk JP, Stewart JW, McGrath PJ, Quitkin FM, Tenke CE (2008) Electroencephalographic alpha measures predict therapeutic response to a selective serotonin reuptake inhibitor antidepressant: pre- and post-treatment findings. Biol Psychiatry 63(12):1171–1177. https://doi.org/10.1016/j.biopsych.2007.10.009
96. Tenke CE, Kayser J, Manna CG, Fekri S, Kroppmann CJ, Schaller JD et al (2011) Current source density measures of electroencephalographic alpha predict antidepressant treatment response. Biol Psychiatr 70(4):388–394. https://doi.org/10.1016/j.biopsych.2011.02.016
97. Kandilarova SS, Terziyski KV, Draganova AI, Stoyanov DS, Akabaliev VH, Kostianev SS (2017) Response to pharmacological treatment in major depression predicted by

electroencephalographic alpha power - a pilot naturalistic study. Folia Med (Plovdiv) 59(3): 318–325. https://doi.org/10.1515/folmed-2017-0040

98. Knott VJ, Telner JI, Lapierre YD, Browne M, Horn ER (1996) Quantitative EEG in the prediction of antidepressant response to imipramine. J Affect Disord 39(3):175–184. https://doi.org/10.1016/0165-0327(96)00003-1
99. Iosifescu DV, Greenwald S, Devlin P, Mischoulon D, Denninger JW, Alpert JE et al (2009) Frontal EEG predictors of treatment outcome in major depressive disorder. Eur Neuropsychopharmacol 19(11): 772–777. https://doi.org/10.1016/j.euroneuro.2009.06.001
100. Arns M, Drinkenburg WH, Fitzgerald PB, Kenemans JL (2012) Neurophysiological predictors of non-response to rTMS in depression. Brain Stimulation 5(4):569–576. https://doi.org/10.1016/j.brs.2011.12.003
101. Cook IA, Leuchter AF, Witte E, Abrams M, Uijtdehaage SH, Stubbeman W et al (1999) Neurophysiologic predictors of treatment response to fluoxetine in major depression. Psychiatry Res 85(3):263–273
102. Leuchter AF, Cook IA, Lufkin RB, Dunkin J, Newton TF, Cummings JL et al (1994) Cordance: a new method for assessment of cerebral perfusion and metabolism using quantitative electroencephalography. NeuroImage 1(3):208–219. https://doi.org/10.1006/nimg.1994.1006
103. Leuchter AF, Cook IA, Uijtdehaage SH, Dunkin J, Lufkin RB, Anderson-Hanley C et al (1997) Brain structure and function and the outcomes of treatment for depression. J Clin Psychiatry 58(Suppl 16):22–31
104. Leuchter AF, Cook IA, Witte EA, Morgan M, Abrams M (2002) Changes in brain function of depressed subjects during treatment with placebo. Am J Psychiatry 159(1):122–129
105. Cook IA, Leuchter AF, Morgan M, Witte E, Stubbeman WF, Abrams M et al (2002) Early changes in prefrontal activity characterize clinical responders to antidepressants. Neuropsychopharmacology 27(1):120–131. https://doi.org/10.1016/S0893-133X(02)00294-4
106. Stubbeman WF, Leuchter AF, Cook IA, Shurman BD, Morgan M, Gunay I et al (2004) Pretreatment neurophysiologic function and ECT response in depression. J ECT 20(3): 142–144. https://doi.org/10.1097/00124509-200409000-00004
107. Cook IA, Leuchter AF, Morgan ML, Stubbeman W, Siegman B, Abrams M (2005) Changes in prefrontal activity characterize clinical response in SSRI nonresponders: a pilot study. J Psychiatr Res 39(5): 461–466. doi: S0022-3956(04)00166-9 [pii]. https://doi.org/10.1016/j.jpsychires.2004.12.002
108. Bares M, Brunovsky M, Kopecek M, Stopkova P, Novak T, Kozeny J et al (2007) Changes in QEEG prefrontal cordance as a predictor of response to antidepressants in patients with treatment resistant depressive disorder: a pilot study. J Psychiatr Res 41(3–4):319–325. doi: S0022-3956(06)00118-X [pii]. https://doi.org/10.1016/j.jpsychires.2006.06.005
109. Bares M, Brunovsky M, Novak T, Kopecek M, Stopkova P, Sos P et al (2010) The change of prefrontal QEEG theta cordance as a predictor of response to bupropion treatment in patients who had failed to respond to previous antidepressant treatments. Eur Neuropsychopharmacol 20(7): 459–466. doi: S0924-977X(10)00067-2 [pii]. https://doi.org/10.1016/j.euroneuro.2010.03.007
110. Bares M, Brunovsky M, Kopecek M, Novak T, Stopkova P, Kozeny J et al (2008) Early reduction in prefrontal theta QEEG cordance value predicts response to venlafaxine treatment in patients with resistant depressive disorder. Eur Psychiatry 23(5): 350–355. doi: S0924-9338(08)01510-1 [pii]. https://doi.org/10.1016/j.eurpsy.2008.03.001
111. Bares M, Novak T, Brunovsky M, Kopecek M, Stopkova P, Krajca V et al (2012) The change of QEEG prefrontal cordance as a response predictor to antidepressive intervention in bipolar depression. A pilot study. J Psychiatr Res 46(2):219–225. doi: S0022-3956(11)00192-0 [pii]. https://doi.org/10.1016/j.jpsychires.2011.09.006
112. Bares M, Brunovsky M, Novak T, Kopecek M, Stopkova P, Sos P et al (2014) QEEG theta Cordance in the prediction of treatment outcome to prefrontal repetitive transcranial magnetic stimulation or venlafaxine ER in patients with major depressive disorder. Clin EEG Neurosci 46:73. https://doi.org/10.1177/1550059413520442
113. Bares M, Novak T, Vlcek P, Hejzlar M, Brunovsky M (2019) Early change of prefrontal theta cordance and occipital alpha asymmetry in the prediction of responses to antidepressants. Int J Psychophysiol 143:1–8. https://doi.org/10.1016/j.ijpsycho.2019.06.006

Chapter 7

Quantitative EEG in Patients with Schizophrenia

Georgi Panov

Abstract

Schizophrenia is a chronic mental illness with a diverse clinical picture. Diagnosis is based on clinical symptoms as well as their duration over time. The use of routine EEG in the evaluation of patients with schizophrenia did not achieve the expected results—to give a pattern corresponding to the clinical picture. In the search for specific EEG findings, the use of quantitative methods of EEG analysis shows encouraging results in the study of intimate mechanisms underlying clinical symptoms. These provide hope for future studies to determine diagnostic specificity and associated therapeutic effectiveness.

Key words Electroencephalography, Schizophrenia, Quantitative EEG, Absolute power, Relative power, Coherence, Treatment

1 Introduction

Quantitative electroencephalography (EEG) as a technique for topographic evaluation of electrophysiological data of the brain is useful in the analysis of the status and prognosis of certain clinical conditions such as concussion, schizophrenia, depression, and substance abuse, as well as the effect of treatment [1].

Analysis of quantitative EEG in patients with schizophrenia can be considered in several aspects: technical provision, the data from the studies in patients without treatment, and analysis of patients against the background of therapy.

In the technical provision of the study, 16-, 19-, or 21-channel EEG recordings with monopolar mounting (averaged reference electrode or averaged lead from ear electrodes) can be used when studying epochs of the EEG recording with a certain duration. Epochs used for EEG evaluation are recommended to be 20 s, and increasing the duration does not usually lead to greater accuracy of the study [2]. Shorter epochs can also be used, as a comparative analysis by a group of authors found no difference in accuracy when using different lengths of estimated time frames [3].

Drozdstoy Stoyanov et al. (eds.), *Computational Neuroscience*, Neuromethods, vol. 199,
https://doi.org/10.1007/978-1-0716-3230-7_7,

EEG spectral features have been found to be highly heritable [4, 5]. These data show that the EEG is state-dependent (varies depending on the state of wakefulness, relaxation, sleep). Each person has the equivalent of an EEG tuning, a natural spontaneous rhythm that the individual exhibits when tested under approximately the same conditions. This relative stability of the spectral characteristics of the EEG provides opportunities for detecting changes related to the development of a certain pathology, as well as for tracking the established changes over time. An age of the patient should also be taken into account when conducting research and especially in comparative studies. With advancing age, there is a shift to slower frequencies of alpha activity with a decrease of 1 Hz every 10 years after age 50 [6].

In search of these individually specific EEG characteristics, authors logically concluded that longer recordings were necessary to extract and record processing epochs without the presence of artefacts [7]. One should not forget the fact that excessively long recordings in the awake state during which the patient stands with eyes closed can logically lead to periods of abduction. In this way, physiologically slower spectra in the EEG can be increased and this can lead to erroneous results. This is also a "trap" to avoid when conducting these longer studies. The presented data show that the analysis of "artifact-free" EEG epochs is of the greatest importance, as is the duration of the study itself and the number of electrodes that are used in the registration of brain activity.

Numerous studies have been conducted in patients with schizophrenia who are not being treated with antipsychotics or other medications. They unequivocally show that certain deviations are present when compared with studies conducted in the general population. Given the fact that schizophrenia is not a homogeneous disorder, the data also show the presence of certain differences in the analyses performed.

The analysis of 30 patients with right-sided functional lateralization showed increased spectral power of delta, theta, and alpha-1 activity when compared with healthy controls. The authors found faster activity within the beta-1 range in the occipital regions as well as a decrease in alpha-2 activity [8]. A study of only 20 treatment-naive patients showed an increase in slow alpha (8–9 Hz) activity and less fast alpha activity (9–13 Hz) [9]. Different results are also available. They found higher mean spectral power in most frequency ranges in patients with schizophrenia [10].

These differences are particularly pronounced in the delta band, in the fast alpha and beta bands, particularly in the left frontal region. Delta activity is most manifest in F7 electrode [11]. These studies at first glance show a different profile of quantitative EEG changes compared to any of the previous results, finding higher power for the high frequencies of the alpha spectrum. The authors of the study indicate that they are correlated with the anxiety and depression scales, which provides an explanation for the increased power in these ranges. These differences may be explained in the context of the differences in the observed clinical cohorts.

Interesting data were presented by another study, where 48 schizophrenic patients were observed who were without treatment at the time of the study. Both patients with predominantly negative symptoms and those with a florid clinical picture were examined. The authors found differences between the two groups. In those with a florid picture, a decrease in alpha-1 and an increase in beta power were registered, which gives the authors the right to conclude that a form of hyperarousal underlying the positive symptomatology process in active psychosis is observed. Conversely, in the negative form of schizophrenia, they found an increase in delta and theta activity bifrontally and bitemporally and a reduction in alpha-1 and an increase in beta activity. Beta activity was more pronounced in those in the florid phase of the disease [12]. Similar results with increased activity in slow-wave spectra in patients with predominantly negative symptoms were also recorded by other authors in more recent studies [13].

In addition to the differences in activity in patients with pronounced negative and positive symptoms, other authors analyze and compare patients with a disorganized type and those with a paranoid type of schizophrenia. In the analyzed groups of patients, no difference was found in terms of gender and age. In all 50 acute and untreated schizophrenic patients, they found higher power for delta, theta, alpha-1, and beta-1 activity and less pronounced alpha-2 activity compared to the control group. These results confirm the studies of the above-described results on increased spectral power in the slow-wave ranges in patients with schizophrenia. Compared to controls, patients with disorganized schizophrenia had significant increases in theta and beta-1 and decreases in alpha-2 activities. They found a significant increase in delta and alpha-1 activity in the occipitoparietal regions and beta-2 activity in the frontal areas of the brain. Patients with paranoid schizophrenia showed significantly increased delta activity in posterior regions, increased alpha-1 activity in anterior regions, and decreased alpha-2 activity in both regions. In the paranoid type, however, there were no significant differences for theta, beta-1, and beta-2 frequency ranges. Patients with disorganized schizophrenia show more increased theta and decreased alpha-2 activity than patients with paranoid schizophrenia [14]. These studies again emphasize the fact that it cannot be considered very correct to talk about EEG studies in patients with schizophrenia in general, and it is appropriate to evaluate changes in neurophysiological indicators in different subtypes of schizophrenia when quantitative EEG studies are conducted.

The analysis of other authors in patients with schizophrenia draws other conclusions. The researchers found no difference in the power of the individual frequency spectra as well as in terms of the average frequency. They found a difference in coherence between the two brain hemispheres. High interhemispheric coherence was found in schizophrenic patients (not undergoing treatment at the time of the study) compared to healthy controls. The data from their research show that brain functions are less lateralized in patients with schizophrenia. The authors evaluated a relatively small group of patients—11, but still questions remain regarding their findings [15].

The data of these authors on the reduced lateralization of brain activity in patients with schizophrenia is also confirmed by the studies of other researchers. They indicated that data on the lack of lateralization in schizophrenic patients changed after treatment, and they found evidence of building lateralization more pronounced for the beta frequency range [16]. Confirmation of the presence of changes in the coherence in the functional organization of the brain in patients with schizophrenia has also been established by other researchers. They observed impaired centroparietal coherence, particularly pronounced in the left hemisphere when performing certain motor tasks [17].

On the basis of the established changes in the quantitative parameters of the EEG, authors conclude that the quantitative EEG is useful for the correct classification and differentiation of patients with schizophrenia from other forms of psychosis. They found that in schizophrenic patients, a higher power for the frontal slow spectra was observed, which was also associated with a greater expression of negative symptoms [18]. These data are confirmed by EEG analysis in 39 patients with untreated schizophrenia, who clearly demonstrate increased delta activity in the frontal regions, which the authors associate with hypoactivity and hypofrontality, results that the authors did not find in patients with schizotypal disorder [19].

Other studies have suggested that a characteristic EEG profile can be found in patients with schizophrenia, but not in patients with other psychotic conditions, who do not meet the criteria for the diagnosis of schizophrenia [20].

A large body of data supports the conclusion that the slow-wave abnormality (mainly delta spectrum) is more or less localized to frontal lobe regions in both untreated and treated schizophrenic patients [11, 18, 21–30].

Other studies find different results. They indicated that EEG spectral abnormalities in patients with schizophrenia are localized in the occipital regions of the brain [8, 31].

The work of John et al. (1994) strongly suggests that patients showing different patterns of topographical distribution of abnormalities in EEG spectral analysis represent different biological subtypes of the disorder. They suggest for a more accurate assessment the examination of five main indicators: the absolute power, the relative power, the coherence, the asymmetry, and the frequency characteristic of the individual spectra in the EEG [22].

Most often, an increase in slow-wave changes in the EEG reflects a disturbed subcortical synchronizing system involving the mesencephalic reticular formation, the reticular formation as a whole, and the thalamus. The analysis of correlations between P300 recordings in patients with 25 schizophrenic patients who were without drug therapy was made. A positive correlation of P300 amplitude with increased theta and delta activity as well as reduced peak frequency of the alpha rhythm was found. After treatment, an increase in the alpha spectral peak was found, which the authors interpreted as impaired thalamocortical regulation in schizophrenic disorders [4].

There are other results that provide other directions in research. Seventeen schizophrenic patients were tested before and 6 weeks after clozapine treatment. Results were compared with a control group of healthy volunteers. The results showed that before the drug intervention, patients had increased spectral power for theta, reduced peak for alpha, increased global power, and increased delta power in the frontal regions. Further slowing of theta and delta rhythms and slowing of the alpha peak frequency were found after clozapine treatment [23].

These results raise the question of whether these are not intermediate data regarding the dynamics of power expression in individual frequency ranges and how these indicators would change with catamnesis follow-up. This question is hardly logical based on the fact that, for example, the provocation of obsessive–compulsive symptoms in schizophrenic patients treated with clozapine appears after about 6 months of therapy [32]. These data indicate that we will most likely observe additional changes in the individual frequency spectra over the course of clozapine treatment.

Sponheim et al. (1994) conducted a comparative study of 102 schizophrenic patients, 44 of whom had a first episode of the disease, and the rest they classified as chronic schizophrenia (58 patients) [33]. They found that in the EEG of patients with schizophrenia, there are more delta and theta waves and less alpha rhythms when compared to healthy controls. No significant difference is found regarding the fast-wave spectra. The authors also found no significant difference between those with a first psychotic episode and those with chronic schizophrenia. No differences were found in the power distribution of individual frequency bands between those on treatment and those not receiving therapy. The authors concluded that the influence of medication does not significantly affect the characteristic of frequency spectra in patients with schizophrenia.

Sponheim and colleagues found that within a population of schizophrenics, those with more pronounced negative symptoms and larger ventricles showed increased spectral power in slow-wave bands [34]. Other authors have found a relationship between EEG findings and treatment effectiveness, finding that patients who respond to treatment tend to have faster activity and less increase in slow activity than patients who do not respond to treatment [35].

Comparison of resting state quantitative EEG of schizophrenic patients with healthy controls found a large difference. When compared with patients with affective disorders, this difference was reduced but remained significant, confirming data that increased delta wave activity in frontal regions was observed in patients with schizophrenia [18].

These findings are confirmed by other studies that also show a reduction in peak spectral power for the alpha frequency range in patients with chronic schizophrenia [36]. The same authors also analyzed the relationship between increased delta activity in the frontal regions and the presence of artifactual activity related to eyeball movements. The eyeballs can be thought of as an electric dipole that changes direction when the eyes are opened and closed. The extent to which this artifactual activity superimposes on the EEG activity varies in the studies presented, and different techniques are presented to remove them [36]. In some cases, it is extremely difficult to distinguish between delta activity and eye movement artifacts. In the same way, the artifactual activity coming from the tongue, the muscles of the scalp, and especially the influence of the ECG on the brain's bioelectrical activity can be considered [37].

When comparing patients with schizophrenia and healthy controls, one should not forget the fact that this artifact is observed in both groups. On the other hand, it is important to note that in patients with active psychotic production, it is possible to observe a low degree of cooperation in carrying out the study, which emphasizes the fact that these artifacts should be detected by examining time epochs without such artifacts (in other words, to analyze short epochs of the record).

The question remains whether the research results are reliable enough. Most studies have shown that retesting is not necessary because test reliability is sufficient, and this has been established in retesting [38, 40]. A study showed that the correlation coefficient was higher in schizophrenic patients $r = 0.94$ compared to healthy controls with $r = 0.70$. Test–retest reliability of $r = 0.9$ can be obtained in both schizophrenic and healthy subjects when artifact-free 8-s epochs are used [41].

For the development of quantitative EEG and its wider incorporation into clinical practice, it is necessary to recommend procedures used in most studies that seek to translate quantitative EEG findings in schizophrenia into clinical diagnostic tests. We believe that it is appropriate to use the frequency ranges already established in practice in conducting the quantitative analysis of EEG in order to be able to ensure a connection with the methods and practices imposed in clinical practice.

Examinations should be carried out in a "resting state" with the eyes closed. The "state of rest" should be clearly defined. By definition, the state of rest is the absence of specific mental activity. Instructions may be limited to "lie still and be awake"; closed eyes in a muted light and sound room.

In the analysis of schizophrenic patients, and especially during the florid stage, it is difficult to define the state of rest due to the fact that we are not able to analyze the patient's mental activity and his cooperation in the study. If a greater cognitive activity is suspected, which would lead to the attenuation of the alpha activity, respectively, the spectral power of it, it is appropriate to conduct a longer recording in order to silence the "engaged attention." Another "trap" in the study is conducting a very long study (as said above), which at the end of the recording can lead to relaxation and the appearance of entrainment, which would artificially activate theta activity. These possible errors in conducting EEG and especially in the use of quantitative analyses should be avoided.

EEG analysis in the schizophrenic patient provides additional opportunities for evaluation in the differential diagnostic and therapeutic approach in the schizophrenic patient. The ease of carrying out the research, the good tolerability by the patient, and the relatively low cost are an opportunity for easy conduction in psychiatric units, ward, and clinics.

In patients with schizophrenia, it is recommended to carry out consecutive studies during the development of the psychotic episode in order to track its dynamics over time and to follow the correlation with neurophysiological indicators. On the other hand, there are currently also therapeutic attempts to influence the negative symptoms in schizophrenia. This makes it possible to follow the effect and search for certain neurophysiological correlates of these promising therapeutic interventions. Despite the presence of heterogeneity in the clinical picture of patients with schizophrenia, it can be concluded that there are established relationships with quantitative EEG findings, and even in the divergent data of the results, a neurophysiological structure corresponding to this heterogeneity can be found.

References

1. McVoy M, Lytle S, Fulchiero E, Aebi ME, Adeleye O, Sajatovic M (2019) A systematic review of quantitative EEG as a possible biomarker in child psychiatric disorders. Psychiatry Res 279:331–344. https://doi.org/10.1016/j.psychres.2019.07.004. Epub 2019 Jul 9
2. Gudmundsson S et al (2007) Reliability of quantitative EEG features. Clin Neurophysiol 118:2162–2171
3. Yan MM, Xu XH, Huang ZL et al (2011) Selection of optimal epoch duration in assessment of rodent sleep-wake profiles. Sleep Biol Rhythms 9:46–55. https://doi.org/10.1111/j.1479-8425.2010.00484.x

4. Kirino E (2004) Correlation between P300 and EEG rhythm in schizophrenia. Clin EEG Neurosci 35(3):137–146. https://doi.org/10.1177/155005940403500306
5. van Beijsterveldt CE, van Baal GC (2002) Twin and family studies of the human electroencephalogram: a review and a meta-analysis. Biol Psychol 61:111–138. https://doi.org/10.1016/S0301-0511(02)00055-8
6. Mankovsky NB, Belonog RP (1971) Aging of the human nervous system in the electroencephalographic aspect. Geriatrics 26(8):100. passim
7. Clements BA, Sponheim SR et al (1994) Resting EEG in first-episode schizophrenia patients, bipolar psychosis patients, and their first-degree relatives. Psychophysiology 31: 486–494. https://doi.org/10.1111/j.1469-8986.1994.tb01052.x
8. Miyauchi T, Tanaka K, Hagimoto H, Miura T, Kishimoto H, Matsushita M (1990) Computerized EEG in schizophrenic patients. Biol Psychiatry 28(6):488–494, ISSN 0006-3223. https://doi.org/10.1016/0006-3223(90)90482-H
9. Omori M, Koshino Y, Murata T, Murata I, Nishio M, Sakamoto K, Tan H, Isaki K (1995) Quantitative EEG in never-treated schizophrenic patients. Biol Psychiatry 38(5): 303–309., ISSN 0006-3223. https://doi.org/10.1016/0006-3223(95)00300-6
10. Nagase Y, Okubo Y, Toru M (1996) Electroencephalography in schizophrenic patients: comparison between neuroleptic-naive state and after treatment. Biol Psychiatry 40(6): 452–456, ISSN 0006-3223. https://doi.org/10.1016/0006-3223(96)00304-6
11. Gattaz WF, Mayer S, Ziegler P, Platz M, Gasser T (1992) Hypofrontality on topographic EEG in schizophrenia. Correlations with neuropsychological and psychopathological parameters. Eur Arch Psychiatry Clin Neurosci 241(6): 328–332. https://doi.org/10.1007/BF02191956. PMID: 1504108
12. Saletu B, Küfferle B, Anderer P, Grünberger J, Steinberger K (1990) EEG-brain mapping in schizophrenics with predominantly positive and negative symptoms: comparative studies with remoxipride/haloperidol. Eur Neuropsychopharmacol 1(1):27–36., ISSN 0924-977X. https://doi.org/10.1016/0924-977X(90)90007-W
13. Gschwandtner U, Zimmermann R, Pflueger MO, Riecher-Rössler A, Fuhr P (2009) Negative symptoms in neuroleptic-naïve patients with first-episode psychosis correlate with QEEG parameters. Schizophrenia Res 115(2–3):231–236. ISSN 0920-9964
14. Miyauchi T, Endo S, Kajiwara S, Ishii M, Okajima J (1996) Computerized electroencephalogram in untreated schizophrenics: a comparison between disorganized and paranoid types. Psychiatry Clin Neurosci 50(2): 71–78. https://doi.org/10.1111/j.1440-1819.1996.tb01666.x
15. Nagase Y, Okubo Y, Matsuura M, Kojima T, Toru M (1992) EEG coherence in unmedicated schizophrenic patients: topographical study of predominantly never medicated cases. Biol Psychiatry 32(11):1028–1034. https://doi.org/10.1016/0006-3223(92)90064-7
16. Merrin EL, Floyd TC (1991) Average reference EEG lateralization in schizophrenic patients. J Neuropsychiatry Clin Neurosci 3(3):307–314. https://doi.org/10.1176/jnp.3.3.307
17. Michelogiannis S, Paritsis N, Trikas P (1991) EEG coherence during hemispheric activation in schizophrenics. Eur Arch Psychiatry Clin Neurosci 241(1):31–34. https://doi.org/10.1007/BF02193751
18. Winterer G, Ziller M, Dorn H, Frick K, Mulert C, Wuebben Y, Herrmann WM (2000) Frontal dysfunction in schizophrenia — a new electrophysiological classifier for research and clinical applications. Eur Arch Psychiatr Clin Neurosci 250:207–214. https://doi.org/10.1007/s004060070026
19. Wuebben Y, Winterer G (2001) Hypofrontality -- a risk-marker related to schizophrenia? Schizophr Res 48(2–3):207–217. https://doi.org/10.1016/s0920-9964(00)00047-5
20. Etevenon P, Pidoux B, Rioux P, Peron-Magnan P, Verdeaux G, Deniker P (1979) Intra- and interhemispheric EEG differences quantified by spectral analysis. Comparative study of two groups of schizophrenics and a control group. Acta Psychiatr Scand 60(1): 57–68. https://doi.org/10.1111/j.1600-0447.1979.tb00265.x
21. Schellenberg R, Knorr W, Schindler M, Kropf S (1990) EEG-power spectral components of schizoaffective disorders. Schizophr Res 3(5–6):357–359. https://doi.org/10.1016/0920-9964(90)90023-Z
22. John ER, Prichep LS, Alper KR, Mas FG, Cancro R, Easton P, Sverdlov L (1994) Quantitative electrophysiological characteristics and subtyping of schizophrenia. Biol Psychiatry 36(12):801–826
23. Knott V, Mahoney C, Kennedy S, Evans K (2001) EEG power, frequency, asymmetry and coherence in male depression. Psychiatry Res Neuroimaging 106(2):123–140, ISSN 0925-4927. https://doi.org/10.1016/S0925-4927(00)00080-9

24. Wada Y, Takizawa Y, Kitazawa S, Jiang ZY, Yamaguchi N (1994) Quantitative EEG analysis at rest and during photic stimulation in drug-naive patients with first-episode paranoid schizophrenia. Eur Arch Psychiatry Clin Neurosci 244(5):247–251
25. Kahn EM, Weiner RD, Coppola R, Kudler HS, Schultz K (1993) Spectral and topographic analysis of EEG in schizophrenic patients. Biol Psychiatry 33(4):284–290
26. Saletu B, Kufferle B, Anderer P, Grunberger J, Steinberger K (1990) EEG-brain mapping in schizophrenics with predominantly positive and negative symptoms. Comparative studies with remoxipride/haloperidol. Eur Neuropsychopharmacol 1(1):27–36
27. Mukundan CR (1986) Computed EEG in schizophrenics. Biol Psychiatry 21(12): 1225–1228
28. Serafetinides EA (1984) EEG lateral asymmetries in psychiatric disorders. Biol Psychol 19(34):237–246
29. Takeuchi K, Takigawa M, Fukuzako H, Hokazono Y, Hirakawa K, Fukuzako T, Ueyama K, Fujimoto T, Matsumoto K (1995) Correlation of third ventricular enlargement and EEG slow wave activity in schizophrenic patients. Psychiatry Res 55(1):1–11
30. Omori M, Koshino Y, Murata T, Murata I, Nishio M, Sakamoto K, Horie T, Isaki K (1995) Quantitative EEG in never-treated schizophrenic patients. Biol Psychiatry 38(5): 305–309
31. Clementz BA, Sponheim SR, Iacono WG, Beiser M (1994) Resting EEG in first-episode schizophrenia patients, bipolar psychosis patients, and their first-degree relatives. Psychophysiology 31(5):486–494
32. Scheltema Beduin AA, Swets M, Machielsen M, Korver N, genetic risk and outcome of psychosis investigators (2012) Obsessive-compulsive symptoms in patients with schizophrenia: a naturalistic cross-sectional study comparing treatment with clozapine, olanzapine, risperidone, and no antipsychotics in 543 patients. J Clin Psychiatry 73(11):1395–1402. https://doi.org/10.4088/JCP.11m07164. PMID: 23218156
33. Sponheim SR, Clementz BA, Iacono WG, Beiser M (1994) Resting EEG in first-episode and chronic schizophrenia. Psychophysiology 31(1):37–43. https://doi.org/10.1111/j.1469-8986.1994.tb01023.x
34. Sponheim SR, Clementz BA, Iacono WG, Beiser M (2000) Clinical and biological concomitants of resting state EEG power abnormalities in schizophrenia. Biol Psychiatry 48(11): 1088–1097
35. Czobor P, Volavka J (1992) Level of haloperidol in plasma is related to electroencephalographic findings in patients who improve. Psychiatry Res 42(2):129–144. https://doi.org/10.1016/0165-1781(92)90077-g
36. Harris A, Melkonian D, Williams L, Gordon E (2006) Dynamic spectral analysis findings in first episode and chronic schizophrenia. Int J Neurosci 116(3):223–246
37. Kendirli MT, Aparci M, Kendirli N, Tekeli H, Karaoglan M, Senol MG, Togrol E (2015) Diagnostic role of ECG recording simultaneously with EEG testing. Clin EEG Neurosci 46(3):214–217. https://doi.org/10.1177/1550059414551554. Epub 2014 Sep 23
38. Lifshitz K, Lee KL, Susswein S (1987) Long-term replicability of EEG spectra and auditory evoked potentials in schizophrenic and normal subjects. Neuropsychobiology 18(4):205–211
39. McGuire KA, Katsanis J, Iacono WG, McGue M (1998) Genetic influences on the spontaneous EEG: an examination of 15-year-old and 17-year-old twins. Dev Neuropsychol 14(1):7–18
40. Gao Y, Tuvblad C, Raine A, Lozano DI, Baker LA (2009) Genetic and environmental influences on frontal EEG asymmetry and alpha power in 9-10-year-old twins. Psychophysiology 46(4):787–796. https://doi.org/10.1111/j.1469-8986.2009.00815.x
41. Lund TR, Sponheim SR, Iacono WG, Clementz BA (1995) Internal consistency reliability of resting EEG power spectra in schizophrenic and normal subjects. Psychophysiology 32(1): 66–71

Part IV

Neuroimaging Methods

Chapter 8

Computational Anatomy Going Beyond Brain Morphometry

Bogdan Draganski, Rositsa Paunova, Adeliya Latypova, and Ferath Kherif

Abstract

In this chapter, we provide a comprehensive historical overview of algorithms for feature extraction of structural magnetic resonance imaging data and the corresponding analytical frameworks. We then focus on the use of T1-weighted images, which still represent the working horse of surface- and voxel-based morphometry to elaborate on the complex relationships between magnetic resonance imaging contrast and underlying brain tissue properties. This critical point is the motivation for embarking on novel structural imaging protocols based on biophysical models, which are sensitive to tissue myelin, iron, and water content. We expand on the concept of voxel-based quantification to demonstrate the added value to existing methods using T1-weighted data—both in terms of robust brain tissue classification and of straightforward neurobiological interpretation of the obtained results. We do not stop short of considering the unresolved issues in voxel-based quantification currently implemented in the data processing and analysis framework of Statistical Parametric Mapping (SPM12). We conclude with an outlook to the future developments and perspectives in computational anatomy, particularly the need for integration of the available magnetic resonance imaging contrasts at the level of statistical analysis without hampering the interpretability of our findings.

Key words Voxel-based quantification, Magnetization transfer saturation, Effective transverse relaxation rate, Longitudinal relaxation rate, Proton density, Relaxometry

1 Introduction

The advent of magnetic resonance imaging (MRI) almost half a century ago helped in addressing one of the fundamental questions in neuroscience about the relationship between brain structure and function. Computational anatomy represents one of the imaging neuroscience domains with a focus on studying brain anatomy using high-resolution structural MRI data and state-of-the-art statistical analyses in three-dimensional space (Fig. 1). Since the mid-1990s, the community is witnessing a continuous expansion of computational anatomy studies covering the wide span between basic and clinical neurosciences. Consequently, advances in MRI technology and computer sciences are pushing the boundaries of

Drozdstoy Stoyanov et al. (eds.), *Computational Neuroscience*, Neuromethods, vol. 199,
https://doi.org/10.1007/978-1-0716-3230-7_8,

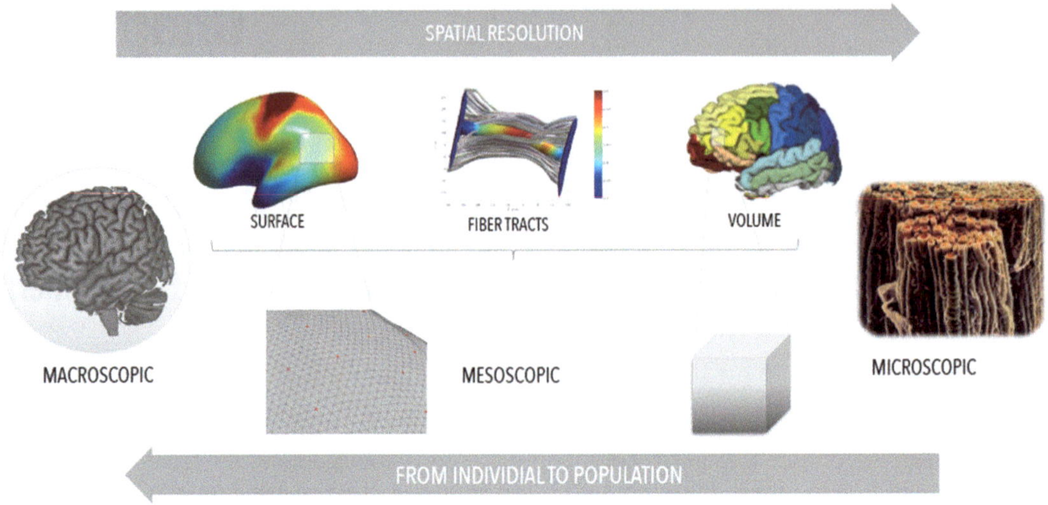

Fig. 1 Schematic representation of surface-, tract- and volume-based computational anatomy methods covering the mesoscopic scale of observation

imaging data diversity, the complexity of algorithms for feature extraction, and the sophistication of dedicated statistical frameworks.

From historical perspective, computational anatomy, like functional MRI, has passed through different stages of maturation including mapping of structural differences across the plethora of brain disorders, looking for causality in longitudinal studies or applying machine learning techniques that address questions about diagnosis and prognosis of a particular brain condition. Most recently, researchers started using deep learning methods not only for computer-assisted diagnosis but also for data augmentation, which increases the number of available MR images through data transforms.

Since the inception of computational anatomy, T1-weighted MR images have been its working horse that provides robust brain morphometry estimates of local volume, cortical thickness, and surface area. In the last decade, MR physics challenged the field and offered a viable alternative to the simplistic morphometric view on anatomy by developing imaging protocols based on biophysical models that tap into the origins of MR contrast in brain tissues. The new techniques of relaxometry, myelin water imaging, and susceptibility mapping, to name but a few, provide not only an exclusive access to previously considered as "difficult" structures, e.g., basal ganglia and thalamus, but carry the promise to deliver quantitative data—the perfect case for unbiased multi-center studies.

The computational anatomy community advanced the sophistication of the frameworks and algorithms for feature extraction to keep the pace with the incremental developments in technology and neuroscience. In the spirit of open science, the surface-based (e.g., FreeSurfer) and voxel-based (e.g., SPM) tools are freely available to the scientific community. There are numerous toolboxes that enhance the functionality and provide in an agile way solutions for dealing with data generated with new MRI protocols.

In this chapter, we offer a comprehensive overview of the new developments in computational anatomy to then discuss their added value and unused potential for studying in vivo the human brain. We deliberately exclude diffusion-weighted imaging from our presentation given its additional complexity from MR physics and feature extraction point of view that goes far beyond the focus of this chapter.

2 Historical Overview

Just a few years after the first publications using manual delineation of regions of interest followed by volume calculation [1, 2], the community swiftly adopted the idea of voxel-based analysis borrowed from functional positron emission tomography and MRI studies [3]. Voxel, in analogy to pixel, represents a value on a three-dimensional regular grid. The implementation of voxel-based analysis was made possible by methodological developments for automated brain tissue classification [3–5] and spatial registration to standardized Cartesian coordinate system [6–8]. The benefits of a voxel-based analysis based on automated feature extraction over a selective manual or semi-automated region-of-interest approach were evident—it enabled the state-of-the-art statistical analysis over the whole-brain offering insight in structure–function relationships going beyond a limited number of predefined regions. Given the growing interest within the basic and clinical neuroscience community, the technique of voxel-based morphometry quickly gained a great deal of popularity and became a household name in imaging neuroscience research laboratories.

Not much time after the first voxel-based morphometry studies, another methodological development, surface-based analysis, offered complementary insights into brain anatomy. Here, the principal prerequisites of brain tissue classification and spatial registration remain the same [9–12]. The complementarity comes from the features projected on the brain surface, Euclidian distance representing cortical thickness at a given point and cortical surface area, and from the statistical analysis performed within each vertex rather than across three-dimensional voxels. One would assume that there is a simple relationship between these metrics, such that the product of cortical thickness and surface area equals the volume of the cortical mantle. However, the methodological differences in tissue classification and spatial registration for the voxel-based method (as implemented in Statistical Parametric Mapping software: www.fil.ion.ucl.ac.uk/spm, Wellcome Centre for Human

Neuroimaging, UCL London, UK) and the surface-based approach (FreeSurfer: www.surfer.nmr.mgh.harvard.edu) preclude any meaningful comparisons and interpretational inferences. In this context, the method for voxel-based estimation of cortical thickness within the Statistical Parametric Mapping framework brought the needed empirical evidence [13].

Over time, there were major developments in the two leading software suits, SPM and FreeSurfer, that brought higher accuracy of tissue classification and spatial registration, additionally to the continuous expansion of statistical analysis sophistication with implementations from the multi-variate, machine learning, and deep learning domains. In our presentation, we deliberately do not touch upon the steadily growing number of other existing methods and software solutions and refer to previous publications on the topic [14–16].

3 Computational Anatomy in Basic and Clinical Neuroscience

There is cumulating empirical evidence to suggest an added value of computational anatomy study of the brain for questions from basic and clinical neuroscience (Fig. 2). This is also supported by the unprecedented success of open data repositories, the Alzheimer's Disease Neuroimaging Initiative (ADNI) (www.adni.loni.usc.edu), the Parkinson's Progression Markers Initiative (PPMI) (www.ppmi-info.org), Imagen (www.imagen-project.org/), and the UK Biobank (www.ukbiobank.ac.uk). The provided structural MRIs alongside rich sets of behavioral, genetic, cognitive, and other relevant variables have made these repositories a unique source for studying the healthy and diseased human brain [17–20]. Taking an

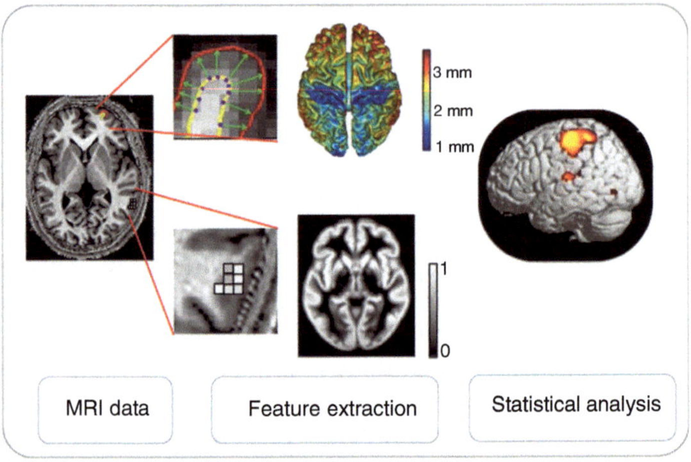

Fig. 2 Schematic representation of MRI data processing and analysis in surface- and volume-based computational anatomy

alternative approach, the ENIGMA initiative filled a major gap in clinical neurosciences by following a strategy close to "pragmatic" trials with uniform data processing followed by statistical analysis of center-specific aggregates rather than funding large-scale data acquisition across different sites. The success of ENIGMA builds not only on the parsimonious data acquisition model but much more on the straightforward way to address a neurobiological question using large-scale data in a truly collaborative manner, which paved the way for major discoveries particularly in the psychiatry domain [21]. In the open spirit of science, there are also critical voices that claim the need for even bigger numbers of observations than the one provided by the aforementioned initiatives [22].

Early computational anatomy studies focused on the mapping of behavioral and cognitive traits, clinical signs, and symptoms to brain structure. The plenitude of investigations covers the wide span between neurodevelopmental disorders, paroxysmal diseases of the brain, chronic neurodegeneration and neuroinflammation, psychiatric conditions, systemic diseases with impact on the brain up to brain maturation, and ageing. The significant part of these studies is cross-sectional—where a cohort of individuals with a particular disorder is compared to presumed healthy volunteers and differences are attributed to the pathological condition of interest. The proof of concept in the clinical domain came from studying brain anatomy differences in autosomal dominant disorders—here, on the example of consistency between region-of-interest and whole-brain results in preclinical and mild forms of Huntington's disease [23, 24]. Similarly, the wealth of evidence for sensitivity of computational anatomy methods toward volume or cortical thickness loss in neurodevelopmental disorders cannot be overestimated. The availability of large-scale genetic and structural MRI data allowed for the successful detection of copy number variations resulting in either loss, i.e., deletion, or gain, i.e., duplication of genes in individuals that result in "mirror" anthropometric phenotypes and associated psychiatric conditions [25]. Most recent results underscore the complexity of the discovered gene–brain structure–behavior relationships demonstrating that CNVs show divergent effects on brain anatomy, which however result in similar clinical phenotypes along the axis of schizophrenia and autism spectrum disorder [26]. Computational anatomy followed the far-reaching attempts in imaging neuroscience to go beyond the mapping of disease signatures toward computer-assisted diagnosis and prognosis in brain disorders. The methods cover the wide span between multi-variate voxel-based approaches [27] via (un)-supervised machine learning techniques [28] and deep learning approaches [29].

Longitudinal studies represent the Holy Grail in computational anatomy when it comes to causality inferences and tracking brain structure patterns of brain maturation and ageing. Against the odds of logistic difficulties, known and unknown data acquisition bias in the temporal and spatial domains, studies with multiple time points brought ample evidence supporting the notion of use-dependent brain plasticity [30], individual variation of cortical maturation linked with cognitive abilities, and risk for schizophrenia [31, 32]. With the very same goal, the open data repositories, ADNI, PPMI, and the UK Biobank, have already acquired or will acquire longitudinal MRI imaging data. The importance and added value of the dynamic aspect of data acquisition at multiple time points is undisputed, particularly for studying progressive neurodegeneration [33]. The intricacies of feature extraction and data preparation of multiple time point images for statistical analysis led to skepticism about the validity of previous findings [34], which were addressed in light of the shortcomings of morphometry studies using T1-weighted data and the need to embrace novel imaging protocols informative about the nature of (use-dependent) MR contrast changes over time [35]. Event-based modeling of cross-sectional data provides viable alternative to the multi-facetted complexity of longitudinal MRI data acquisition [36]. Subsequent studies confirmed the validity of the approach by demonstrating plausible sequence of events matching the clinical progression of Alzheimer's and Huntington's disease [37–39].

Another promising analytical method that can be applied to both cross-sectional and longitudinal MRI data is the analysis of structural covariance. Stemming from the assumption that connected cortical and subcortical areas share similar morphometry characteristics [40], the concept of structural covariance was validated in a number of computational anatomy studies (for review, see [41]). The winning point here is the departure from the "winner takes all" type of segregative mass-univariate analysis toward integrative look at the brain from a network perspective [41]. Along these lines, we draw parallels to the recently popular lesion network mapping approach in the functional connectivity domain [42] that successfully bridges the anatomical study in small-size cohorts with covariance analysis of large-scale resting state MRI data [43, 44].

4 Limitations of Computational Anatomy Using T1-Weighted Data

There is an ongoing controversy about the validity of past and present computational anatomy studies aiming at in vivo characterization of brain's structure. The much-needed progress in understanding the link between brain structure and function strongly depends on the potential to quantify in vivo the neurobiological properties of the human brain anatomy [45, 46]. Advances in the field validated the unique role of post mortem investigations introduced by Brodmann, Voigt, von Economo, and Koskinas at the beginning of the last century for achieving observer-independent delineation of cortical areas based on cyto- or myeloarchitectonic characteristics [47, 48]; however these methods are extremely laborious and not fully applicable in vivo. This limits their use in large-scale studies that aim at unbiased assessment of intersubject variability or at longitudinal investigations of trajectories of change over time.

Despite computational anatomy research spanning over nearly three decades, the echo of a recent study partially focusing on cortical thickness metrics, concluding that "reproducible brain-wide association studies require thousands of individuals" [22], gives rise to legitimate skepticism not only outside but also inside the imaging neuroscience field. One of the main issues here, besides considerations on statistical significance thresholds and interpretability in the context of large-scale studies, is the very nature of brain imaging protocols used in computational anatomy that do not necessarily follow recent advances in the field. With reference to this, we see the widespread use of T1-weighted imaging as the major source of "spurious" morphometric findings [49]. By design, T1-weighted protocols cannot disambiguate the differential contributions of brain tissue properties, myelin, iron, or tissue water, to MR contrast. This renders features assessed from T1-weighted data susceptible to the effects of brain maturation, ageing, or pathological processes on MR contrast. Therefore, the unknown relationship between brain tissue property changes, MR contrast and derived metrics of volume, or cortical thickness makes the interpretation in the context of morphometry findings as demonstrated empirically [49, 50] (Fig. 3).

Novel MRI approaches designed to overcome the critical issue of T1-weighted imaging in computational anatomy introduce imaging protocols that are measuring the unique contribution of myelin, iron, and MR-visible tissue water [51–54]. Relaxometry is one of the well-established in the field quantitative MRI (qMRI) approaches that has shown potential benefits at different levels of data processing and analysis within the voxel-based and surface-based frameworks of SPM and FreeSurfer [49, 55]. In the next

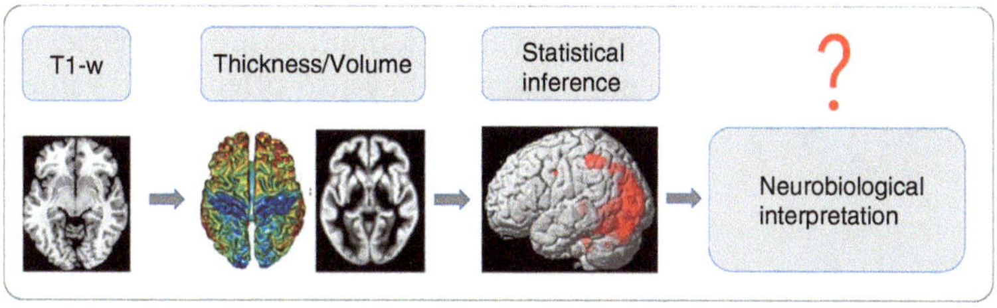

Fig. 3 Computational anatomy using T1-weighted images—from data through analysis and interpretation

paragraphs, we will summarize the scientific evidence for improved tissue classification, neurobiology-informed interpretation of morphometry results, and added value of qMRI in longitudinal studies.

5 Improved Brain Tissue Classification Using qMRI

The obvious benefit of the meanwhile established qMRI relaxometry protocol was the improved contrast-to-noise ratio for "difficult" structures, e.g., iron-rich basal ganglia. The automated tissue classification using maps of magnetization transfer saturation (MTsat) and the default SPM12 parameters provided not only reliable and robust detection of the dorso-lateral putamen but also allowed for delineating the pallidum, substantia nigra, subthalamic nucleus, and dentate nucleus that are usually assigned to brain's white matter when using T1-weighted images [56]. Given the probabilistic Bayesian framework of SPM12 with tissue priors derived from T1-weighted images, the introduction of enhanced tissue priors that allocate pallidum, substantia nigra, subthalamic nucleus, and dentate nucleus to the gray matter brought additional detection accuracy of these structures when using qMRI data [57].

The improved tissue classification of basal ganglia structures when using MTsat maps is explained by their high iron content that in the case of T1-weighted data shorten the T1 time, which results in loss of gray-white matter contrast with increasing iron content. Considering the empirical evidence about the exponential increase of iron in the basal ganglia with age [58], it becomes clear that the interpretation of basal ganglia volume differences in the context of ageing studies should be interpreted with caution [55]. This line of argumentation can be extrapolated to different anatomical areas and to different brain tissue properties, e.g., changes in intracortical myelin content during brain maturation [50] and differences in intracortical free water with ageing [49], to name but a few.

6 "In Vivo Histology" Using qMRI

Meanwhile, there is an extant body of literature showing that relaxometry-based quantitative MRI maps with sensitivity to microstructural brain tissue properties not only provide improved contrast, but they also offer added value to claim a straightforward neurobiological interpretation of MR contrast differences (Fig. 4). We must clarify here that the interpretation of the qMRI maps as indicative for myelin, iron, and tissue unbound water is an oversimplification of the underlying biophysical models. Indeed, there are first studies being published that correlate individual qMRI maps with state-of-the-art histology or established means for iron content measurements [59–61]; however, there should be a great

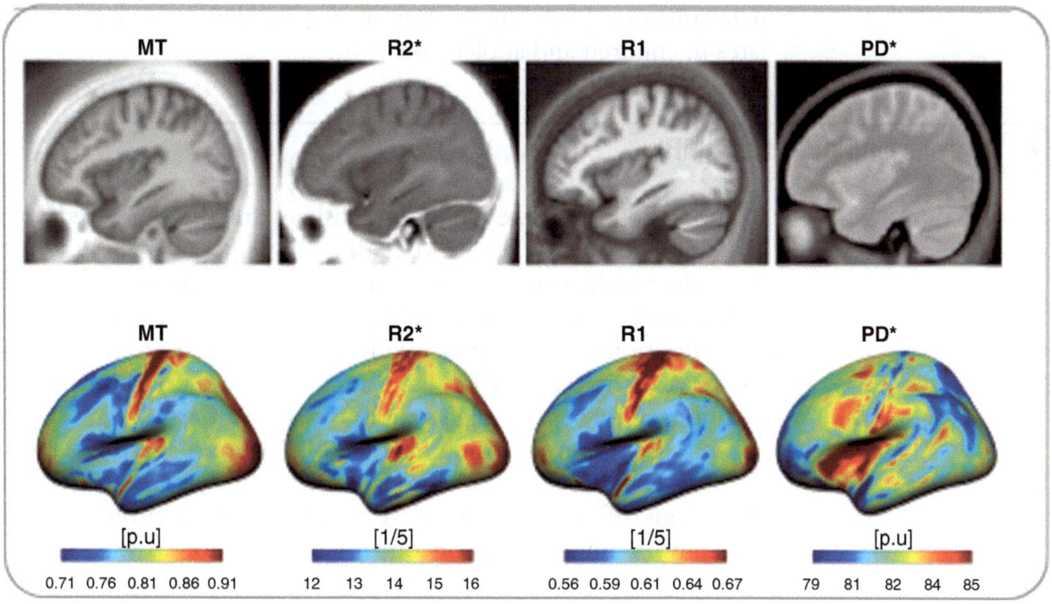

Fig. 4 Quantitative relaxometry-based MRI with maps of magnetization transfer (MT) saturation, effective transverse relaxation rate (R2*), longitudinal relaxation rate (R1) and effective proton density (PD*)

portion of caution when generalizing these to interpretation over the whole brain and in certain pathological conditions.

The field of relaxometry-based qMRI applications covers already studies of brain maturation [62–66], inflammatory brain disorders [67], neurodegeneration [61], ageing [53, 68–71], spinal cord injury [72, 73], modulation of brain plasticity [74, 75] and beyond. One of the promising developments is the combination of qMRI-derived indicators of myelin content with estimates of axonal density to obtain an index close to the g-ratio describing the axons' diameter in relation to its myelin sheath [76, 77]. A reliable g-ratio estimate opens new opportunities to link brain structure and function via the known physiological parameters of signal conduction velocity and area/system-specific g-ratio. This has been recently demonstrated in the domain of inter-individual differences in autobiographic memory recall [78].

A potentially game-changing characteristic of qMRI is its quantitative character, which precludes cite- or vendor-related bias in the longitudinal setting or in multi-centric studies. The field is moving toward validation of this claim for the case of multi-centric studies; however there are some controversies to be addressed [52, 73, 79, 80]. Our recent study on the effects of electro-convulsive stimulation for pharmacotherapy refractory major depression suggested a qMRI-tailored data processing algorithm and a multi-variate statistical analysis strategy for longitudinal relaxometry multi-parameter maps (MPMs) [74]. This was a development of a previous concept applied to longitudinal MPMs for studying the socioeconomic determinants of myelination trajectories and their behavioral correlates in children and adolescents [66].

7 Current Limitations of qMRI

Compared with the long-established methods based on T1-weighted images, the qMRI developments are still suffering from the insufficient maturity of the tailored data processing and analytical framework within SPM [81]. The proposed voxel-based quantification strategy with a weighted averaging step fulfils the aim to preserve the total qMRI signal after linear and non-linear interpolations during spatial registration to standard Cartesian space but also has limitations of losing regional MR contrast specificity [53].

Currently, there are major efforts to address one of the main limitations of our established relaxometry qMRI protocol—the difficulty of implementation on different vendor MRI scanners. This confines the number of laboratories benefitting from this technique to only a few, which hampers the progress of open science and contributions from a wider community. We have created a dedicated SPM toolbox [81] for data processing of qMRI data following the VBQ approach [53] coupled with exemplary data sets [82].

8 Outlook

The qMRI "in vivo histology" approach holds a great promise to brain computational anatomy to another interpretational level—both for single-center and multi-center studies and for cross-sectional and longitudinal study designs. The near future should bring more empirical evidence in support of the "in vivo histology" interpretational claims. An excellent example are recent studies in Parkinson's disease showing correlation between pre-synaptic dopamine transporter quantity and basal ganglia's iron content [83] additionally to their functional impact [84]. The fine-grained correlative studies of substantia nigra's nigrosome using histology in parallel with qMRI are paving the way forward [61].

Despite a decade of developments and some tooting problems of its large-scale application, qMRI has delivered in various imaging neuroscience domains. It is a matter of short time until we see more studies with a broader field of applications in basic and clinical neuroscience.

Acknowledgments

BD is supported by the Swiss National Science Foundation (grants 32003B_159780, CRSK-3_190185, 324730_192755, and the Sinergia VascX project), the EU ERA-NET iSee project, the SPHN SACR project, the CLIMACT project, the Leenaards Foundation, and the HeadFirst Innosuisse project. LREN is grateful to the ROGER DE SPOELBERCH and Partridge Foundations for their generous financial support.

References

1. Coffey CE, Wilkinson WE, Parashos IA, Soady SA, Sullivan RJ, Patterson LJ et al (1992) Quantitative cerebral anatomy of the aging human brain: a cross-sectional study using magnetic resonance imaging. Neurology 42(3 Pt 1):527–536
2. Coffey CE, Wilkinson WE, Weiner RD, Parashos IA, Djang WT, Webb MC et al (1993) Quantitative cerebral anatomy in depression. A controlled magnetic resonance imaging study. Arch Gen Psychiatry 50(1):7–16
3. Wright IC, McGuire PK, Poline JB, Travere JM, Murray RM, Frith CD et al (1995) A voxel-based method for the statistical analysis of gray and white matter density applied to schizophrenia. NeuroImage 2(4):244–252
4. Bullmore E, Brammer M, Rouleau G, Everitt B, Simmons A, Sharma T et al (1995) Computerized brain tissue classification of magnetic resonance images: a new approach to the problem of partial volume artifact. NeuroImage 2(2):133–147
5. Ashburner J, Friston KJ (2000) Voxel-based morphometry--the methods. NeuroImage 11(6 Pt 1):805–821
6. Ashburner J, Neelin P, Collins DL, Evans A, Friston K (1997) Incorporating prior knowledge into image registration. NeuroImage 6(4):344–352
7. Ashburner J, Andersson JL, Friston KJ (1999) High-dimensional image registration using symmetric priors. NeuroImage 9(6 Pt 1): 619–628
8. Ashburner J, Friston KJ (1999) Nonlinear spatial normalization using basis functions. Hum Brain Mapp 7(4):254–266
9. Dale AM, Fischl B, Sereno MI (1999) Cortical surface-based analysis. I Segmentation and surface reconstruction. NeuroImage 9(2): 179–194
10. Fischl B, Sereno MI, Dale AM (1999) Cortical surface-based analysis. II: Inflation, flattening, and a surface-based coordinate system. NeuroImage 9(2):195–207
11. Fischl B, Sereno MI, Tootell RB, Dale AM (1999) High-resolution intersubject averaging and a coordinate system for the cortical surface. Hum Brain Mapp 8(4):272–284
12. Fischl B, Dale AM (2000) Measuring the thickness of the human cerebral cortex from magnetic resonance images. Proc Natl Acad Sci 97(20):11050–11055
13. Hutton C, Draganski B, Ashburner J, Weiskopf N (2009) A comparison between voxel-based cortical thickness and voxel-based morphometry in normal aging. NeuroImage 48(2): 371–380
14. Ashburner J, Csernansky JG, Davatzikos C, Fox NC, Frisoni GB, Thompson PM (2003) Computer-assisted imaging to assess brain structure in healthy and diseased brains. Lancet Neurol 2(2):79–88
15. Eickhoff SB, Constable RT, Yeo BTT (2018) Topographic organization of the cerebral cortex and brain cartography. NeuroImage 170: 332–347
16. Eickhoff SB, Yeo BTT, Genon S (2018) Imaging-based parcellations of the human brain. Nat Rev Neurosci 19(11):672–686
17. Weber CJ, Carrillo MC, Jagust W, Jack CR, Shaw LM, Trojanowski JQ et al (2021) The Worldwide Alzheimer's Disease Neuroimaging Initiative: ADNI-3 updates and global perspectives. Alzheimers Dement N Y N 7(1):e12226
18. Tafuri B, Lombardi A, Nigro S, Urso D, Monaco A, Pantaleo E et al (2022) The impact of harmonization on radiomic features in Parkinson's disease and healthy controls: a multicenter study. Front Neurosci 16:1012287
19. Miller KL, Alfaro-Almagro F, Bangerter NK, Thomas DL, Yacoub E, Xu J et al (2016) Multimodal population brain imaging in the UK Biobank prospective epidemiological study. Nat Neurosci 19(11):1523

20. Mascarell Maričić L, Walter H, Rosenthal A, Ripke S, Quinlan EB, Banaschewski T et al (2020) The IMAGEN study: a decade of imaging genetics in adolescents. Mol Psychiatry 25(11):2648–2671
21. Thompson PM, Jahanshad N, Ching CRK, Salminen LE, Thomopoulos SI, Bright J et al (2020) ENIGMA and global neuroscience: a decade of large-scale studies of the brain in health and disease across more than 40 countries. Transl Psychiatry 10(1):100
22. Marek S, Tervo-Clemmens B, Calabro FJ, Montez DF, Kay BP, Hatoum AS et al (2022) Reproducible brain-wide association studies require thousands of individuals. Nature 603(7902):654–660
23. Rosas HD, Liu AK, Hersch S, Glessner M, Ferrante RJ, Salat DH et al (2002) Regional and progressive thinning of the cortical ribbon in Huntington's disease. Neurology 58(5): 695–701
24. Aylward EH, Brandt J, Codori AM, Mangus RS, Barta PE, Harris GJ (1994) Reduced basal ganglia volume associated with the gene for Huntington's disease in asymptomatic at-risk persons. Neurology 44(5):823–828
25. Modenato C, Martin-Brevet S, Moreau CA, Rodriguez-Herreros B, Kumar K, Draganski B et al (2021) Lessons learned from neuroimaging studies of copy number variants: a systematic review. Biol Psychiatry 90(9):596–610
26. Kopal J, Kumar K, Saltoun K, Modenato C, Moreau CA, Martin-Brevet S et al (2022) Rare CNVs and phenome-wide profiling: a tale of brain-structural divergence and phenotypical convergence [Internet]. bioRxiv [cited 2022 Nov 27]. p. 2022.04.23.489093. Available from: https://www.biorxiv.org/content/10.1101/2022.04.23.489093v1
27. Kawasaki Y, Suzuki M, Kherif F, Takahashi T, Zhou SY, Nakamura K et al (2007) Multivariate voxel-based morphometry successfully differentiates schizophrenia patients from healthy controls. NeuroImage 34(1):235–242
28. Klöppel S, Stonnington CM, Chu C, Draganski B, Scahill RI, Rohrer JD et al (2008) Automatic classification of MR scans in Alzheimer's disease. Brain J Neurol 131 (Pt 3):681–689
29. Yang Z, Nasrallah IM, Shou H, Wen J, Doshi J, Habes M et al (2021) A deep learning framework identifies dimensional representations of Alzheimer's Disease from brain structure. Nat Commun 12(1):7065
30. Draganski B, Gaser C, Busch V, Schuierer G, Bogdahn U, May A (2004) Neuroplasticity: changes in grey matter induced by training. Nature 427(6972):311–312
31. Sowell ER, Thompson PM, Leonard CM, Welcome SE, Kan E, Toga AW (2004) Longitudinal mapping of cortical thickness and brain growth in normal children. J Neurosci 24(38):8223–8231
32. Thompson PM, Vidal C, Giedd JN, Gochman P, Blumenthal J, Nicolson R et al (2001) Mapping adolescent brain change reveals dynamic wave of accelerated gray matter loss in very early-onset schizophrenia. Proc Natl Acad Sci U S A 98(20):11650–11655
33. Dicks E, Vermunt L, van der Flier WM, Visser PJ, Barkhof F, Scheltens P et al (2019) Modeling grey matter atrophy as a function of time, aging or cognitive decline show different anatomical patterns in Alzheimer's disease. NeuroImage Clin 22:101786
34. Thomas C, Baker CI (2013) Teaching an adult brain new tricks: a critical review of evidence for training-dependent structural plasticity in humans. NeuroImage 73:225–236
35. Draganski B, Kherif F (2013) In vivo assessment of use-dependent brain plasticity--beyond the 'one trick pony' imaging strategy. NeuroImage 73:255–259; discussion 265–267
36. Young AL, Oxtoby NP, Daga P, Cash DM, Fox NC, Ourselin S et al (2014) A data-driven model of biomarker changes in sporadic Alzheimer's disease. Brain J Neurol 137(Pt 9): 2564–2577
37. Fonteijn HM, Modat M, Clarkson MJ, Barnes J, Lehmann M, Hobbs NZ et al (2012) An event-based model for disease progression and its application in familial Alzheimer's disease and Huntington's disease. NeuroImage 60(3):1880–1889
38. Oxtoby NP, Young AL, Cash DM, Benzinger TLS, Fagan AM, Morris JC et al (2018) Data-driven models of dominantly-inherited Alzheimer's disease progression. Brain J Neurol 141(5):1529–1544
39. Archetti D, Ingala S, Venkatraghavan V, Wottschel V, Young AL, Bellio M et al (2019) Multi-study validation of data-driven disease progression models to characterize evolution of biomarkers in Alzheimer's disease. NeuroImage Clin 24:101954
40. Mechelli A, Friston KJ, Frackowiak RS, Price CJ (2005) Structural covariance in the human cortex. J Neurosci 25(36):8303–8310
41. Evans AC (2013) Networks of anatomical covariance. NeuroImage 80:489–504
42. Fox MD (2018) Mapping symptoms to brain networks with the human connectome. N Engl J Med 379(23):2237–2245

43. Fox MD, Buckner RL, Liu H, Chakravarty MM, Lozano AM, Pascual-Leone A (2014) Resting-state networks link invasive and noninvasive brain stimulation across diverse psychiatric and neurological diseases. Proc Natl Acad Sci U S A 111(41):E4367–E4375
44. Padmanabhan JL, Cooke D, Joutsa J, Siddiqi SH, Ferguson M, Darby RR et al (2019) A human depression circuit derived from focal brain lesions. Biol Psychiatry 86(10):749–758
45. Amunts K, Hawrylycz MJ, Van Essen DC, Van Horn JD, Harel N, Poline JB et al (2014) Interoperable atlases of the human brain. NeuroImage 99:525–532
46. Amunts K, Zilles K (2015) Architectonic mapping of the human brain beyond Brodmann. Neuron 88(6):1086–1107
47. Amunts K, Schleicher A, Zilles K (2007) Cytoarchitecture of the cerebral cortex--more than localization. NeuroImage 37(4): 1061–1065; discussion 1066–1068
48. Annese J, Pitiot A, Dinov ID, Toga AW (2004) A myelo-architectonic method for the structural classification of cortical areas. NeuroImage 21(1):15–26
49. Lorio S, Kherif F, Ruef A, Melie-Garcia L, Frackowiak R, Ashburner J et al (2016) Neurobiological origin of spurious brain morphological changes: a quantitative MRI study. Hum Brain Mapp 37(5):1801–1815
50. Natu VS, Gomez J, Barnett M, Jeska B, Kirilina E, Jaeger C et al (2019) Apparent thinning of human visual cortex during childhood is associated with myelination. Proc Natl Acad Sci U S A
51. Helms G, Dathe H, Kallenberg K, Dechent P (2008) High-resolution maps of magnetization transfer with inherent correction for RF inhomogeneity and T1 relaxation obtained from 3D FLASH MRI. Magn Reson Med 60(6):1396–1407
52. Weiskopf N, Suckling J, Williams G, Correia MM, Inkster B, Tait R et al (2013) Quantitative multi-parameter mapping of R1, PD*, MT, and R2* at 3T: a multi-center validation. Front Neurosci 7:95
53. Draganski B, Ashburner J, Hutton C, Kherif F, Frackowiak RSJ, Helms G et al (2011) Regional specificity of MRI contrast parameter changes in normal ageing revealed by voxel-based quantification (VBQ). NeuroImage 55(4):1423–1434
54. Weiskopf N, Edwards LJ, Helms G, Mohammadi S, Kirilina E (2021) Quantitative magnetic resonance imaging of brain anatomy and in vivo histology. Nat Rev Phys [Internet] [cited 2021 Nov 1];3(8):570–88. Available from: https://www.nature.com/articles/s42254-021-00326-1
55. Lorio S, Lutti A, Kherif F, Ruef A, Dukart J, Chowdhury R et al (2014) Disentangling in vivo the effects of iron content and atrophy on the ageing human brain. NeuroImage 103: 280–289
56. Helms G, Draganski B, Frackowiak R, Ashburner J, Weiskopf N (2009) Improved segmentation of deep brain grey matter structures using magnetization transfer (MT) parameter maps. NeuroImage 47(1): 194–198
57. Lorio S, Fresard S, Adaszewski S, Kherif F, Chowdhury R, Frackowiak RS et al (2016) New tissue priors for improved automated classification of subcortical brain structures on MRI. NeuroImage 130:157–166
58. Hallgren B, Sourander P (1958) The effect of age on the non-haemin iron in the human brain. J Neurochem 3(1):41–51
59. Morawski M, Kirilina E, Scherf N, Jäger C, Reimann K, Trampel R et al (2018) Developing 3D microscopy with CLARITY on human brain tissue: towards a tool for informing and validating MRI-based histology. NeuroImage 182:417–428
60. Kirilina E, Helbling S, Morawski M, Pine K, Reimann K, Jankuhn S et al (2020) Superficial white matter imaging: contrast mechanisms and whole-brain in vivo mapping. Sci Adv 6(41):eaaz9281
61. Brammerloh M, Morawski M, Friedrich I, Reinert T, Lange C, Pelicon P et al (2021) Measuring the iron content of dopaminergic neurons in substantia nigra with MRI relaxometry. NeuroImage 239:118255
62. Natu VS, Rosenke M, Wu H, Querdasi FR, Kular H, Lopez-Alvarez N et al (2021) Infants' cortex undergoes microstructural growth coupled with myelination during development. Commun Biol 4(1):1191
63. Grotheer M, Yeatman J, Grill-Spector K (2021) White matter fascicles and cortical microstructure predict reading-related responses in human ventral temporal cortex. NeuroImage 227:117669
64. Grotheer M, Rosenke M, Wu H, Kular H, Querdasi FR, Natu VS et al (2022) White matter myelination during early infancy is linked to spatial gradients and myelin content at birth. Nat Commun 13(1):997
65. Ziegler G, Hauser TU, Moutoussis M, Bullmore ET, Goodyer IM, Fonagy P et al (2019) Compulsivity and impulsivity traits linked to attenuated developmental frontostriatal

myelination trajectories. Nat Neurosci 22(6): 992–999

66. Ziegler G, Moutoussis M, Hauser TU, Fearon P, Bullmore ET, Goodyer IM et al (2020) Childhood socio-economic disadvantage predicts reduced myelin growth across adolescence and young adulthood. Hum Brain Mapp 41(12):3392–3402
67. Granziera C, Wuerfel J, Barkhof F, Calabrese M, De Stefano N, Enzinger C et al (2021) Quantitative magnetic resonance imaging towards clinical application in multiple sclerosis. Brain J Neurol 144(5):1296–1311
68. Slater DA, Melie-Garcia L, Preisig M, Kherif F, Lutti A, Draganski B (2019) Evolution of white matter tract microstructure across the life span. Hum Brain Mapp 40(7):2252–2268
69. Taubert M, Roggenhofer E, Melie-Garcia L, Muller S, Lehmann N, Preisig M et al (2020) Converging patterns of aging-associated brain volume loss and tissue microstructure differences. Neurobiol Aging 88:108–118
70. Callaghan MF, Freund P, Draganski B, Anderson E, Cappelletti M, Chowdhury R et al (2014) Widespread age-related differences in the human brain microstructure revealed by quantitative magnetic resonance imaging. Neurobiol Aging 35(8):1862–1872
71. Trofimova O, Loued-Khenissi L, DiDomenicantonio G, Lutti A, Kliegel M, Stringhini S et al (2021) Brain tissue properties link cardio-vascular risk factors, mood and cognitive performance in the CoLaus|PsyCoLaus epidemiological cohort. Neurobiol Aging 102: 50–63
72. David G, Pfyffer D, Vallotton K, Pfender N, Thompson A, Weiskopf N et al (2021) Longitudinal changes of spinal cord grey and white matter following spinal cord injury. J Neurol Neurosurg Psychiatry 92(11):1222–1230
73. Seif M, Leutritz T, Schading S, Emmengger T, Curt A, Weiskopf N et al (2022) Reliability of multi-parameter mapping (MPM) in the cervical cord: a multi-center multi-vendor quantitative MRI study. NeuroImage 264: 119751
74. Gyger L, Ramponi C, Mall JF, Swierkosz-Lenart K, Stoyanov D, Lutti A et al (2021) Temporal trajectory of brain tissue property changes induced by electroconvulsive therapy. NeuroImage 232:117895
75. Lazari A, Salvan P, Verhagen L, Cottaar M, Papp D, van der Werf OJ et al (2022) A macroscopic link between interhemispheric tract myelination and cortico-cortical interactions during action reprogramming. Nat Commun 13(1):4253
76. Stikov N, Campbell JS, Stroh T, Lavelée M, Frey S, Novek J et al (2015) In vivo histology of the myelin g-ratio with magnetic resonance imaging. NeuroImage 118:397–405
77. Campbell JSW, Leppert IR, Narayanan S, Boudreau M, Duval T, Cohen-Adad J et al (2018) Promise and pitfalls of g-ratio estimation with MRI. NeuroImage 182:80–96
78. Clark IA, Mohammadi S, Callaghan MF, Maguire EA (2022) Conduction velocity along a key white matter tract is associated with autobiographical memory recall ability. elife 11:e79303
79. Gracien RM, Maiworm M, Brüche N, Shrestha M, Nöth U, Hattingen E et al (2020) How stable is quantitative MRI? - Assessment of intra- and inter-scanner-model reproducibility using identical acquisition sequences and data analysis programs. NeuroImage 207:116364
80. Karakuzu A, Biswas L, Cohen-Adad J, Stikov N (2022) Vendor-neutral sequences and fully transparent workflows improve inter-vendor reproducibility of quantitative MRI. Magn Reson Med 88(3):1212–1228
81. Tabelow K, Balteau E, Ashburner J, Callaghan MF, Draganski B, Helms G et al (2019) hMRI - a toolbox for quantitative MRI in neuroscience and clinical research. NeuroImage 194: 191–210
82. Callaghan MF, Lutti A, Ashburner J, Balteau E, Corbin N, Draganski B et al (2019) Example dataset for the hMRI toolbox. Data Brief 25: 104132
83. Depierreux F, Parmentier E, Mackels L, Baquero K, Degueldre C, Balteau E et al (2021) Parkinson's disease multimodal imaging: F-DOPA PET, neuromelanin-sensitive and quantitative iron-sensitive MRI. NPJ Park Dis 7(1):57
84. Drori E, Berman S, Mezer AA (2022) Mapping microstructural gradients of the human striatum in normal aging and Parkinson's disease. Sci Adv 8(28):eabm1971

Chapter 9

Nonlinear Methods for the Investigation of Psychotic Disorders

Alexandra Korda, Marina Frisman, Christina Andreou, and Stefan Borgwardt

Abstract

The basic aspects of nonlinear dynamics theory's utility and impact on understanding psychotic disorders are reviewed and discussed in this chapter. A different view of nonlinear approaches as a metaphor, analogy, and mathematical/physical concept is presented, with an emphasis on how the latter can be applied to empirical knowledge of psychotic disorders. The typical behavior of patients with psychotic disorders and general chaotic systems is discussed, as well as the dynamical analysis of time series extracted from such systems. The original question is rephrased in chaos theory terminology: "Are psychotic disorders described as dynamical processes?" There are some implications to viewing psychotic disorders as dynamic processes which enable the chaos theory approach to give deeper insights.

Key words Chaos, Fractals, Nonlinear dynamics, Psychotic disorders

1 Introduction

Psychotic disorders affect brain function through various combinations of psychopathological symptoms that include formal thought disorders, delusions, hallucinations, disorganized speech, and bizarre behavior. About 3% of people will experience a psychotic episode at some point in their lives [1]. Especially for adolescents under the age of 18, access to diagnosis, prevention/intervention, and care for mental health problems are still relative compared to limited physical problems in general [2]. Psychotic disorders place a considerable burden on the people affected and on society, which leads to a paradigm shift toward the use of objective indicators to assess disease severity and disease progression and the widespread use of new methodologies for prevention and prediction based on digital measurements. Despite mounting scientific data supports research in psychosis, the gap between research evidence and clinical practices is still wide due to several important challenges: The

Drozdstoy Stoyanov et al. (eds.), *Computational Neuroscience*, Neuromethods, vol. 199,
https://doi.org/10.1007/978-1-0716-3230-7_9,

unspecific nature of psychosis and related psychosocial difficulties in transitional age youth makes it difficult to identify specific risks and protective factors for psychotic disorders. The identification of psychotic disorders is often challenging because of their complex interactions with an individual's environment.

Additionally, psychosis is a highly heterogeneous, dynamic, and multifactorial phenomenon, despite widespread recognition of their inherently complex nature (including genetic environment and psychosomatic interactions, as well as lifelong developmental and other experiential changes); progress in our understanding of mental disorders remains slow [3]. Many potential predictors for psychosis are time-dependent in that they are measured at various points throughout, e.g., the assessment process, and may change over time. Most research studies in psychosis and transition to psychosis have not used the dynamic nature of time-dependent measures, i.e., using assessments, EEG, fMRI, and speech, despite the progress of the nonlinear dynamical systems theory, which increases our understanding of complex system dynamics over the last 50 years [4]. By applying linear methods on time-dependent data, important information of the nature of the data, psychotic disorders in this case, remains hidden. The introduction of nonlinear dynamical systems in psychosis research will give insights to the identification of the disorder and its pathophysiology, making potential progress on the establishment of biomarkers in psychosis.

The formalism of this framework is broad enough to be applied in related fields, such as biology or psychology, where complex systems are the norm rather than the exception. This chapter aims to improve the understanding of nonlinear dynamical systems in the medical community, help clinicians understand nonlinear approaches, and put their findings into context. We aim to demonstrate that this framework can be useful in scientific fields such as neuroscience and psychiatry, where objects have natural time or spatial dependency (i.e., dynamical properties) and nonlinear properties. Nonlinear dynamics concepts applied to these topics are precise and can shed new light on mental functioning and dysfunction. Focusing on psychotic disorders, we aim to (1) evaluate possible fields of application for these methods and (2) examine how psychotic disorders can be seen in a nonlinear dynamical context. A brief introduction on nonlinear dynamical systems theory and how can be evaluated is presented followed by case studies in psychosis research.

2 How to Evaluate Nonlinear Dynamical Systems?

In psychosis research, the term "time" can be referred to (a) naturally perceived time (e.g., change of symptoms), (b) subunits of time (e.g., neurobiological processes, EEG, and fMRI), and (c) series resulting from the conversion of other units (e.g., spatial series from structural MRI). Dynamic models represent mental processes and phenomena that change over time, more like a film than a photograph, where multiple variables interact over time. Systems are collections of components/variables that interact with one another. A dynamical system is a system in which the coordinates of the points, which evolve over time, undergo a fixed rule and are analyzed using differential equations—an equation with one or more derivatives of a function. Systems have the propensity to look for equilibrium—a behavior state that uses as little energy as possible—where they can stay if they are not forced to change. Systems with multiple levels of equilibrium occasionally switch between states. Nonlinear dynamical systems are distinguished mathematically by the observation that the input is not always proportional to the output, i.e., increased input does not always result in proportionally increased output. As a result, minor changes in the input can result in an entirely different sequence of outputs which causes chaotic behavior. A lawful but exquisite sensitivity to initial conditions characterizes chaotic dynamical systems. Initial conditions are the starting values of the variables investigated in the dynamical system that affects the evolution of the system. Because of this sensitivity to initial conditions, we can see how initially similar behaviors evolve into a striking divergence of behavioral patterns over time. For example, negative mood in one period is related to mood in the prior period, while internal and external influences affect the last of the negative mood. The sensitivity on the initial conditions in mood disorders is characterized by inertia in persons diagnosed with depression [5], which means that the negative mood in patients with depression cannot change rapidly in time evolution. Computational models based on nonlinear dynamics and systems theory have recently emerged as a promising tool in neuroscience for describing the large-scale, neural system-level dynamics present in brain electrical activity in patients with schizophrenia [6]. Rather than focusing on microscale neuronal activity, methods that characterize large-scale, system-level dynamics may aid in understanding emergent phenomena of behavior and cognition [7]. Below, the most common used methods on dynamical system's analysis (a) chaos and (b) fractals are presented.

3 Methods

In medicine and biology, nonlinear dynamical systems analysis have been reported as effective computational methods for analyzing complex data [8]. Several mental disorders are considered as dynamical systems that follow nonlinear patterns, resulting in complex emotional/cognitive, interpersonal interactions and phenomenological presentations [9]. A deeper investigation at the nonlinear estimators characterizes these nonlinear dynamical systems is needed. For example, while in Euclidean geometry, objects are described with simple formulas (e.g., volume, thickness, etc.), for the description of nonlinear dynamical systems, it is necessary to introduce new estimators that may not have one precise size [10, 11], i.e., correlation dimension, Lyapunov exponent, fractal dimension, entropy, etc. All these estimators capture the chaos and fractality of the examined dynamical systems; the meaning of these two definitions will be explained here.

Chaos Nonlinear phenomena are responsible for many complex and fascinating natural phenomena. The theory of nonlinear dynamical systems, also known as "chaos theory" [4], has now advanced to the point where it is possible to study self-organization and pattern formation in the brain's complex neuronal networks. One approach to nonlinear time series analysis is to reconstruct an attractor—a set of states toward which a system tends to evolve, for a wide variety of initial conditions of the system—of the underlying dynamical system from EEG, fMRI time series, or brain MRI and characterize it in terms of its dimension (an estimate of the system's degrees of freedom) or its Lyapunov exponents which characterizes the rate of separation of the initial neighbor points, and entropy (reflecting unpredictability of the dynamics due to the sensitive dependence on initial conditions). Natural systems, including the brain, frequently appear chaotic because complex nonlinear dynamical processes typically drive them. For example, disruption in the fluid coordination of multiple brain regions contributes to impairments in information processing and the constellation of symptoms seen in neuropsychiatric disorders. Psychosis, one of the most incapacitating mental disorders, is thought to be caused in part by such complex networks, which leads to impaired auditory information processing as well as cognitive and psychosocial deficits. Understanding the seamless orchestration of many interconnected brain regions required to produce the integrated cognition experience is one of neuroscience's fundamental challenges.

Fractals Fractal dimension (FD) is a single non-integer value that enables us to describe how an object fills the space more correctly than possible with the "classical" dimension from Euclidean geometry. More complex objects tend to have a higher FD, while less complex objects have a lower FD. It is essential to notice that the complexity of an object doesn't characterize this object completely. Two objects may have the same FD but appear completely different (like it is with the Euclidean dimension). Nevertheless, the FD provides a good overview and measurement for the overall complexity and structure of any object's shape, which can also be a medical signal like a heart rate, a time series, or the brain cortex.

The FD seems to be an innovative way to describe the complex morphology of different brain structures, including the cortex, gray matter, and white matter. In the past decades, interest in FD has been growing, and more and more studies have investigated the usability of FD for the description of brain morphology in several neurological diseases (i.e., FD was able to explain neurodegeneration, neoplasms, and even epilepsy [11]).

Many studies in psychosis research were employed nonlinear dynamical system's theory on different modalities. In this chapter, we examine the application of all possible estimators of chaos and fractals in psychosis. A brief overview of the use cases is presented below.

4 Use Cases

fMRI We argue that the temporal architecture of schizophrenia is characterized by bursts of complex, nonlinear phenomena interspersed with genuinely random events. Using nonlinear methods to analyze these patterns of molecular (e.g., calcium channel activity) to molar (e.g., symptom level) phenomena can provide new approaches to understanding complex temporal and sequential shifts in neural substrate activity and pathophysiology. Different regions of the human brain functionally connect with one another, forming a brain functional network, and the time evolution of functional connectivity between different brain regions exhibits complex nonlinear dynamics.

A class of neural networks (recurrent neural networks) is an example of dynamical system implementation and provides models of cognitive processes [12]. Recurrent neuronal networks have been used to propose the interpretation of several mental dysfunctions. The network's state of activity is represented in its state space—the Euclidean space in which the variables on the axes are the state variables that evolve over time—and a trajectory in this space represents the time evolution of this state. After some time, those networks reach an equilibrium (a kind of attractor). The strength of connections between neurons determines the number and relationships of those attractors. In the case of schizophrenia, it has been proposed that problems with cortical pruning during development [13] may result in a decrease in neural network storage ability and the formation of spurious attractors. These attractors do not correspond to stored memories and attract many initial conditions; thus, they were linked to the reality distortion observed in schizophrenia. In a study [14], the dynamic functional connectivity is built by analyzing resting-state functional magnetic resonance imaging data, and its nonlinear properties are defined by sample entropy (SampEn, a commonly used nonlinear measure of the degree of regularity of time series), with higher SampEn values indicating greater complexity. It has been demonstrated that the patients have significantly higher SampEn at various brain levels and that the patients' significantly higher SampEn primarily causes this phenomenon in the visual cortex. Furthermore, SampEn of the visual cortex is shown to correlate with illness duration or symptom severity scores significantly and positively. In a study on psychedelics drugs [15], FD of cortical functional connectivity and BOLD time series was investigated to identify differences in the effect of psychedelic drugs, LSD, and psilocybin in healthy controls. The authors identified changes in the FD of BOLD signals in brain regions associated with the dorsal attention network. Psychedelic drugs significantly increased the FD of the functional connectivity network, LSD significantly increased the FD of the BOLD signal, and psilocybin showed a non-significant trend in the same direction. Another study [16] implemented nonlinear dynamic causal modeling (DCM) that models neurobiological processes at the neuronal population level in healthy controls, method that has not been applied in psychosis research which would inform us about the causality between not direct connected networks.

EEG Current approaches to neurophysiological biomarker analysis rely primarily on linear methods, which may fail to capture the wealth of information in the entire EEG signals, including nonlinear dynamics. Most event-related potential (ERP) studies in neuropsychiatry have relied on linear EEG signal analysis. ERP peaks and latencies [17, 18] or time–frequency analyses [19, 20] and cross-frequency coupling are examples of linear methods [21]. While highly informative, such conventional linear methods focus on a priori determined time-locked events or frequency ranges (i.e., delta, theta, gamma bands) and thus fail to capture important underlying nonlinear system dynamics. Therefore, characterizing general dynamical components of broadband data without such constraints may reveal information about altered system-level states associated with schizophrenia. For example, in Lainscsek et al. [22], delay differential analysis (DDA), a nonlinear method based on embedding theory from theoretical physics which measures the chaos, was applied to EEG recordings from 877 patients with schizophrenia and 753 nonpsychiatric comparison subjects who participated in the Consortium on the Genetics of Schizophrenia (COGS-2) study. In both groups, DDA revealed significant nonlinear dynamical architecture related to auditory information processing. Notably, significant DDA changes occurred before those observed using traditional linear methods. Patients with schizophrenia had substantial abnormalities in both linear and nonlinear features. These findings highlight the advantages of nonlinear brain signal analysis and the need for future research into the relationship between DDA features and information processing pathophysiology. The results show that brain signals can be analyzed unusually to extract important large-scale dynamical information that is not visible using conventional methods commonly used to study neuropsychiatric disorders. The largest Lyapunov exponent, a measure of chaos, was also showed to be higher in patients with schizophrenia vs. healthy controls in [23]. In Korda et al. [19], EEGs from patients with schizophrenia were analyzed with continuous wavelet transformation. The output time–frequency images were fed into a convolutional neural network able to distinguish patients with schizophrenia against healthy controls and identify differences in specific rhythm's bands not predetermined. Fractal dimension indicated statistically significant differences in the complex structure of the EEG signal between patients with schizophrenia and healthy controls [24, 25]. In Raghavendra et al. [26], the fractal dimension on recent onset psychosis patients with positive symptoms was higher than both healthy controls and patients with negative symptoms.

Speech Pathophysiological studies of schizophrenia must consider one of its most distinguishing features: disorganization and impoverishment in verbal behavior. Patients with schizophrenia are frequently described as having monotonous, flat, and emotionless speech. Distinctive speech patterns are evaluated qualitatively during the diagnostic process and significantly impact the quality of everyday social interactions. Many characteristics of speech, such as prosody and inter-words time intervals, can be measured objectively. Speaking differently can reveal cognitive and communicative difficulties, which are the hallmark symptoms of schizophrenia and other psychotic disorders. Although difficult for clinicians to identify, these language changes can occur prior to more obvious psychotic symptoms, delaying diagnosis and lowering the standard of care. To give clinicians a simple-to-collect, scalable, and objective marker for earlier identification of psychotic disorders like schizophrenia, a study funded partly by the National Institute of Mental Health set out to identify speech patterns associated with psychosis [27]. The duration of pauses in patients with schizophrenia with predominantly negative symptoms was related to flat affect and alogia. Pauses were context-dependent in healthy subjects and patients with schizophrenia who did not have a thought disorder but not in thought-disordered patients with schizophrenia [28]. These linear methods for quantifying pauses in schizophrenia can only provide a limited amount of information about the dynamical system that underpins the phenomenon of pausing. In [29], 15 patients with schizophrenia and 15 healthy volunteers are recorded speaking spontaneously for 10–15 min. In normal speech, the series of inter-word time intervals has the characteristics of a low-dimensional chaotic attractor. The attractor is deconstructed in psychosis and re-established after antipsychotic treatment. Shannon entropy, a measure of time series complexity calculated from symbolic dynamics, is higher in psychotic speech, which is also characterized by higher levels of phase coupling: higher bicoherence, as determined by bi-spectral analysis. In [30], voice dynamics allowed discrimination of patients with schizophrenia from healthy controls with a balanced accuracy of 85.68% using nonlinear time series analysis techniques. Applying nonlinear dynamical methods to speech recognition enables a content-independent, pure measure of the form of normal thought, its distortion in psychosis, and its restoration during treatment.

Facial Expressions Alterations in emotional facial expressions are a hallmark of psychopathology and may be present before the onset of mental disorders. Facial expression is widely used to evaluate emotional impairment in neuropsychiatric disorders [31]. Patients with various psychiatric conditions, particularly psychotic disorders in which an individual's sense of reality is impaired, may exhibit abnormal facial expressions [32]. The most recent reviews on the topic are almost two decades old, and even though impairment of facial expressions is recognized as a major diagnostic feature in schizophrenia, research in this field has been scarce. Automatic face tracking, feature extraction, and the production of temporal profiles of each facial muscle activity are all calculated using nonlinear methods. These profiles are quantified to calculate the frequency of individual and combined action units (AUs) in videos, which can help with statistical research on widespread illnesses that impact facial expression. The forehead, nose, and right cheek significantly differed between moderately and markedly ill patients with schizophrenia using support vector machine (SVM) classification algorithm [33]. The nature of facial expression deficits among individuals at clinical high risk (CHR) for developing psychosis using a comprehensive approach, incorporating clinical interview ratings and automated facial expression coding analysis, has been presented [34]. The interpersonal coordination (synchrony) of head movement during patient–therapist clinical interviews was investigated in patients with depression [35] at 7-week intervals over a period of 21 weeks. Interpersonal coordination of head movement between patients and therapists was measured using windowed cross-correlation, which showed no change over the course of treatment with a change in depression severity. Further studies using facial expressions in diagnosis, prevention, and prediction of psychosis should be investigated in this direction.

Assessments Several studies on the time evolution of symptoms have shown that nonlinear dynamical processes may be involved in the recurrence of symptoms in conditions such as schizophrenia [36]. These findings may contribute to a more cautious interpretation of the time course of these illnesses than conventional theories. When performing physiological functions or suffering from diseases, the human brain, a complex nonlinear system with hierarchical levels, exhibits complex spatiotemporal behaviors at different hierarchical levels. Understanding the human brain requires identifying nonlinear dynamics of spatiotemporal behaviors that dynamically evolve over time. Given the complex and seemingly random pattern of thought and behavior of schizophrenia patients, it seems reasonable to ask whether nonlinear dynamics methods, which aim to quantify the evolution of nonlinear systems over time, may be useful in understanding schizophrenia. Introducing dynamics concepts such as complexity and dynamical changes (i.e., bifurcations) allows for a new perspective on mind–brain function and dysfunction and the time evolution of symptoms. Association disturbances (i.e., the disruption of normal links between thoughts) are easily observed over time and were proposed by Bleuler as one of the key features of schizophrenia [37]. Events are organized in a complex but rule-driven manner across time and sequences of observations on multiple levels, from neural substrates to symptoms and social behavior [38]. The relationships between symptom dimensions and cognition have primarily been studied in non-affective psychosis. Different psychopathological indicators, such as anxiety, depression, and positive and negative symptoms, change over time and are measured at several assessment time points. In [39], linear and nonlinear associations between cognition and four symptom dimensions (reality distortion, negative symptoms, disorganization, and depression) in non-affective psychosis were examined in a population-based cohort of 166 patients with first-onset psychosis using regression analyses. The associations of negative symptoms with cognition include both linear and nonlinear elements, suggesting that this dimension is not a unified concept.

MRI Abnormalities in brain gray matter (GM) topology has been consistently observed in schizophrenia; these appear to be present even during the first episode of psychosis (FEP) and in patients at clinical high risk (CHR) [40]. Cortical surface area and thickness abnormalities have been consistently observed in schizophrenia, including in FEP [41]. GM morphology is inherently complex, and nonlinear dynamics analyses of these spatial data are appropriate mathematical techniques for extracting their statistical properties. Converting images into sequences with time series analysis tools has been used to solve various image data mining problems [42]. In previous research [40], the largest Lyapunov exponent was applied to distinguish patients with psychosis against clinical high-risk and healthy controls. Between-group differences were identified by (a) comparing the largest Lyapunov exponent brain maps, which resulted in statistically significant differences in FEP and CHR compared to HC, and (b) matching the largest Lyapunov exponent series with the Morlet wavelet, which resulted in statistically significant differences in the scalograms of FEP against CHR and HC. In another study [43], authors investigated the difference in the complexity of gray matter brain structures between pathological MRI images and the images of healthy controls. They used T1-weighted MRI images of 39 HC, 25 SCZ, and 11 bipolar disorders (BPD) and applied the 2D box-counting method for the FD calculation. They used a global and a local approach to identify subtle brain alterations. They found reduced FD values in BPD and SCZ patients, especially in regions often affected by these disorders, like the temporal and frontal lobes. Zhao et al. [44] also used a box-counting-based method but calculated the FD in 3D images. They used MRI images of 19 SCZ patients and 19 HC. While they also demonstrated a reduction of FD in patients affected with schizophrenia, the regions were different from the group of Squarcina et al. (2017). The main regions investigated were the hippocampus and thalamus. Non-segmented brain MR images were analyzed in [45] using radiomics texture features for the identification of first-episode psychosis (FEP) vs. controls and FEP vs. clinical high risk with no later transition. The difference of entropy and contrast features calculated on image intensities showed to enhance the subject's clinical status prediction.

5 Summary and Outlook

Nonlinear dynamical systems can identify a set of variables (e.g., psychotic symptoms) which control the system transition from one state to another, revealing the variables that force the dynamics of psychotic states. However, psychopathological phenomena, like many other processes in nature, are complex processes that evolve over time and are well described and understood within deterministic chaotic and nonlinear dynamical systems theory. The purpose of this chapter was to highlight the potential benefits of applying nonlinear dynamics approaches to the field of psychiatry, especially in this increased digitalized era. Since, clinical research and practice are using objective digital assessments of patient behavior at a rapid rate. Digital measuring instrument development and validation in psychiatry present considerable opportunities. Several behavioral aspects of psychotic disorders, including alogia (poor speech) and affective flattening (diminished emotional expression or emotional withdrawal), can be quantitatively assessed using standardized tasks and coding schemas, which can be automated by using computer vision and vocal acoustic machine learning models. The involvement of nonlinear dynamic systems theory is now urgent to be implemented in the quantization of the digital information received from naturally occurring data in psychosis research.

We hope to inform researchers and clinicians that nonlinear dynamic systems applications can provide an additional useful longitudinal framework for understanding psychotic disorders. They may also provide a new, more descriptive, and ultimately more predictive approach, leading to a better understanding of the interrelationship between psychotic disorder's psychological, neurobiological, and genetic underpinnings.

References

1. Moreno-Küstner B, Martín C, Pastor L (2018) Prevalence of psychotic disorders and its association with methodological issues. A systematic review and meta-analyses. PLoS One 13(4):e0195687
2. Saunders NR, Gandhi S, Chen S et al (2020) Health care use and costs of children, adolescents, and young adults with somatic symptom and related disorders. JAMA Netw Open 3(7): e2011295–e2011295
3. Roessner V, Rothe J, Kohls G et al (2021) Taming the chaos?! Using eXplainable Artificial Intelligence (XAI) to tackle the complexity in mental health research. Eur Child Adolesc Psychiatry 30(8):1143–1146
4. Strogatz SH (2015) Nonlinear dynamics and chaos: with applications to physics, biology, chemistry, and engineering, 2nd edn. CRC Press
5. Hosenfeld B, Bos EH, Wardenaar KJ et al (2015) Major depressive disorder as a nonlinear dynamic system: bimodality in the frequency distribution of depressive symptoms over time. BMC Psychiatry 15(1):222
6. Breakspear M (2017) Dynamic models of large-scale brain activity. Nat Neurosci 20(3): 340–352
7. Paulus MP, Braff DL (2003) Chaos and schizophrenia: does the method fit the madness? Biol Psychiatry 53(1):3–11
8. Pham TD, Abe T, Oka R et al (2015) Measures of morphological complexity of gray matter on magnetic resonance imaging for control age grouping. Entropy 17(12):1099–4300
9. Huber MT, Braun HA, Voigt K et al (2001) Some computational aspects of the kindling model for neuropsychiatric disorders. Neurocomputing 38(40):1297–1306
10. Fernández A, Gómez C, Hornero R et al (2013) Complexity and schizophrenia. Prog Neuro-Psychopharmacol Biol Psychiatry 45: 267–276
11. John AM, Elfanagely O, Ayala CA et al (2015) The utility of fractal analysis in clinical. Neuroscience 26(6):633–645
12. Pezard L, Nandrino JL (2001) Dynamic paradigm in psychopathology: "chaos theory", from physics to psychiatry. Encéphale 27(3): 260–268
13. Hoffman RE, Dobscha SK (1989) Cortical pruning and the development of schizophrenia: a computer model. Schizophr Bull 15(3): 477–490
14. Jia Y, Gu H (2019) Identifying nonlinear dynamics of brain functional networks of patients with schizophrenia by sample entropy. Nonlinear Dyn 96(4):2327–2340
15. Varley TF, Carhart-Harris R, Roseman L et al (2020) Serotonergic psychedelics LSD & psilocybin increase the fractal dimension of cortical brain activity in spatial and temporal domains. NeuroImage 220:117049
16. Stephan KE, Kasper L, Harrison LM et al (2008) Nonlinear dynamic causal models for fMRI. NeuroImage 42(2):649–662
17. Oribe N, Hirano Y, Kanba S et al (2015) Progressive reduction of visual P300 amplitude in patients with first-episode schizophrenia: an ERP Study. Schizophr Bull 41(2):460–470
18. Turetsky BI, Dress EM, Braff DL et al (2015) The utility of P300 as a schizophrenia endophenotype and predictive biomarker: clinical and socio-demographic modulators in COGS-2. Schizophr Res 163(1):53–62
19. Korda AI, Ventouras E, Asvestas P et al (2022) Convolutional neural network propagation on electroencephalographic scalograms for detection of schizophrenia. Clin Neurophysiol 139: 90–105
20. Light GA, Zhang W, Joshi YB et al (2017) Single-dose Memantine improves cortical oscillatory response dynamics in patients with schizophrenia. Neuropsychopharmacology 42(13):2633–2639
21. Kirihara K, Rissling AJ, Swerdlow NR et al (2012) Hierarchical Organization of Gamma and Theta Oscillatory Dynamics in Schizophrenia. Biol Psychiatry 71(10):873–880
22. Lainscsek C, Sampson AL, Kim R et al (2019) Nonlinear dynamics underlying sensory processing dysfunction in schizophrenia. Proc Natl Acad Sci 116(9):3847–3852
23. Kutepov IE, Dobriyan VV, Zhigalov MV et al (2020) EEG analysis in patients with

schizophrenia based on Lyapunov exponents. Inform Med Unlock 18:100289
24. Dick OE, Murav'eva SV, Lebedev VS et al (2022) Fractal structure of brain electrical activity of patients with mental disorders. Front Physiol 13. https://doi.org/10.3389/fphys.2022.905318
25. Namazi H, Aghasian E, Ala TS (2019) Fractal-based classification of electroencephalography (EEG) signals in healthy adolescents and adolescents with symptoms of schizophrenia. Technol Health Care 27:233–241
26. Raghavendra BS, Dutt DN, Halahalli HN et al (2009) Complexity analysis of EEG in patients with schizophrenia using fractal dimension. Physiol Meas 30(8):795
27. Alpert M, Kotsaftis A, Pouget ER (1994) At issue: speech fluency and schizophrenic negative signs. Schizophr Bull 23:171–177
28. Spitzer M, Beuckers J, Beyer S et al (1994) Contextual insensitivity in thought-disordered schizophrenic patients: evidence from pauses in spontaneous speech. Lang Speech 37(2): 171–185
29. Todder D, Avissar S, Schreiber G (2013) Non-linear dynamic analysis of inter-word time intervals in psychotic speech. IEEE J Transl Eng Health Med 1:2200107
30. Fusaroli R, Simonsen A, Weed E et al (2014) Poster #M197 non-linear dynamics of speech in schizophrenia: a machine-learning approach. Schizophr Res 153:S262
31. Hamm J, Kohler CG, Gur RC et al (2011) Automated Facial Action Coding System for dynamic analysis of facial expressions in neuropsychiatric disorders. J Neurosci Methods 200(2):237–256
32. Gupta T, Haase CM, Strauss GP et al (2022) Alterations in facial expressions of emotion: determining the promise of ultrathin slicing approaches and comparing human and automated coding methods in psychosis risk. Emotion 22:714–724
33. Jian B-L, Chen C-L, Chu W-L et al (2017) The facial expression of schizophrenic patients applied with infrared thermal facial image sequence. BMC Psychiatry 17(1):229
34. Gupta T, Haase CM, Strauss GP et al (2019) Alterations in facial expressivity in youth at clinical high-risk for psychosis. J Abnorm Psychol 128(4):341–351
35. Bhatia S, Goecke R, Hammal Z et al (2019) Automated measurement of head movement synchrony during dyadic depression severity interviews. In 2019 14th IEEE International Conference on Automatic Face & Gesture Recognition 14–18 May 2019, p 1–8
36. Breakspear M (2006) The nonlinear theory of schizophrenia. Australian New Zealand J Psychiatry 40(1):20–35
37. Bleuler E (1950) Dementia praecox or the group of schizophrenias. Dementia praecox or the group of schizophrenias. International Universities Press, Oxford, p 548
38. Paulus MP, Geyer MA, Braff DL (1996) Use of methods from chaos theory to quantify a fundamental dysfunction in the behavioral organization of schizophrenic patients. Am J Psychiatr 153(5):714–717
39. Kravariti E, Russo M, Vassos E et al (2012) Linear and non-linear associations of symptom dimensions and cognitive function in first-onset psychosis. Schizophr Res 140(1): 221–231
40. Korda AI, Andreou C, Avram M et al (2022) Chaos analysis of the brain topology in first-episode psychosis and clinical high risk patients. Front Psych 13. https://doi.org/10.3389/fpsyt.2022.965128
41. Schultz CC, Koch K, Wagner G et al (2010) Increased parahippocampal and lingual gyrification in first-episode schizophrenia. Schizophr Res 123(2):137–144
42. Réjichi S, Chaabane F (2016) Brain tumor extraction using graph based classification of MRI time series for diagnostic assistance. In: 2016 International Symposium on Signal, Image, Video and Communications 21–23 November 2016
43. Squarcina L, Castellani U, Bellani M et al (2017) Classification of first-episode psychosis in a large cohort of patients using support vector machine and multiple kernel learning techniques. NeuroImage 145:238–245
44. Zhao G, Denisova K, Sehatpour P et al (2016) Fractal dimension analysis of subcortical gray matter structures in schizophrenia. PLoS One 11(5):e0155415
45. Korda AI, Andreou C, Rogg HV et al (2022) Identification of texture MRI brain abnormalities on first-episode psychosis and clinical high-risk subjects using explainable artificial intelligence. Transl Psychiatry 12(1):481

Chapter 10

Carving Out the Path to Computational Biomarkers for Mental Disorders

Ronald Sladky, Anna Todeva-Radneva, and Claus Lamm

Abstract

For the last 20 years, there has been impressive progress in cognitive neuroscience and neuroimaging. Unfortunately, the translation from basic research to real world clinical applications in the domain of mental disorders has been limited. A major contribution of functional MRI would be to provide computational biomarkers for precision medicine and treatment optimization. Before this is possible, a number of open conceptual and methodological challenges need to be addressed, and some of them will be outlined in this chapter. For illustration purposes, we focus on the amygdala and its relevance in anxiety disorders. We cover a few central aspects on how certain pitfalls can preclude reliable empirical work on the amygdala and, thus, robust computational biomarkers. Overcoming them requires multidisciplinary collaboration on the level of psychiatric and neuroscientific theories on the one hand, and on fMRI acquisition, signal processing, and analysis on the other. While we are optimistic that neuroimaging will provide a substantial contribution for the discovery of computational biomarkers in psychiatry, psychopathologies and their neuropathophysiology are complex so we cannot propose an easy solution. Instead, we focus on some challenges that emerge in the overlap of theory and application.

Key words Emotions, Anxiety, Mental disorders, Amygdala, Functional MRI

1 Introduction

1.1 The Complexity of Understanding Brain Function and Dysfunction

In his lectures in 1890, the Viennese psychiatry professor Theodor Meynert deplored that, in contrast to other fields in medicine, psychiatry has not made the transition from a descriptive discipline toward an *explaining science*. In fact, over the last 130 years, psychiatry has made tremendous progress in the diagnostics and treatment of mental illnesses. However, compared to other fields in medicine, a precise understanding of the etiology, pathophysiology, and progression is insufficient for many diseases. Despite the availability and success of a wide range of treatment options, core brain mechanisms and neural principles that indicate or explain efficacious treatments are unclear. Admittedly, this is not surprising considering that the brain is the most complex organ in the

Drozdstoy Stoyanov et al. (eds.), *Computational Neuroscience*, Neuromethods, vol. 199,
https://doi.org/10.1007/978-1-0716-3230-7_10,

human body and constantly shaped by biological, psychological, and social factors in a complex, nonlinear fashion. Appreciating this complexity, Theodor Meynert already proposed that psychopathologies and their treatment should be addressed by an interdisciplinary and mechanistic perspective on the brain, just like anatomy has evolved toward an explanatory science by relating its observations from histology, developmental, and comparative biology.

While none of the factors relevant for development and progression of psychopathologies appear monocausal, it is only reasonable to assume that if mental illnesses are (at least in part) brain illnesses there have to be some observable biomarkers in the brain for the presence or absence of these illnesses based on the computational level of pathophysiological mechanisms. Functional neuroimaging has been very successful in mapping brain functions to brain areas, unveiled quantifiable differences between study populations, and in so doing informed more and more refined theories about brain mechanisms. Proper theorizing and application of these theories are important. The history of neuroscience has taught us that these mechanistic concepts have always been inspired or explained by referring to metaphors and narratives that reflected the cultural and technological state of the art. With varying degree of seriousness, the brain has been illustrated as operating using cogwheels, hydraulic pumps, steam engines, electric cables, or, most recently, computer hardware that performs computations, implements the mind as a software, or instantiates some form of virtual reality. Is it organized like a manufacturing line in a factory, an office building with several different departments, a homunculus sitting in a Cartesian theater, like a democratic parliament, or a social network?

Thinking about these questions is not an intellectual pastime activity. They provide the conceptual framework for how hypotheses are formed and experiments are designed. However, that is not the only motivation as pragmatic considerations also play a role in experimental design and in doing so influence the way brain functions are operationalized. The empirical results are then used to update theories, yet they are not independent of hidden assumptions, methodological (quasi) standards, and narratives taken for granted. This means, methodological choices—may they be well-founded and necessary or naïve and arbitrary—can change the direction of how theories evolve over time. On the other hand, not paying attention to pragmatic methodological constraints is not fruitful, either. An effect cannot be studied if it is not operationalized for the empirical methods at hand. In sum, theoretical background, methodological choices, empirical results, and their interpretation conditionally depend on each other.

We will discuss how these considerations apply to the identification of computational biomarkers in psychiatry. In doing so, this chapter will focus mainly on the amygdala and anxiety disorders.

Yet, many of the principles described herein generalize to other psychopathologies and brain areas and illustrate general principles of empirical neuroscientific research. First, we will investigate the justification of using the amygdala as a potential target of interest by discussing its function and operationalization in experiments. Second, we address potential challenges in measuring neural activation in the amygdala both because of its complex anatomy and methodological challenges specific to functional MRI. We argue that finding optimal solutions for these two issues are of central importance for the identification and practical application of amygdala-based biomarkers.

1.2 The Role of Emotions in Anxiety and Other Mental Disorders

On a symptomatic level, the experience of negative emotions and impairments in emotion regulation are well-known phenotypes of many psychiatric disorders [1–3]. In anxiety disorders, in particular, excessive fear and anxiety are hallmark symptoms (ICD-11, https://icd.who.int). Modifying the experience of these negative emotions or regulating their intensity are targets or outcomes of successful therapeutic interventions [4]. Experiencing negative emotions biases the perception of the world [5] and executive functions [6, 7], making anxiety disorders a high-risk factor for major depressive disorders [8] and disability [9]. There is also a well-established relationship between negative emotional experience and chronic stress [10], which in turn leads to negative somatic and psychological downstream effects [11].

While symptoms and consequences of anxiety and other mental disorders are multifaceted, we can assume that there is a relationship to disordered brain functions that might be observable using neuroimaging. In fact, applications of fMRI have shown that deficits in motor areas can be informative to evaluate motor impairments [12], language areas for language impairments [13], and visual areas for vision impairments [14]. The same should hold true for brain areas that are involved in perceiving and experiencing emotions given that emotions play a key role in anxiety and other mental disorders.

1.3 The Role of the Amygdala in Emotions

Historically, the amygdala has been considered one of the key regions for experiencing and perceiving emotions [15]. Since then, robust empirical evidence has led to the popular notion that the amygdala is the emotion center of the brain. Amygdala's role in fear and anxiety is likely the most well-studied topic due to its high relevance for clinical and social neuroscience as well as because of the comparably simple experimental implementations and strong behavioral and neural effects. After all, there is a high ecological significance and value to efficient and effective responses toward danger [16], which are understood as a set of stereotypical and automatic threat responses [17, 18]. Beyond animal research and human neuroimaging studies, there is also strong evidence from humans with amygdala lesions that exhibit strongly reduced threat responses and subjective feelings of fear [19] and, e.g., marked impairments in detecting deceit in social decision making [20, 21]. Later on, we will discuss some challenges to this simplified notion of the amygdala's involvement in emotions and highlight a more nuanced role in how emotions are processed in the brain.

On a neural level, amygdala reactivity during threat perception is assumed to be implemented by input from sensory cortices and the thalamus to the amygdala in partly independent processing streams [22]. This traditional dual road model has been extended to a more complex multi-road model [23] strengthening the notion that emotion perception is not implemented in the amygdala alone but in a more distributed brain network. Likewise, the experience of emotions is also considered to rely on distributed brain networks [24]. While the amygdala appears to be central to emotional experience, it is neither sufficient nor necessary. This is most convincingly evidenced in a person with amygdala lesions, who can experience strong fear in response to CO_2 inhalation, and was unaffected in other tested fear-induction settings, such as fear-inducing emotional images and videos, a visit to a house of horrors, and life-threatening situations in real life [25]. On the other hand, there is strong evidence from a controlled amygdala stimulation study in epilepsy patients using acute monopolar electrical stimulation that did not affect the subjective experience of emotions despite robustly observable physiological changes [26].

In sum, the present evidence shows clearly that the amygdala is involved in the perception and experience of emotions. Yet, there is neither a specialization for fear alone nor is it the only brain area relevant for emotion and fear processing. Its undisputed central role in a variety of emotional processes makes it a highly valued research target in clinical neuroscience and an ideal biomarker candidate.

1.4 Amygdala Activation and Connectivity in Anxiety and Other Mental Disorders

A typical neuroimaging finding in anxiety disorders is an increased amygdala response to fear-inducing stimuli [27, 28]. This hyperactivation is in contrast to an amygdala hypoactivation toward rewarding stimuli, which is observed in the context of depression-related anhedonia symptoms [29, 30] or initial hyperactivation followed by hypoactivation in schizophrenia [31]. This means, in these and other conditions, amygdala activation can be an informative biomarker taking into account the difference in BOLD response [32].

Amygdala hyperactivation in anxiety disorders can be explained in part by an increased reactivity toward threat and uncertainty in anxiety disorders [33]. On the other hand, there is also a reduction in prefrontal regulation that leads to hypervigilance (i.e., increased attentional focus toward threat), impaired safety learning, and dysfunctional cognitive control [34]. In particular, the orbitofrontal cortex plays a relevant role in amygdala regulation [35, 36], and reduced resting state and task connectivity has been shown, e.g., in social anxiety disorder [37–39]. In contrast, e.g., selective serotonin reuptake inhibitors, a pharmacological intervention with demonstrated applicability in anxiety and affective disorders, were shown to increase amygdala downregulation in healthy volunteers [40] consistent with connectivity changes following successful psychotherapeutic interventions [41] and theories of reappraisal success [42]. Additionally, amygdala activation and connectivity changes during emotion processing have been observed for the NMDA receptor antagonist ketamine [43] and the mixed serotonin receptor agonist psilocybin [44], which have both demonstrated the potential of therapeutic efficacy in anxiety and other mental disorders [45, 46].

By and large, the present literature suggests some involvement of the amygdala in anxiety and other mental disorders [47]. Connectivity studies and changes after therapeutic interventions suggest that it plays a role in the pathophysiological mechanisms providing further justification for its usefulness as a biomarker.

1.5 Structural and Functional Alterations in the Amygdala as a Possible Differential Diagnostic Biomarker for Mental Disorders

Considering its role in emotion processing and the impairment of the emotional response being a consistent phenomenological state across the majority of mental disorders, the amygdala could also be a morphological substrate with differential diagnostic value in psychiatry. For instance, a reduction of the bilateral amygdala volume has been established in major depressive disorder, bipolar disorder, and also first-episode schizophrenia [48]. In addition, the volume reduction was significantly more pronounced in the latter two than in major depressive disorder.

A meta-analysis by McTeague et al. demonstrated convergent amygdala hyperactivation during emotion-processing tasks in non-psychotic (e.g., anxiety, unipolar depression, bipolar depression, obsessive-compulsive disorder, etc.) versus psychotic disorders (e.g., schizophrenia, schizoaffective disorder, etc.) [2]. Furthermore, in the same meta-analysis an overlapping increased activation of the left amygdala and the hippocampus was observed in anxiety and major depressive disorders whereas aberrant activity of the right amygdala and right ventrolateral prefrontal cortex was associated with bipolar disorder.

Differences in amygdala connectivity have also been implicated in various mental disorders. Resting state functional connectivity of the amygdala and the right middle cingulate has been shown to be reduced in patients with first episode schizophrenia as compared to individuals at high-risk for schizophrenia and healthy controls [49]. Decreased resting state functional connectivity between the right amygdala and the left anterior hippocampus has been noted as a possible discriminating alteration in unipolar major depressive disorder compared to bipolar disorder and healthy individuals [50].

Moreover, bipolar disorder type II has been differentiated from borderline personality disorder via a lower resting state functional connectivity between the amygdala and the middle frontal gyrus, which was further associated with affective shifts, higher impulsivity, functional impairment, etc. [51]. All of the hitherto described findings suggest the possibility of the amygdala being a possible differentiating biomarker for mental illnesses.

1.6 Real World Challenges for Amygdala-Based Biomarkers

So far, we have discussed a number of reasons why the amygdala could be an excellent biomarker for psychiatric conditions such as anxiety disorders. However, despite the robust theoretical basis, there are a number of challenges regarding the practical applications. In the following, we will focus on a few key issues that influence the outcome of this endeavor. Admittedly, there is a strong interaction of these issues, which creates the impression that any research in this domain is hopelessly stuck in a vicious circle. To overcome this deadlock, we aim at making these challenges explicit and unmask some of the mutual dependencies. We argue that there is not one overall grand solution that magically dissolves all challenges. There is not the one ideal scanner or sequence parameter that provides the optimal data at the press of a button. There is not the one ideal analysis method that will provide answers to complex research problems given just some arbitrary data. Instead, the challenges are multi-faceted and follow the Anna Karenina principle named after a quote from Leo Tolstoy's novel: *All happy families are alike; each unhappy family is unhappy in its own way.*

Emphasizing the complexities as well as the various sources of limitations is not meant to frustrate the reader. Instead, we want to raise awareness on the reasons for these complexities and that there is not a single monolithic solution or investment that can solve all issues at the same time. Instead, we encourage to investigate the challenges from an engineering perspective and applying the successful paradigm that goes back to Julius Caesar: *I divide and conquer.* This means to appreciate the complexity of an unsurmountable task and then to break it apart into independent subtasks as far as needed to make them easy to handle. Having smaller tasks also allows for feasible outcome evaluation, i.e., to verify that a specific goal has been met, that the protocol is working as expected, and that a device is actually functioning. In practice this can mean, e.g., the following: Is my task valid for my target brain region, do I have a problem with the timing and synchronization of my scripts, and is it actually possible to measure the brain response with my imaging sequence settings? This kind of verification and troubleshooting is usually only possible if the subtasks and preliminary goals have been operationalized properly before setting course for novel scientific discoveries.

2 Materials

2.1 Magnetic Resonance Imaging Hardware

To date, magnetic resonance imaging (MRI) provides the best way for in vivo neuroimaging of the amygdala and other subcortical brain regions. Not only does it allow for high resolution anatomical imaging, functional MRI (fMRI) can be used to acquire time-resolved images that are sensitive to local changes in blood oxygenation, which is a well-established proxy for clustered neural activity. MRI scanners are becoming more available and affordable as scanner manufacturers are able to optimize their hardware and software designs to reduce construction, material, and consumable costs. However, while the versatility of the fMRI method allows for many compromises, there is no such thing as a free lunch. In fact, some of the methodological challenges that will be discussed below can be mitigated by investing in research grade MRI scanners with field strengths of 3 Tesla and more, a sensitive multi-channel head coil for high signal to noise ratio (SNR), and an optimal gradient system (i.e., fast rise time and high maximum gradient strength) to allow for the use of fast echo planar imaging (EPI) methods. In detail, compromises in field strengths entail a linear reduction of overall MRI signal intensity impacting factors such as specificity (e.g., the ability to distinguish between tissue types). High sensitivity of the receiver coil, on the other hand, can improve the amount of detected original signals despite higher levels of noise.

If these hardware investments are not an option, increasing the number of data points provides a solution to improve data quality by suppressing the effect of noise through averaging, following the law of large numbers. One way to achieve this is by generating more data points by increasing the sampling rate, i.e., reducing the repetition time of the image acquisition. While this option also achieves a more accurate sampling of the BOLD signal, its possibilities are limited by the sluggishness of the BOLD effect.

Another solution for noise suppression is to increase the repetition of task trials, which entails either longer experiments and measurements or fewer individual task conditions. Increasing the repetitions of task trials is in many cases undesired as patients in particular will struggle with long experimental runs leading to less task compliance and excessive head motion. Beyond that, prolonged exposure to the task stimulus will create learning effects resulting in neuronal adaptation processes, such as habituation, a phenomenon particularly relevant in the amygdala. A reduction of individual task conditions, on the other hand, can be achieved by formulating precise hypotheses and experimental interventions. This requires a precise understanding of the effects of interest and their neural mechanisms. We will discuss these issues in detail in the Subheading 3.

Overall, these explanations demonstrate the consequences of several investment and experimental choices that require careful deliberation, setup, and ongoing empirical evaluation of the setup, experimental protocol, and data. The complexity of the imaging part additionally highlights the need for dedicated research support by personnel with experience in neuroimaging and MR physics—and that hardware investments alone are not sufficient.

2.2 Computing Hardware

For data analysis and modeling dedicated computing hardware is required. In comparison to an MRI scanner, however, the costs will be negligible. Based on previous experience, modern prosumer and small enterprise hardware allow for successful data processing, as long as fast data access and long-term data security demands can be met. This means, e.g., that a state-of-the-art desktop workstation with sufficiently large internal disk space and RAM attached to an external redundant backup system could be a cost-efficient solution. While multi-core servers and high-performance computing clusters allow for parallel data processing and high data throughput, they also require experienced research support personnel and end-users with high levels of technical expertise. With this in mind, the computing infrastructure should be designed to address the end-users' skill level and demands. Investing time and resources in usability and ongoing domain-general training are central requirements for planning new computing infrastructure. Overlooking or ignoring these human factors inevitably degrades the usefulness and cost-effectiveness of the infrastructure provided.

3 Methods

3.1 Experimental Design: Amygdala Function

Typical tasks to investigate amygdala reactivity involve presentation of emotional stimuli. Amygdala function is not restricted to only one sensory domain, in fact it is known to receive multimodal input and perform multisensory integration [52]. Consequently, a wide range of experimental paradigms are used, e.g., *visual*, such as negatively valanced (e.g., shocking or disgusting) [53, 54] or positively valanced (e.g., humorous or erotic) emotional images [55] and faces [56], *auditory*, such as dissonant audio cues [57], *olfactory*, such as pleasant or unpleasant odors [58], and *somatic*, such as painful stimulation [59].

Detailed work in rodent models has demonstrated that amygdala output [60] is used to regulate and initiate defensive behaviors [61–63] and plays a central role in fear conditioning [64, 65]. A direct translation from rodent to primate research is challenging due to phylogenetic changes specifically in the amygdala region [66]. While there is good cross-species evidence for similar defensive behavior initiation and regulation [67], there is less consensus if mere anticipation of negative stimuli robustly increases amygdala activation [68]. Yet, it was observable for highly anxious participants [69–71] as a consequence of negative expectations and biases during anticipation, e.g., in anxiety disorders [34] and depression [72]. On the other hand, there is also some risk that elevated feelings of anxiety and stress when undergoing fMRI experiments could be an important confound [73], in particular at the beginning of the scanning [74]. This means, an increase in amygdala activation in people with anxious predispositions could be an artifact of the experimental setup and less specific for psychopathologies per se [75]. This demonstrates that translational neuroscience can be complex and task demands do not necessary translate across different species, i.e., what might be ecologically relevant for rodents does not necessarily entail the same relevance for human participants.

In fact, the amygdala has been proposed to play an important role for relevance and novelty detection to guide attention and learning [76]. Given that understanding the feelings and intentions of conspecifics has a high ecological relevance for human beings, it is not surprising that recognizing [77, 78], distinguishing [79], passively viewing [80], and even subliminal presentation of emotional faces increase amygdala activity [81, 82]. In this context, amygdala hyperactivation has been observed in anxiety disorders [27, 28] and increased amygdala reactivity toward negatively contextualized neutral face images is a risk factor for impaired mental well-being [83]. Besides different forms of hypo- or hyperactivation [29–31], another relevant biomarker could be delayed amygdala recovery from an emotional stimulus, which has been linked with trait neuroticism [84] and inhibited temperament [85]. This means, other than peak amygdala activation, which is the focus of conventional fMRI analysis, there are informative signal properties that could be identified using a time course analysis, such as time to peak, slope, recovery to baseline time, area under the curve, and features observable using frequency domain analyses [86, 87].

Besides inter-trial temporal dynamics, there is also an ongoing debate about the temporal reliability and stability of the amygdala response to emotional faces across trials, runs, and sessions. While some studies report high stability across task sessions [88] and trials [89–91], other studies demonstrate amygdala habituation in healthy participants [92–94]. In the context of social anxiety disorders, however, amygdala habituation could also be seen as a distinctive reaction to the exposure of an inherently aversive stimulus [91]. This means, in addition to amygdala reactivity, signal changes across task trials and within an individual task trial could be informative.

Another lesson from animal models is the central role of the amygdala in various forms of learning, especially in volatile or vicarious contexts [95]. In support of this notion, it has been shown in humans that participants with (basolateral) amygdala lesions perform worse in a social learning task [20, 21], where they had to repeatedly decide whether to trust another player in an economic decision-making game. This finding has been replicated in neurotypical participants [96], highlighting the amygdala's important role in prolonged instrumental learning. In this task, in contrast to face stimuli, no habituation was observed, indicating a longer-lasting involvement to maintain successful task performance. This means that ecologically valid, social, and possibly non-social probabilistic learning tasks could have particular potential for identifying pathology-specific amygdala and behavioral biomarkers.

In sum, we conclude that the amygdala, in theory, would be a good biomarker candidate given its central role in a variety of neuronal systems, cognitive processes, and experiential phenomena. However, conceptual flaws in the study and task design lead to suboptimal amygdala responses or even a complete absence of task-related activation. Careful planning and validation of the experimental setup and model will be essential.

3.2 Experimental Design: Amygdala Anatomy

In addition to the amygdala's functional intricacies, another level of complexity is added by its anatomical and regional functional heterogeneity. In human neuroimaging, the amygdala is typically treated as a singular structure [97], which might be an invalid oversimplification [98]. Two important subnuclei are the basolateral (BLA) and the central part of the amygdala (CeA), which possibly extends to the bed nucleus of the stria terminalis (BST) [99], which is heavily connected to the CeA but anatomically separated by a white matter structure known as the internal capsule in primates [52]. On a functional level, the BLA is considered a primary input region for multimodal sensory data, while the CeA, on the other hand, projects strongly to brainstem and hypothalamus to facilitate automatic responses to threat [52]. Previous studies have found highly relevant fear- and anxiety-related CeA responses [65, 69], which highlight that this region, despite being small and some of its effects overshadowed by the larger BLA, should be addressed in a differentiated way. While the CeA could be related to fast threat responses, the BST appears to play a role in more prolonged states of anxiety [100]. Taken together, the BLA, CeA, and BST could provide the necessary architecture to form a computational network that hierarchically encodes associations between states of the world and their affective value on different time scales and levels of abstractness. It can be assumed that there are functional dissociations between the amygdaloid subregions that are reflected by a difference in the temporal profiles of the task response. This can lead to potentially destructive signal mixing, as time delayed or anticorrelated signals that are sampled and averaged together can lead to attenuation of the overall measured BOLD response.

Besides the BLA, CeA, and BST, there are more subnuclei with different cytoarchitecture, suggesting an even more complex involvement in the neural computation. A recent review suggests that the optimal strategy for amygdala parcellation is not even clear on an anatomic level (Ufer et al., in preparation). It can be expected that the missing consensus might lead to future revisions of how subregions should be defined anatomically. The debate on amygdala segmentation might appear only of theoretical value, yet there are important pragmatic implications. Many amygdala interneurons are GABAergic [101] leading to a strong inhibitory effect. This means that, e.g., amygdala regulation by prefrontal regions is achieved by long range excitatory glutaminergic projections, which target amygdala GABAergic neural populations that implement local inhibition [102]. Amygdala output, in turn, is frequently implemented by a neural mechanism known as disinhibition, i.e., neural activation that is facilitated by suppressing an ongoing inhibition. Output suppression and regulation plays a major role in amygdala's function leading to the notion that it is a *silent region* as local spontaneous neural activation does not easily propagate to other brain areas [52, 103], suppressing behavioral downstream consequences. As a consequence, uncovering the actual neural mechanisms that caused an observable effect on the fMRI BOLD signal is not straightforward as relevant amygdala output could have been suppressed as a consequence of neurotypical or pathological physiology. Likewise, the absence of an observable difference in the fMRI BOLD signal alone does not necessarily mean that there were no differences in neural computation between different study populations or conditions. The reason for this is that local metabolic demands that cause BOLD signal fluctuations can either be caused by the activation of excitatory or inhibitory neural subpopulations. To some extent, apparent amygdala hyperactivation observed in fMRI could reflect the opposite of the actual neural activation. In this case, the relatively increased BOLD signal could be caused by more local inhibition, leading to less amygdala neural activation but more metabolic demand.

We list a number of challenges for amygdala fMRI originating from the anatomical, cytoarchitectonic, and functional heterogeneity of the amygdaloid cluster. As countermeasure, we propose (a) the implementation of validated experimental protocols that avoid destructive signal mixing resulting from the temporal differences in the neural activation of the individual subpopulations, (b) the use of a high temporal and spatial resolution in fMRI to distinguish the different functional contributions during task performance (discussed in more detail in [1, 2]), and (c) taking into account amygdala connectivity and regulation by other brain regions (discussed in more detail in Subheading 3.3).

3.3 Experimental Design: Connectivity and Amygdala Regulation

The amygdala is extensively connected to a wide range of subcortical and cortical regions that are involved in a diverse set of different brain functions. On one hand, this is the reason for the importance of the amygdala and functions as an important hub in many domains. On the other hand, the amygdala might only be one aspect of a compound biomarker that involves a network of functionally connected brain areas. When considering the brain as a self-organized dynamic system, the causal interaction of these brain areas could be pivotal for understanding the mechanisms leading to regional activation as well as for behavioral, computational, and experiential phenotypes and how they relate to psychopathologies. For example, fast amygdala reactivity to an aversive stimulus might not necessarily lead to automatic threat responses or subjectively experienced distress, if efficient downregulation by other brain regions occurs or if attentional resources are utilized elsewhere [104]. In fact, the context-dependency of prefrontal-amygdala regulation has been shown [38] and successful reappraisal alters prefrontal network connectivity [42], which could be an explanation for amygdala connectivity changes after successful psychotherapy [41]. In this context, with fMRI-based neurofeedback of the amygdala gaining popularity, regulation success could be a promising tool as it reflects the participants' ability to voluntarily control their amygdala activation on demand.

3.4 Functional MRI of the Amygdala

So far, we have addressed issues pertaining mainly to the experimental design and already mentioned that amygdala fMRI also requires imaging with high spatial and temporal resolution. Even though fMRI can be an excellent method for subcortical functional imaging, it faces a number of challenges that need to be addressed.

Historically, there were serious concerns in the neuroimaging community that the amygdala and ventral prefrontal regions might be inaccessible for reliable BOLD fMRI [105], due to the air cavities formed by the paranasal sinuses that cause magnetic field inhomogeneities leading to signal dropouts in the ventral brain. As BOLD fMRI relies on detecting susceptibility-related signal differences (i.e., the difference between oxygenated and deoxygenated hemoglobin), the imaging sequences used are particularly affected by these susceptibility artefacts, e.g., gradient echo planar imaging sequence with relatively long echo times. Noteworthy, these issues are a lesser concern for other imaging modalities (e.g., anatomical MRI) or for fMRI in other species with different cranial anatomy. For example, in dogs the most profound signal dropouts are observed in medial dorsofrontal and not ventral brain areas such as the amygdala or hypothalamus [106].

The effects of signal loss due to spin dephasing can be compensated by using a high spatial resolution with small voxel sizes [107], which is also indicated to distinguish the functional contributions of the individual amygdala subregions to efficiently avoid signal mixing. Two factors limit the use of arbitrarily high resolutions.

First, decreasing the slice thickness reduces the covered acquisition volume. To compensate, more slices need to be acquired, which leads to longer repetition times reducing the temporal resolution. A whole-brain coverage may not be essential when using optimized experimental protocols that only target the amygdala and other more ventral areas such as the ventral striatum, midbrain regions, anterior insula, ventral temporal lobe (e.g., the fusiform face), as well as ventrolateral and ventromedial prefrontal areas. All these regions have been implicated in a wide range of mental disorders before, which makes them an ideal focus for an optimized imaging protocol. The disadvantage of this approach is that the experimental hypotheses need to be clear and explicit and there are less possibilities for post-hoc troubleshooting. For example, to verify the accuracy of the experimental timing and synchronization of the stimulus computer and MRI scanner clocks, the motor response in the hand area of the primary motor cortex provides a robust signal that can be related to button presses. This means that careful validation of the experiment is key before applying it to the actual research context. Finally, if support by the MRI scanner is available, novel optimized acquisition methods such as multiband echo planar imaging [108] can provide a good solution to achieve both high temporal and spatial resolution.

Second, the image signal to noise ratio degrades as voxel sizes decrease. The common way to reduce the signal noise during preprocessing is to apply spatial smoothing. Finding a good tradeoff between sensitivity (i.e., detecting the BOLD response despite the noise) and specificity (i.e., avoiding signal mixing) can be an empirical question that is tested with an independent pilot dataset, ideally. Smoothing that respects the tissue boundaries can be a good compromise. This can be achieved, e.g., in a volume of interest analysis, by extracting all the individual time courses within a given tissue boundary and averaging over these time courses. Another solution for increasing the SNR is to conduct longer experiments with more task trials [109]. Besides the longer scan times that can be intolerable for patients, the disadvantage is that there are learning and exposure effects that can alter the consistency of the amygdala signal, as we discussed in Subheading 3.1. Agreeably, the least practical but best solution to increase SNR in the amygdala is to move to higher field strengths such as 7 Tesla, which will be discussed in more detail in Subheading 3.5.

With respect to imaging parameters, besides the temporal and spatial resolution, the echo time (TE) plays an important role. Image quality can be improved by using short echo times, as distortions and other susceptibility-related artifacts are visibly reduced. Unsurprisingly, a reduction in susceptibility-related signal effects also reduces the BOLD-sensitivity, leading to less image contrast between deoxygenated and oxygenated hemoglobin. This means that longer echo times that approximately optimize for the hemoglobin-specific relaxation time contrasts while preventing the dominance of imaging artifacts is key. Validation of optimal BOLD sensitivity in the amygdala, given the experimental setup, should occur before the actual measurements. A straightforward approximation is to calculate and systematically compare the temporal signal to noise ratio SNR_t for each voxel based on its time course y defined as:

$$\mathrm{SNR}_t(y) = \frac{\mu_y}{\sigma_y} = \frac{\mathrm{mean}(y)}{\mathrm{stddev}(y)}$$

In practice, to quantify BOLD sensitivity in the amygdala based on the SNR_t, resting state or task datasets can be used. Robust tasks such as the emotion discrimination task [79] would be a good option as the they can be used both to quantify the size of a contrast of interest and as an independent functional localizer for a more specific task. To quantify BOLD sensitivity on a more fundamental level, a breath holding paradigm can be used as it leads to well-observable whole brain changes in blood oxygen levels [110].

3.5 Ultra-High Field Functional MRI of the Amygdala

Increasing the number of time points by faster or longer measurements increases SNR_t by reducing the impact of noise (σ_y). Another possibility to increase the SNR_t is by increasing the amount of detectable BOLD signal (μ_y). Higher magnetic field strengths lead to an almost linear increase in image SNR [111], which only in part translates to an increase in temporal SNR_t as the reduction of thermal noise is counter acted by the increase of physiological noise [112]. Consequently, there could be concerns that increases in field strengths do not provide benefits for BOLD fMRI in the amygdala and other ventral brain areas, as they are particularly susceptible to physiological artefacts.

Using an emotional face discrimination task [79], we demonstrated that amygdala BOLD sensitivity was indeed increased at 7 T compared to 3 T in the amygdala [113]. A limitation of this study was the use of two different study populations, minor differences in the task implementations, and differences in signal parameters. To overcome this limitation, we conducted an additional study [114], where all these factors were matched between the scanner, except the echo times, to account for the different field strength specific $T2^*$ relaxation effects.

Sixteen healthy volunteers (11f/5 m, 24.56 ± 3.1 years) underwent two sessions of fMRI in a Siemens 7 T MAGNETOM and a Siemens 3 T TIM Trio whole-body MRI scanner using a single-shot echo planar sequence with the following sequence parameters: matrix size = 128 × 128 px × 20 slices, repetition time of TR = 1.8 s, $TE_{7T}/TE_{3T} = 23/40$ ms, FoV = 190 × 190 mm^2, 3 mm slice thickness. We employed an emotional face (Faces) and an object discriminated task (Objects) as introduced by Hariri et al. [79]. Data were preprocessed (slice timing correction, realignment,

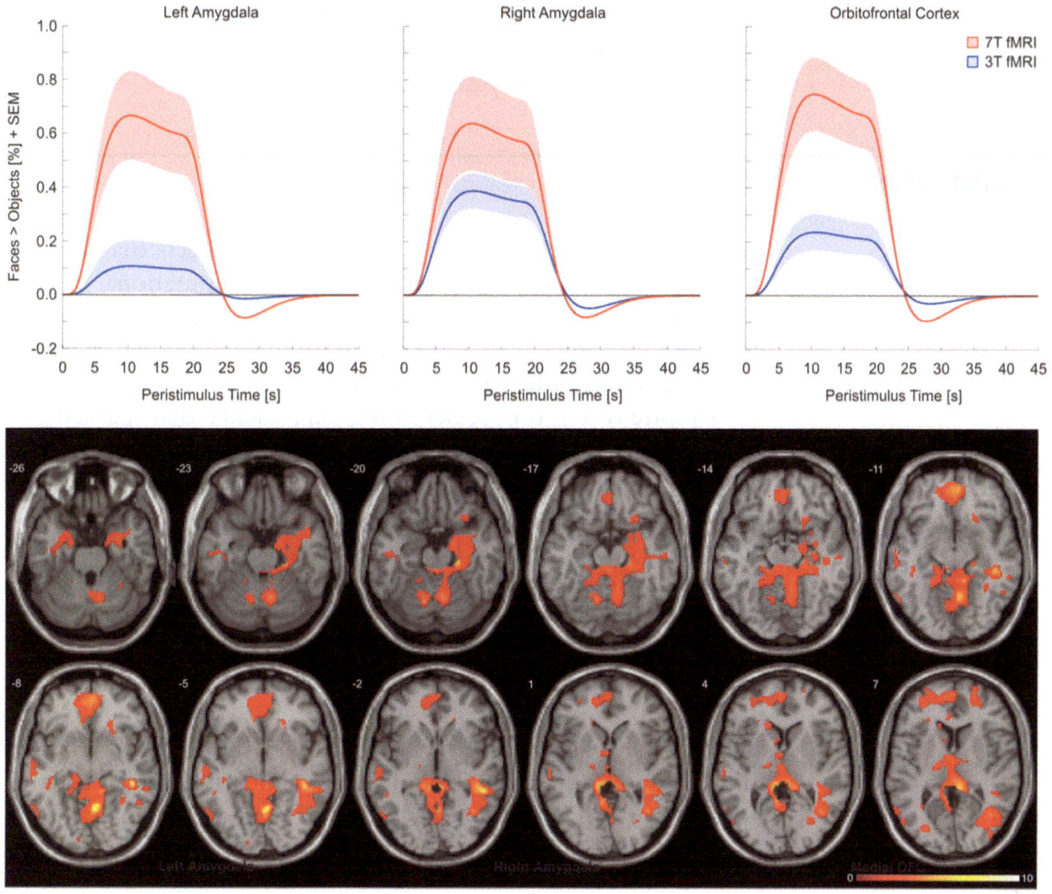

Fig. 1 Comparison of 3T and 7T amygdala fMRI. Amygdala activation was higher at 7 Tesla ($t_{\text{right}} = 6.00$, $t_{\text{left}} = 4.81$) compared to 3 Tesla ($t_{\text{right}} = 2.59$, $t_{\text{left}} = 2.71$). Activation was also increased in other ventral areas, such as the medial orbitofrontal cortex, which plays an important role in amygdala regulation. The upper panel depicts the estimated percent signal change differences between the faces and the objects condition at 3T and 7T. The lower panel depicts the paired t-test t-map with the significant increases of 7T vs. 3T

normalization, 8 mm FWHM spatial smoothing) and analyzed in SPM8 (FIL Methods Group, University College London, UK). A paired t-test was used for comparison between the 3 T and 7 T scans and revealed a significant increase in amygdala activation at 7 T (Fig. 1).

These findings are in line with theoretical predictions and support similar comparison studies [111], in particular those that targeted subcortical brain areas [113, 115]. In practice, the most important benefits of amygdala fMRI at 7 T are the possibility to increase image resolution, while maintaining optimal levels of image and temporal SNR, leading to an increased BOLD sensitivity (i.e., increasing the chance of detecting weak signal differences) and localization specificity (i.e., increasing the chance of spatially discriminating the source of the BOLD signal) [116]. Both of these positive effects are particularly relevant for detecting and discriminating activation in amygdala subnuclei [100].

4 Conclusion

In this chapter we explained multiple crucial aspects that are relevant for discovering and using fMRI-based computational biomarkers. For illustration, we focused on the amygdala's role in anxiety disorders but argue that most of the principles will generalize to other mental disorders and target regions.

Despite recent progresses in big data science that sparked promises of data-driven biomarker discovery, we would like to emphasize the benefits of hypothesis-driven research. It aims for conception of mechanistic models of psychopathologies that are intelligible and applicable for humans, in particular clinical professionals and, ideally, patients and their families as well. Additionally, well-founded theories allow to constrain the search space, given that even successful data-driven discovery of brain–behavioral phenotype associations are complex and require large datasets of more than 50,000 individuals [117]. Finally, as we explained in this chapter, there are many diverse pitfalls on the different levels of functional neuroimaging, from biased implicit theoretical assumptions to inadequate operationalization of brain functions to insufficient data quality. Using adequate theoretical models that provide testable predictions allows for avoiding these pitfalls and uncover their reasons. Ultimately, overcoming the challenges can allow for explanatory computational biomarkers that can transform clinical practice and neuroscientific theories of psychopathologies.

References

1. Aldao A, Nolen-Hoeksema S, Schweizer S (2010) Emotion-regulation strategies across psychopathology: a meta-analytic review. Clin Psychol Rev 30:217–237. https://doi.org/10.1016/j.cpr.2009.11.004
2. McTeague LM, Rosenberg BM, Lopez JW et al (2020) Identification of common neural circuit disruptions in emotional processing across psychiatric disorders. Am J Psychiatry 177:411–421. https://doi.org/10.1176/appi.ajp.2019.18111271
3. Werner K, Gross JJ (2010) Emotion regulation and psychopathology: a conceptual framework. In: Emotion regulation and psychopathology: a transdiagnostic approach to etiology and treatment. The Guilford Press, New York, pp 13–37
4. Newman MG, Fisher AJ (2013) Mediated moderation in combined cognitive behavioral therapy versus component treatments for generalized anxiety disorder. J Consult Clin Psychol 81:405–414. https://doi.org/10.1037/a0031690
5. Vinograd M, Williams A, Sun M et al (2020) Neuroticism and interpretive bias as risk factors for anxiety and depression. Clin Psychol Sci J Assoc Psychol Sci 8:641–656. https://doi.org/10.1177/2167702620906145
6. Forbes CE, Poore JC, Krueger F et al (2014) The role of executive function and the dorsolateral prefrontal cortex in the expression of neuroticism and conscientiousness. Soc Neurosci 9:139–151. https://doi.org/10.1080/17470919.2013.871333

7. Mineka S, Kelly KA (1989) The relationship between anxiety, lack of control and loss of control. In: Stress, personal control and health. John Wiley & Sons, Oxford, pp 163–191
8. Kaufman J, Charney D (2000) Comorbidity of mood and anxiety disorders. Depress Anxiety 12:69–76. https://doi.org/10.1002/1520-6394(2000)12:1+<69::AID-DA9>3.0.CO;2-K
9. Stein MB (2004) Public Health Perspectives on Generalized Anxiety Disorder 5
10. Patriquin MA, Mathew SJ (2017) The neurobiological mechanisms of generalized anxiety disorder and chronic stress. Chronic Stress 1: 2470547017703993. https://doi.org/10.1177/2470547017703993
11. Blackburn-Munro G, Blackburn-Munro RE (2001) Chronic Pain, Chronic Stress and Depression: Coincidence or Consequence?: HPA axis involvement in comorbidity of chronic pain and depression. J Neuroendocrinol 13:1009–1023. https://doi.org/10.1046/j.0007-1331.2001.00727.x
12. Zemke AC, Heagerty PJ, Lee C, Cramer SC (2003) Motor cortex organization after stroke is related to side of stroke and level of recovery. Stroke 34:e23–e26. https://doi.org/10.1161/01.STR.0000065827.35634.5E
13. Perani D, Cappa SF, Tettamanti M et al (2003) A fMRI study of word retrieval in aphasia. Brain Lang 85:357–368. https://doi.org/10.1016/S0093-934X(02)00561-8
14. Ritter M, Hummer A, Ledolter AA et al (2019) Correspondence between retinotopic cortical mapping and conventional functional and morphological assessment of retinal disease. Br J Ophthalmol 103:208–215. https://doi.org/10.1136/bjophthalmol-2017-311443
15. Klüver H, Bucy PC (1939) Preliminary analysis of function of the temporal lobes in monkeys. Arch Neurol Psychiatr 42:979–1000. https://doi.org/10.1001/archneurpsyc.1939.02270240017001
16. LeDoux J (2003) The emotional brain, fear, and the amygdala. Cell Mol Neurobiol 23: 727–738. https://doi.org/10.1023/A:1025048802629
17. Öhman A (2005) The role of the amygdala in human fear: automatic detection of threat. Psychoneuroendocrinology 30:953–958. https://doi.org/10.1016/j.psyneuen.2005.03.019
18. Dolan RJ, Vuilleumier P (2003) Amygdala automaticity in emotional processing. Ann N Y Acad Sci 985:348–355
19. Feinstein JS, Adolphs R, Damasio A, Tranel D (2011) The human amygdala and the induction and experience of fear. Curr Biol 21:34–38. https://doi.org/10.1016/j.cub.2010.11.042
20. van Honk J, Eisenegger C, Terburg D et al (2013) Generous economic investments after basolateral amygdala damage. Proc Natl Acad Sci 110:2506–2510. https://doi.org/10.1073/pnas.1217316110
21. Rosenberger LA, Eisenegger C, Naef M, et al (2019) The human basolateral amygdala is indispensable for social experiential learning. Curr Biol 29:3532–3537.e3. https://doi.org/10.1016/j.cub.2019.08.078
22. Ledoux J (1998) The emotional brain: the mysterious underpinnings of emotional life. Simon and Schuster
23. Pessoa L, Adolphs R (2011) Emotion and the brain: multiple roads are better than one. Nat Rev Neurosci 12:425–425. https://doi.org/10.1038/nrn2920-c2
24. Siegel EH, Sands MK, Van den Noortgate W et al (2018) Emotion fingerprints or emotion populations? A meta-analytic investigation of autonomic features of emotion categories. Psychol Bull 144:343–393. https://doi.org/10.1037/bul0000128
25. Amaral DG, Adolphs R (2016) Living without an Amygdala. Guilford Publications
26. Inman CS, Bijanki KR, Bass DI et al (2020) Human amygdala stimulation effects on emotion physiology and emotional experience. Neuropsychologia 145:106722. https://doi.org/10.1016/j.neuropsychologia.2018.03.019
27. Phan KL, Fitzgerald DA, Nathan PJ, Tancer ME (2006) Association between amygdala hyperactivity to harsh faces and severity of social anxiety in generalized social phobia. Biol Psychiatry 59:424–429. https://doi.org/10.1016/j.biopsych.2005.08.012
28. Brühl AB, Delsignore A, Komossa K, Weidt S (2014) Neuroimaging in social anxiety disorder—a meta-analytic review resulting in a new neurofunctional model. Neurosci Biobehav Rev 47:260–280. https://doi.org/10.1016/j.neubiorev.2014.08.003
29. Keedwell PA, Andrew C, Williams SCR et al (2005) The neural correlates of anhedonia in major depressive disorder. Biol Psychiatry 58: 843–853. https://doi.org/10.1016/j.biopsych.2005.05.019
30. Suslow T, Konrad C, Kugel H et al (2010) Automatic mood-congruent amygdala responses to masked facial expressions in major depression. Biol Psychiatry 67:155–

160. https://doi.org/10.1016/j.biopsych.2009.07.023
31. Suslow T, Lindner C, Dannlowski U et al (2013) Automatic amygdala response to facial expression in schizophrenia: initial hyperresponsivity followed by hyporesponsivity. BMC Neurosci 14:140. https://doi.org/10.1186/1471-2202-14-140
32. Swartz JR, Knodt AR, Radtke SR, Hariri AR (2015) A neural biomarker of psychological vulnerability to future life stress. Neuron 85: 505–511. https://doi.org/10.1016/j.neuron.2014.12.055
33. Rosen JB, Donley MP (2006) Animal studies of amygdala function in fear and uncertainty: relevance to human research. Biol Psychol 73: 49–60. https://doi.org/10.1016/j.biopsycho.2006.01.007
34. Grupe DW, Nitschke JB (2013) Uncertainty and anticipation in anxiety: an integrated neurobiological and psychological perspective. Nat Rev Neurosci 14:488–501. https://doi.org/10.1038/nrn3524
35. Sladky R, Hahn A, Karl I-L et al (2021) Dynamic causal modeling of the prefrontal/amygdala network during processing of emotional faces. Brain Connect
36. Phillips ML, Drevets WC, Rauch SL, Lane R (2003) Neurobiology of emotion perception I: the neural basis of normal emotion perception. Biol Psychiatry 54:504–514. https://doi.org/10.1016/S0006-3223(03)00168-9
37. Hahn A, Stein P, Windischberger C et al (2011) Reduced resting-state functional connectivity between amygdala and orbitofrontal cortex in social anxiety disorder. NeuroImage 56:881–889
38. Minkova L, Sladky R, Kranz GS et al (2017) Task-dependent modulation of amygdala connectivity in social anxiety disorder. Psychiatry Res Neuroimaging 262:39–46
39. Sladky R, Höflich A, Küblböck M et al (2015) Disrupted effective connectivity between the amygdala and orbitofrontal cortex in social anxiety disorder during emotion discrimination revealed by dynamic causal modeling for fMRI. Cereb Cortex 25:895–903
40. Sladky R, Spies M, Hoffmann A, et al (2015) (S)-citalopram influences amygdala modulation in healthy subjects: a randomized placebo-controlled double-blind fMRI study using dynamic causal modeling. NeuroImage: 243–250
41. Young KS, Burklund LJ, Torre JB et al (2017) Treatment for social anxiety disorder alters functional connectivity in emotion regulation neural circuitry. Psychiatry Res Neuroimaging 261:44–51. https://doi.org/10.1016/j.pscychresns.2017.01.005
42. Morawetz C, Bode S, Baudewig J, Heekeren HR (2017) Effective amygdala-prefrontal connectivity predicts individual differences in successful emotion regulation. Soc Cogn Affect Neurosci 12:569–585. https://doi.org/10.1093/scan/nsw169
43. Scheidegger M, Henning A, Walter M et al (2016) Ketamine administration reduces amygdalo-hippocampal reactivity to emotional stimulation: Amygdalo-Hippocampal Reactivity and Ketamine Administration. Hum Brain Mapp 37:1941–1952. https://doi.org/10.1002/hbm.23148
44. Kraehenmann R, Schmidt A, Friston K et al (2016) The mixed serotonin receptor agonist psilocybin reduces threat-induced modulation of amygdala connectivity. NeuroImage Clin 11:53–60. https://doi.org/10.1016/j.nicl.2015.08.009
45. Costi S, Van Dam NT, Murrough JW (2015) Current status of ketamine and related therapies for mood and anxiety disorders. Curr Behav Neurosci Rep 2:216–225. https://doi.org/10.1007/s40473-015-0052-3
46. Muttoni S, Ardissino M, John C (2019) Classical psychedelics for the treatment of depression and anxiety: a systematic review. J Affect Disord 258:11–24. https://doi.org/10.1016/j.jad.2019.07.076
47. Anand A, Shekhar A (2006) Brain imaging studies in mood and anxiety disorders: special emphasis on the amygdala. Ann N Y Acad Sci 985:370–388. https://doi.org/10.1111/j.1749-6632.2003.tb07095.x
48. Shi J, Guo H, Liu S et al (2022) Subcortical brain volumes relate to neurocognition in first-episode schizophrenia, bipolar disorder, major depression disorder, and healthy controls. Front Psych 12:747386. https://doi.org/10.3389/fpsyt.2021.747386
49. Wang X, Yin Z, Sun Q et al (2021) Comparative study on the functional connectivity of amygdala and hippocampal neural circuits in patients with first-episode schizophrenia and other high-risk populations. Front Psych 12: 627198. https://doi.org/10.3389/fpsyt.2021.627198
50. Ambrosi E, Arciniegas DB, Madan A et al (2017) Insula and amygdala resting-state functional connectivity differentiate bipolar from unipolar depression. Acta Psychiatr Scand 136:129–139. https://doi.org/10.1111/acps.12724

51. Reich DB, Belleau EL, Temes CM et al (2019) Amygdala resting state connectivity differences between bipolar II and borderline personality disorders. Neuropsychobiology 78:229–237. https://doi.org/10.1159/000502440
52. Freese JL, Amaral DG (2009) Neuroanatomy of the primate amygdala. In: The human amygdala. The Guilford Press, New York, pp 3–42
53. Irwin W, Davidson RJ, Lowe MJ et al (1996) Human amygdala activation detected with echo-planar functional magnetic resonance imaging. NeuroReport 7:1765–1769. https://doi.org/10.1097/00001756-199607290-00014
54. Schneider F, Grodd W, Weiss U et al (1997) Functional MRI reveals left amygdala activation during emotion. Psychiatry Res Neuroimaging 76:75–82. https://doi.org/10.1016/S0925-4927(97)00063-2
55. Canli T, Zhao Z, Desmond JE et al (1999) fMRI identifies a network of structures correlated with retention of positive and negative emotional memory. Psychobiology 27:441–452. https://doi.org/10.3758/BF03332139
56. Hariri AR, Tessitore A, Mattay VS et al (2002) The amygdala response to emotional stimuli: a comparison of faces and scenes. NeuroImage 17:317–323. https://doi.org/10.1006/nimg.2002.1179
57. Pallesen KJ, Brattico E, Bailey C et al (2005) Emotion processing of major, minor, and dissonant chords. Ann N Y Acad Sci 1060:450–453. https://doi.org/10.1196/annals.1360.047
58. Patin A, Pause BM (2015) Human amygdala activations during nasal chemoreception. Neuropsychologia 78:171–194. https://doi.org/10.1016/j.neuropsychologia.2015.10.009
59. Simons LE, Moulton EA, Linnman C et al (2014) The human amygdala and pain: evidence from neuroimaging. Hum Brain Mapp 35:527–538. https://doi.org/10.1002/hbm.22199
60. Tovote P, Fadok JP, Lüthi A (2015) Neuronal circuits for fear and anxiety. Nat Rev Neurosci 16:317–331
61. Tovote P, Esposito MS, Botta P et al (2016) Midbrain circuits for defensive behaviour. Nature 534:206–212. https://doi.org/10.1038/nature17996
62. Tye KM, Prakash R, Kim S-Y et al (2011) Amygdala circuitry mediating reversible and bidirectional control of anxiety. Nature 471:358–362. https://doi.org/10.1038/nature09820
63. Fadok JP, Krabbe S, Markovic M et al (2017) A competitive inhibitory circuit for selection of active and passive fear responses. Nature 542:96–100. https://doi.org/10.1038/nature21047
64. Haubensak W, Kunwar PS, Cai H et al (2010) Genetic dissection of an amygdala microcircuit that gates conditioned fear. Nature 468:270–276. https://doi.org/10.1038/nature09553
65. Ciocchi S, Herry C, Grenier F et al (2010) Encoding of conditioned fear in central amygdala inhibitory circuits. Nature 468:277–282. https://doi.org/10.1038/nature09559
66. Janak PH, Tye KM (2015) From circuits to behaviour in the amygdala. Nature 517:284–292. https://doi.org/10.1038/nature14188
67. Terburg D, Scheggia D, Triana del Rio R et al (2018) The basolateral amygdala is essential for rapid escape: a human and rodent study. Cell 175:723–735.e16. https://doi.org/10.1016/j.cell.2018.09.028
68. Visser RM, Bathelt J, Scholte HS, Kindt M (2021) Robust BOLD responses to faces but not to conditioned threat: challenging the Amygdala's reputation in human fear and extinction learning. J Neurosci 41:10278–10292. https://doi.org/10.1523/JNEUROSCI.0857-21.2021
69. Oler J, Fox AS, Shackman AJ, Kalin N (2016) The central nucleus of the amygdala is a critical substrate for individual differences in anxiety. In: Living without an amygdala. The Guilford Press, pp 218–251
70. Shackman AJ, Fox AS, Oler JA et al (2013) Neural mechanisms underlying heterogeneity in the presentation of anxious temperament. Proc Natl Acad Sci 110:6145–6150
71. Fox AS, Oler JA, Shackman AJ et al (2015) Intergenerational neural mediators of early-life anxious temperament. Proc Natl Acad Sci 112:9118–9122
72. Strigo IA, Simmons AN, Matthews SC et al (2008) Association of major depressive disorder with altered functional brain response during anticipation and processing of heat pain. Arch Gen Psychiatry 65:1275–1284. https://doi.org/10.1001/archpsyc.65.11.1275
73. Pripfl J, Robinson S, Leodolter U et al (2006) EEG reveals the effect of fMRI scanner noise on noise-sensitive subjects. NeuroImage 31:332–341. https://doi.org/10.1016/j.neuroimage.2005.11.031

74. Kragel PA, Knodt AR, Hariri AR, LaBar KS (2016) Decoding spontaneous emotional states in the human brain. PLoS Biol 14: e2000106. https://doi.org/10.1371/journal.pbio.2000106
75. Shackman AJ, Tromp DPM, Stockbridge MD et al (2016) Dispositional negativity: an integrative psychological and neurobiological perspective. Psychol Bull 142:1275–1314. https://doi.org/10.1037/bul0000073
76. Sander D, Grafman J, Zalla T (2003) The human amygdala: an evolved system for relevance detection. Rev Neurosci 14:303–316. https://doi.org/10.1515/REVNEURO.2003.14.4.303
77. Staudigl T, Minxha J, Mamelak AN, et al (2022) Saccade-related neural communication in the human medial temporal lobe is modulated by the social relevance of stimuli. Sci Adv 8:eabl6037. https://doi.org/10.1126/sciadv.abl6037
78. Cao R, Li X, Brandmeir NJ, Wang S (2021) Encoding of facial features by single neurons in the human amygdala and hippocampus. Commun Biol 4:1394. https://doi.org/10.1038/s42003-021-02917-1
79. Hariri AR, Mattay VS, Tessitore A et al (2002) Serotonin transporter genetic variation and the response of the human amygdala. Science 297:400–403
80. Guyer AE, Monk CS, McClure-Tone EB et al (2008) A developmental examination of amygdala response to facial expressions. J Cogn Neurosci 20:1565–1582
81. Brooks SJ, Savov V, Allzén E et al (2012) Exposure to subliminal arousing stimuli induces robust activation in the amygdala, hippocampus, anterior cingulate, insular cortex and primary visual cortex: a systematic meta-analysis of fMRI studies. NeuroImage 59:2962–2973. https://doi.org/10.1016/j.neuroimage.2011.09.077
82. Etkin A, Klemenhagen KC, Dudman JT et al (2004) Individual differences in trait anxiety predict the response of the basolateral amygdala to unconsciously processed fearful faces. Neuron 44:1043–1055. https://doi.org/10.1016/j.neuron.2004.12.006
83. Puccetti NA, Schaefer SM, van Reekum CM et al (2021) Linking amygdala persistence to real-world emotional experience and psychological well-being. J Neurosci 41:3721–3730. https://doi.org/10.1523/JNEUROSCI.1637-20.2021
84. Schuyler BS, Kral TRA, Jacquart J et al (2014) Temporal dynamics of emotional responding: amygdala recovery predicts emotional traits. Soc Cogn Affect Neurosci 9:176–181. https://doi.org/10.1093/scan/nss131
85. Blackford JU, Avery SN, Shelton RC, Zald DH (2009) Amygdala temporal dynamics: temperamental differences in the timing of amygdala response to familiar and novel faces. BMC Neurosci 10:145. https://doi.org/10.1186/1471-2202-10-145
86. Küblböck M, Woletz M, Höflich A et al (2014) Stability of low-frequency fluctuation amplitudes in prolonged resting-state fMRI. NeuroImage 103:249–257
87. Woletz M, Hoffmann A, Tik M et al (2019) Beware detrending: optimal preprocessing pipeline for low-frequency fluctuation analysis. Hum Brain Mapp 40:1571–1582
88. Johnstone T, Somerville LH, Alexander AL et al (2005) Stability of amygdala BOLD response to fearful faces over multiple scan sessions. NeuroImage 25:1112–1123. https://doi.org/10.1016/j.neuroimage.2004.12.016
89. Gur RE, Loughead J, Kohler CG et al (2007) Limbic activation associated with misidentification of fearful faces and flat affect in schizophrenia. Arch Gen Psychiatry 64:1356–1366. https://doi.org/10.1001/archpsyc.64.12.1356
90. Infantolino ZP, Luking KR, Sauder CL et al (2018) Robust is not necessarily reliable: from within-subjects fMRI contrasts to between-subjects comparisons. NeuroImage 173: 146–152. https://doi.org/10.1016/j.neuroimage.2018.02.024
91. Sladky R, Höflich A, Atanelov J et al (2012) Increased neural habituation in the amygdala and orbitofrontal cortex in social anxiety disorder revealed by fMRI. PLoS One 7:e50050
92. Nord CL, Gray A, Charpentier CJ et al (2017) Unreliability of putative fMRI biomarkers during emotional face processing. NeuroImage 156:119–127. https://doi.org/10.1016/j.neuroimage.2017.05.024
93. Plichta MM, Schwarz AJ, Grimm O et al (2012) Test-retest reliability of evoked BOLD signals from a cognitive-emotive fMRI test battery. NeuroImage 60:1746–1758. https://doi.org/10.1016/j.neuroimage.2012.01.129
94. Geissberger N, Tik M, Sladky R et al (2020) Reproducibility of amygdala activation in facial emotion processing at 7T. NeuroImage 211:116585
95. Phelps EA (2006) Emotion and cognition: insights from studies of the human amygdala. Annu Rev Psychol 57:27–53. https://doi.

org/10.1146/annurev.psych.56.091103.070234
96. Sladky R, Riva F, Rosenberger LA et al (2021) Basolateral and central amygdala orchestrate how we learn whom to trust. Commun Biol 4: 1–9
97. Gupta R, Koscik TR, Bechara A, Tranel D (2011) The amygdala and decision-making. Neuropsychologia 49:760–766. https://doi.org/10.1016/j.neuropsychologia.2010.09.029
98. Balleine BW, Killcross S (2006) Parallel incentive processing: an integrated view of amygdala function. Trends Neurosci 29:272–279. https://doi.org/10.1016/j.tins.2006.03.002
99. Fox AS, Shackman AJ (2019) The central extended amygdala in fear and anxiety: closing the gap between mechanistic and neuroimaging research. Neurosci Lett 693:58–67. https://doi.org/10.1016/j.neulet.2017.11.056
100. Sladky R, Geissberger N, Pfabigan DM et al (2018) Unsmoothed functional MRI of the human amygdala and bed nucleus of the stria terminalis during processing of emotional faces. NeuroImage 168:383–391
101. Bauer EP, LeDoux JE (2004) Heterosynaptic long-term potentiation of inhibitory interneurons in the lateral amygdala. J Neurosci 24:9507–9512
102. Ray RD, Zald DH (2012) Anatomical insights into the interaction of emotion and cognition in the prefrontal cortex. Neurosci Biobehav Rev 36:479–501
103. Ledoux J, Schiller D (2009) The human amygdala: insights from other animals. In: The human amygdala. Guilford Press, pp 43–60
104. Brosch T, Wieser MJ (2011) The (non)-automaticity of amygdala responses to threat: on the issue of fast signals and slow measures. J Neurosci 31:14451–14452. https://doi.org/10.1523/JNEUROSCI.4089-11.2011
105. Merboldt K-D, Fransson P, Bruhn H, Frahm J (2001) Functional MRI of the human amygdala? NeuroImage 14:253–257. https://doi.org/10.1006/nimg.2001.0802
106. Karl S, Sladky R, Lamm C, Huber L (2021) Neural responses of pet dogs witnessing their caregiver's positive interactions with a conspecific: an fMRI study. Cereb Cortex Commun 2:tgab047
107. Robinson S, Windischberger C, Rauscher A, Moser E (2004) Optimized 3 T EPI of the amygdalae. NeuroImage 22:203–210. https://doi.org/10.1016/j.neuroimage.2003.12.048
108. Moeller S, Yacoub E, Olman CA et al (2010) Multiband multislice GE-EPI at 7 tesla, with 16-fold acceleration using partial parallel imaging with application to high spatial and temporal whole-brain fMRI. Magn Reson Med 63:1144–1153. https://doi.org/10.1002/mrm.22361
109. Murphy K, Bodurka J, Bandettini PA (2007) How long to scan? The relationship between fMRI temporal signal to noise ratio and necessary scan duration. NeuroImage 34:565–574. https://doi.org/10.1016/j.neuroimage.2006.09.032
110. Kastrup A, Krüger G, Glover GH, Moseley ME (1999) Assessment of cerebral oxidative metabolism with breath holding and fMRI. Magn Reson Med 42:608–611. https://doi.org/10.1002/(SICI)1522-2594(199909)42:3<608::AID-MRM26>3.0.CO;2-I
111. van der Zwaag W, Francis S, Head K et al (2009) fMRI at 1.5, 3 and 7 T: characterising BOLD signal changes. NeuroImage 47: 1425–1434
112. Triantafyllou C, Hoge RD, Krueger G et al (2005) Comparison of physiological noise at 1.5 T, 3 T and 7 T and optimization of fMRI acquisition parameters. NeuroImage 26:243–250. https://doi.org/10.1016/j.neuroimage.2005.01.007
113. Sladky R, Baldinger P, Kranz GS et al (2013) High-resolution functional MRI of the human amygdala at 7T. Eur J Radiol 82: 728–733
114. Sladky R, Baldinger P, Kranz G, et al (2012) A quantitative comparison of functional MRI of the human amygdala at 3 and 7 Tesla. Beijing
115. Hahn A, Kranz GS, Seidel E-M et al (2013) Comparing neural response to painful electrical stimulation with functional MRI at 3 and 7T. NeuroImage 82:336–343
116. Wald LL, Triantafyllou C, Hoge RD (2015) Field strength dependence of contrast and noise in fMRI. In: Uludag K, Ugurbil K, Berliner L (eds) fMRI: from nuclear spins to brain functions. Springer US, Boston, pp 793–818
117. Marek S, Tervo-Clemmens B, Calabro FJ et al (2022) Reproducible brain-wide association studies require thousands of individuals. Nature 603:654–660. https://doi.org/10.1038/s41586-022-04492-9

Chapter 11

Neuroimaging Methods Investigating Anterior Insula Alterations in Schizophrenia and Mood Disorders

Sevdalina Kandilarova and Dora Zlatareva

Abstract

Anterior insula (AI) is involved in a variety of functions including compassion, empathy, taste, perception, motor control, self-awareness, cognitive functioning, interpersonal experience, and awareness of homeostatic emotions such as hunger, pain, and fatigue. Since the majority of these brain functions are impaired in major psychiatric disorders, it is not surprising that neuroimaging methods have revealed structural as well as functional and connectivity alterations of this structure. This work gives an overview of the most recent findings of AI changes in schizophrenia (SCZ), bipolar disorder (BD), and major depressive disorder (MDD). In summary, strong meta-analytic evidence supports the association between schizophrenia and reduction of the gray matter volume (GMV) of AI bilaterally. Similar correlation has been found in the first episode psychosis and in subjects with high familial risk. Significant reduction of the gyrification of the AI has been reported as well. On one hand, BD shares common traits with schizophrenia in terms of GMV reductions in AI but the alterations seem to be less extensive. On the other hand, MDD demonstrates shared GMV reductions of AI with BD. Thus, structural alterations of the AI emerge as a common feature of all three major psychiatric disorders though expressed in a different extent. Similarly, functional MRI has demonstrated common AI disruptions across these major mental illnesses when probed with various interoceptive and cognitive control tasks. In addition, connectivity studies in SCZ, BD, and MDD report impaired functional as well as effective connectivity of the AI, especially as part of the salience network (SN). Thus, the evidence reviewed supports the hypothesis that in SCZ the abnormality of the SN results in predominance of externalizing mental representations leading to psychotic symptoms (such as delusions and hallucinations), while in depression it causes an overrepresented internalizing leading to the self-defeating depressive symptoms (ruminations, low self-esteem).

Key words Anterior insula, Functional MRI, Structural MRI, Connectivity, Schizophrenia, Bipolar disorder, Major depressive disorder

1 Introduction

The insular cortex is a portion of the cerebral cortex hiding deep in the Sylvian fissure within each hemisphere of the human brain. It is traditionally divided into anterior and posterior parts with regions of variable cytoarchitecture, ranging from agranular to granular [1]. In terms of function, the insular part of the cortex has been

Drozdstoy Stoyanov et al. (eds.), *Computational Neuroscience*, Neuromethods, vol. 199,
https://doi.org/10.1007/978-1-0716-3230-7_11,

suggested to play a role in compassion, empathy, taste, perception, motor control, self-awareness, cognitive functioning, interpersonal experience, and awareness of homeostatic emotions such as hunger, pain, and fatigue [2]. In his seminal paper, Craig [3] proposed that the insula's role in these diverse phenomena may be accounted for by its involvement in interoception. It was proposed that along the insula's caudal to rostral axis there exist successively higher order re-representations of the bodily interoceptive information. In the mid-insula and posterior insula interoception is represented in its most embodied format, and in the anterior insula in a more abstracted way.

In addition, information about emotional salience is further represented in brain regions connected to the AI, including the amygdala, orbitofrontal cortex, and anterior cingulate cortex (ACC). This is supported by the anatomical and functional connectivity of the AI. There are direct projections from the ventral medial thalamic nucleus and the central nucleus of the amygdala to the anterior insula. In turn, AI projects to the anterior amygdaloid region as well as medial, cortical, accessory basal magnocellular, medial basal, and lateral amygdaloid nuclei [4].

Recent functional studies suggest that the AI mediates the dynamic interactions between large-scale brain networks engaged in externally oriented attention and self-cognition. As mentioned earlier, insular cortex is involved in a variety of cognitive, affective, and regulatory processes, including interoception, emotional reactions, and empathy. These functions are supported by the bottom-up identification of salient events, the shifting between other large-scale networks to promote access to attention and working memory when a salient event is detected, and the involvement of the anterior and posterior parts of the insula to adjust autonomic response to salient stimuli [5].

Thus, the salience network (SN) with its main hubs including AI and ACC enables different brain regions to generate appropriate behavioral reactions to salient stimuli [6]. The perception of both visual, and auditory emotional information, pain, and subjective projections of the self are all insula-related functions disrupted in major psychiatric disorders including psychotic, mood, and anxiety disorders. Recent magnetic resonance imaging (MRI) findings of structural, functional, and connectivity alterations of AI across schizophrenia (SCZ) and mood disorders will be selectively reviewed in the following paragraphs.

2 Structural Changes

Quantitative structural MRI methods can be grossly divided into voxel-based morphometry (VBM) and surface-based morphometry (SBM). VBM can be used for the estimation of the regional volumes of the grey matter (GM), white matter, and the cerebrovascular fluid, while with SBM cortical surfaces can be modelled using cortical thickness, area, volume, and curvature measurements [7, 8]. Both methods have been used for the exploration of the structural changes of the AI related to different psychiatric conditions.

VBM studies have consistently linked schizophrenia to GM volume reductions in various brain regions including the insular cortex bilaterally with a more pronounced effect on the anterior part as identified by Shepherd et al. in their meta-analysis [9]. Another voxel-wise analysis revealed an association between GM reduction and disease duration in the left AI [10]. Interestingly reduced volume in this same region along with left anterior cingulate cortex (as part of the SN) was found to be significantly correlated with the severity of reality distortion [11]. Moreover SBM analysis demonstrated regions with significant reductions in gyrification (hypogyria) predominantly in the left hemisphere, involving the insula [12].

Noteworthy, the first episode schizophrenia as well as chronic schizophrenia was characterized by bilateral insula volume reductions in another meta-analytical study conducted by Ellison-Wright and collaborators [13] and in left insula according to the study published by Fusar-Poli et al. [14]. Among other regions, bilateral insula GM volume reductions have been identified in a recent meta-analysis of structural brain imaging studies in relatives of patients with schizophrenia reflecting the familial risk for the disorder [15].

In bipolar disorder (BD), gray matter reductions were detected in the anterior cingulate and bilateral insula substantially overlapping with areas of loss identified in schizophrenia [16]. As the authors of this meta-analysis noted, GM changes in SCZ were more extensive and involved limbic and neocortical structures along with the paralimbic regions affected in BD. Moreover, the first episode BD as compared to at risk individuals demonstrated more GM loss in the medial superior frontal gyrus and insula [17]. SBM methods revealed reduced surface area of the insula in BP I and BP II patients as compared to healthy individuals [18].

Recent meta-analysis identified common AI GM loss in mood disorders including BD and major depressive disorder [19]. Earlier studies demonstrated clusters of GM reduction in the anterior cingulate cortex and bilateral insula that were shared by the two diagnostic groups [20]. Thinner insular cortex (among other regions) in MDD compared to healthy controls (HC) was reported by the ENIGMA consortium as well [21].

In their seminal paper Goodkind et al. reported a meta-analysis based on more than 15,000 subjects across 6 psychiatric disorders—schizophrenia, bipolar disorder, depression, addiction, obsessive-compulsive disorder, and anxiety. They found three regions with converging gray matter loss across all diagnostic entities. Those regions included dorsal anterior cingulate, and right and left insula [22]. Thus, the hubs of the salience network emerge as a common neuroanatomical site for major mental illnesses.

3 Functional Alterations

Functional task-related MRI is a popular noninvasive method for the investigation of brain function with the advantage of very good spatial resolution and moderate temporal resolution as compared to other methods such as electroencephalography. It is based on the registration of the blood oxygenation level dependent (BOLD) signal changes induced by specific stimuli. Insula function has been probed by a variety of tasks including the use of touch and pain stimulation, auditory, and visual emotional stimuli with or without cognitive components in the test (recognition, decision making, etc.)

Functional aberrations observed in schizophrenic patients, which are likely related to altered insula function, include pain insensitivity, deficits in sensory-emotional integration such as poor recognition of emotions in facial expressions, emotionally blunt speech, impairments in distinguishing self from non-self, and the occurrence of hallucinations [23]. Stubbs et al. reported in their meta-analysis that compared to healthy controls schizophrenic patients consistently demonstrated elevated pain threshold and pain tolerance, as well as reduced physiological response to noxious stimuli and pain intensity/unpleasantness ratings [24]. These findings were not influenced by the antipsychotic medication status of the patients.

In a pilot functional MRI (fMRI) study unmedicated SCZ patients had decreased activation in pain affective-cognitive processing regions (insula, posterior cingulate cortex, brainstem), and an over-activation of the primary sensory-discriminative pain processing region [25]. When receiving electric shocks, the schizophrenic patients displayed significantly lower activation of the middle insula correlated with the magnitude of positive symptoms [26]. Thus, the reduced insula reactivity seems to reflect the diminished pain sensitivity in SCZ.

Poor facial recognition is another well-established finding in schizophrenia research [27]. Earlier fMRI studies demonstrated that SCZ patients failed to activate the right fusiform gyrus which is one of several brain areas involved in facial information processing [28]. Later research confirmed and extended this finding by demonstrating lack of differential activation of this same region in SCZ during the presentation of faces vs chairs [29].

Recent meta-analysis of emotion recognition reported decreased activation in the right ventrolateral prefrontal cortex (PFC), cingulate, insula, and subcortical regions (amygdala, thalamus, caudate, lentiform nucleus, and putamen) in schizophrenic patients as compared to HC [30]. Interestingly, insula and bilateral medial PFC showed decreased activation in the schizophrenia group during theory of mind tasks which suggest the important role of the insula dysfunction in social cognition deficits.

Hallucinations (especially auditory) are considered as a core feature of SCZ. In functional MRI, these experiences have been linked to activation in the right homologue of Broca's area, bilateral insula, bilateral supramarginal gyri, and right superior temporal gyrus [31]. In addition, specific BOLD response of the left AI has been shown to precede patients' awareness of hallucinatory experience [32]. Interestingly, bilateral insula activation was linked to tactile hallucinations [33].

Disturbed interoception has also been reported in depression [34]. For rxample, MDD patients showed significantly higher scores in body perception and reduced signal changes during rest in the left anterior insula, and both were positively correlated with depression severity [35]. In a recent meta-analysis, the activity of the insular cortex at rest has been found to differentiate between MDD and BD with higher values in unipolar depression [36]. On the other hand, BD patients showed significantly less activation in AI along with other brain regions during cognitive interference trials [37]. Interestingly, the same study revealed task-related functional connectivity disruptions.

Meta-analytic evidence suggests that disrupted dorsal mid-insula activation during interoception represents a common feature across a variety of psychiatric disorders including SCZ, BD, MDD, anorexia, and anxiety disorder as explored by fMRI paradigms ranging from pain and hunger to interoceptive attention [38]. Moreover, AI has been identified among the regions demonstrating common disruptions in cognitive control tasks across major psychiatric disorders [39]. Along with the findings of common structural disturbances of the insular cortex in different psychiatric conditions [22], these reports underline the importance of this underestimated structure in neuropsychiatric research.

4 Impaired Connectivity

Brain regions can be interconnected both on the structural and on the functional level. Structural connectivity studies such as diffusion-weighted imaging or diffusion tensor imaging explore the anatomical links between brain regions. Functional connectivity (FC), on the other hand, reflects the correlated activation of brain areas not necessarily anatomically connected. The FC method is based on the analysis of functional MRI time series data using different areas of interest (seeds or voxels). FC analysis can only define which regions are connected. In contrast effective connectivity (EC) provides an estimate of the influence that one neural system exerts over another. EC can inform us about the directionality (from region one to region two or the opposite) and the type (inhibitory or excitatory) of the causal connections [40]. The most popular method used is called dynamic causal modelling (DCM) [41]. Both FC and EC can be measured during resting state (intrinsic connectivity) or during specific task-related conditions.

Schizophrenia patients and individuals with high-risk for psychosis have demonstrated disturbed FC of the nodes of the SN as well as aberrant interactions of the SN with other large brain networks [42, 43]. Moreover, reduced FC of the same network has been reported during information processing and reward prediction [44]. Based on these and other findings, the "aberrant salience" hypothesis was suggested by Kapur [45]. It states that inaccurate assessment of irrelevant stimuli might cause abnormal salience processing in individuals with psychosis. Thus, hallucinations might be a direct consequence of the aberrant salience of internal representations, while delusions might represent cognitive attempt to make sense of the disturbed perceptions.

In support of this understanding are the recent findings of reduced FC between the left ventral AI and other SN regions in subjects with At-Risk-Mental-State along with reduced structural connectivity in terms of fractional anisotropy (FA) and axial diffusivity of white-matter tracts, including frontal-striatal-thalamic circuits and the cingulum. Symptom severity correlated with the FA measures extracted from these disrupted white-matter regions. Moreover, FC between the bilateral insular cortex was further reduced in subjects who transitioned to psychosis [46].

Recent multimodal study combining voxel-based morphometry and resting-state FC in early-stage SCZ patients delineated significantly reduced GM volume in both bilateral AI and ACC [47]. Moreover, significantly reduced FC within the SN was found in the patient group. Both structural and connectivity changes were correlated with hallucinations. On the meta-analytic level, SCZ patients were characterized by converging dysconnectivity and GM volume reductions in insula, lateral postcentral cortex, striatum, and thalamus [48] .

Using the method of DCM, our group found an excitatory connection from the ACC to the AI in the SCZ group during rest [49]. We believe this finding suggests that patients stay in a resting-state of "aberrant salience" which may interfere with the function of the default mode network (DMN). Earlier we reported strong involvement of the DMN during task performance [50]. Since the core function of the SN is to support the dynamic switching between the states of rest and task, this over-engagement with internal stimuli at rest disturbs the appropriate activation within the DMN.

In our most recent transdiagnostic study on effective connectivity in schizophrenia and mood disorders, we have been able to demonstrate the differential diagnostic potential of the brain connectome features involving major nodes of the SN, central executive network (CEN), and the limbic system [51]. Interestingly, the self-inhibitory connection of the AI emerged as a feature of both mood disorders and SCZ and it was capable of distinguishing them both from HC. This led us to the assumption that the disrupted self-regulation of this key hub of the SN might represent a shared mechanism of both affective and psychotic disorders.

Early research on resting state fMRI demonstrated decreased amplitude of low-frequency fluctuations in the left posterior insula and superior parietal lobule and increased amplitude of low frequency fluctuations (ALFF) in the right dorsal AI in BD compared to MDD [52]. Later publication suggested that patients with BD had significantly altered right anterior insula FC with inferior parietal lobule (as part of the central executive network) when compared to both MDD and HC [53]. Moreover, the finding significantly discriminated patients with bipolar depression from the other two groups.

A newer approach called dynamic FC (dFC) explores the changes of the activation correlations over time. Using this method decreased dFC between the right ventral AI (rvAI) and right ventrolateral prefrontal cortex that was common for BD and MDD was reported [54]. The authors delineated some distinct features such as the increases in dFC between the rvAI and right precuneus, temporal pole, and left dorsolateral prefrontal cortex that were MDD-related, and increases in the dFC between the right dorsal AI (rdAI) and left inferior parietal lobule and right middle occipital gyrus associated with BD. On the other hand, shared decreased network switching rate of key hubs in DMN was reported for both BD and MDD groups while only MDD patients had decreased switching rate in salience network and striatum [55].

Another recent study found that MDD was associated with increased FC between rdAI and right superior frontal gyrus, along with increased FC between left posterior insula and right precentral gyrus and right thalamus. Notably, machine learning based on the combined FC of ACC, bilateral insula, and bilateral amygdala seeds with the whole brain discriminated BD from MDD with an accuracy of 91.30% [56]. Similarly, we demonstrated that abnormalities in resting-state neural connectivity of the anterior insula, amygdala, and prefrontal cortex may differentiate the depressive episodes of MDD and BD with an overall accuracy of 80.4% [51].

Various connectivity disturbances have been reported in unipolar depression as well. MDD patients have been characterized by decreased intrinsic FC within the SN and strong correlation between the severity of symptoms and the decreased FC of the right AI [57]. In our previous research, we have found decreased effective connectivity between the AI and the dorsolateral prefrontal cortex (DLPFC) in depression compared to healthy controls, along with an aberrant connection from the amygdala to the AI only in the patient group [58]. In line with these findings, Hwang et al. demonstrated decreased FC between DLPFC and insula in subthreshold depression [59].

Another important aspect is the connectivity between major brain networks. In this regard, the meta-analysis of resting state studies by Mulders et al. confirmed an increased connectivity between the SN and the anterior DMN [60]. However, other authors demonstrated mixed meta-analytical results, concerning the connectivity between the SN and nodes of both the CEN and DMN [61]. They found both hypo- and hyperconnectivity of the SN. Additional studies will be needed to shed some more light on this area of research.

5 Conclusion

In this review we demonstrated that neuroimaging methods such as voxel-based and surface-based morphometry, functional MRI during execution of various tasks and at rest including functional and effective connectivity yielded strong converging evidence for the impaired structure, function, and connectivity of the AI. In summary, strong meta-analytic evidence supports the association between schizophrenia and reduction of the GM volume of AI bilaterally. Similar correlation has been found in first episode psychosis and in subjects with high familial risk. Significant reduction of the gyrification of the AI has been reported as well. On one hand, BD shares common traits with schizophrenia in terms of GMV reductions in AI but the alterations seem to be less extensive. On the other hand, MDD demonstrates shared GMV reductions of AI with BD. Thus, structural alterations of the AI emerge as common features of all three major psychiatric disorders though expressed in a different extent. Similarly, functional MRI has demonstrated common AI disruptions across these major mental illnesses when probed with various interoceptive and cognitive control tasks. In addition, connectivity studies in SCZ, BD, and MDD report impaired functional as well as effective connectivity of the AI. As a key node of the SN this region appears to play an important role in the pathophysiology of both SCZ and mood disorders and provides a plausible mechanism explaining their symptom formation and clinical presentation. Thus, the evidence reviewed supports the hypothesis that in SCZ the abnormality of the SN results in predominance of externalizing mental representations leading to psychotic symptoms (such as delusions and hallucinations), while in depression it causes an overrepresented internalizing leading to the self-defeating depressive symptoms (ruminations, low self-esteem). The well-known role of the AI in interoception and the findings of various interoceptive disturbances in all three diagnostic categories was a less explored aspect of current research warranting further investigation.

References

1. Gogolla N (2017) The insular cortex. Curr Biol 27(12):R580–R5r6. https://doi.org/10.1016/j.cub.2017.05.010
2. Shura RD, Hurley RA, Taber KH (2014) Insular cortex: structural and functional neuroanatomy. J Neuropsychiatry Clin Neurosci 26(4): iv–282. https://doi.org/10.1176/appi.neuropsych.260401
3. Craig AD (2009) How do you feel--now? The anterior insula and human awareness. Nat Rev Neurosci. vol 1. England. p. 59–70
4. Uddin LQ, Nomi JS, Hébert-Seropian B, Ghaziri J, Boucher O (2017) Structure and function of the human insula. J Clin Neurophysiol 34(4):300–306. https://doi.org/10.1097/wnp.0000000000000377
5. Uddin LQ (2015) Salience processing and insular cortical function and dysfunction. Nat Rev Neurosci 1:55–61
6. Menon V, Uddin LQ (2010) Saliency, switching, attention and control: a network model of insula function. Brain Struct Funct 214(5-6):655–67. https://doi.org/10.1007/s00429-010-0262-0.
7. Ashburner J, Friston KJ. Voxel-based morphometry--the methods. NeuroImage 2000;11(6 Pt 1):805–821. doi: https://doi.org/10.1006/nimg.2000.0582
8. Pantazis D, Leahy RM, Nichols TE, Styner M (2004) Statistical surface-based morphometry using a nonparametric approach. 2004 2nd IEEE International Symposium on Biomedical Imaging: Nano to Macro (IEEE Cat No 04EX821), p. 1283–1286, Vol. 2
9. Shepherd AM, Matheson SL, Laurens KR, Carr VJ, Green MJ (2012) Systematic meta-analysis of insula volume in schizophrenia. Biol Psychiatry 72(9):775–784. https://doi.org/10.1016/j.biopsych.2012.04.020
10. Nickl-Jockschat T, Schneider F, Pagel AD, Laird AR, Fox PT, Eickhoff SB (2011) Progressive pathology is functionally linked to the domains of language and emotion: meta-analysis of brain structure changes in schizophrenia patients. Eur Arch Psychiatry Clin Neurosci 261 Suppl 2(Suppl 2):S166–S171. https://doi.org/10.1007/s00406-011-0249-8
11. Palaniyappan L, Mallikarjun P, Joseph V, White TP, Liddle PF (2011) Reality distortion is related to the structure of the salience network in schizophrenia. Psychol Med 41(8): 1701–1708. https://doi.org/10.1017/s0033291710002205
12. Palaniyappan L, Liddle PF (2012) Aberrant cortical gyrification in schizophrenia: a surface-based morphometry study. J Psychiatry Neurosci 37(6):399–406. https://doi.org/10.1503/jpn.110119
13. Ellison-Wright I, Glahn DC, Laird AR, Thelen SM, Bullmore E (2008) The anatomy of first-episode and chronic schizophrenia: an anatomical likelihood estimation meta-analysis. Am J Psychiatry 165(8):1015–1023. https://doi.org/10.1176/appi.ajp.2008.07101562
14. Fusar-Poli P, Smieskova R, Serafini G, Politi P, Borgwardt S (2014) Neuroanatomical markers of genetic liability to psychosis and first episode psychosis: a voxelwise meta-analytical comparison. World J Biol Psychiatry 15(3):219–228. https://doi.org/10.3109/15622975.2011.630408
15. Zhang W, Sweeney JA, Yao L, Li S, Zeng J, Xu M et al (2020) Brain structural correlates of familial risk for mental illness: a meta-analysis of voxel-based morphometry studies in relatives of patients with psychotic or mood disorders. Neuropsychopharmacology 45(8): 1369–1379. https://doi.org/10.1038/s41386-020-0687-y
16. Ellison-Wright I, Bullmore E (2010) Anatomy of bipolar disorder and schizophrenia: a meta-analysis. Schizophr Res 117(1):1–12. https://doi.org/10.1016/j.schres.2009.12.022
17. Long J, Qin K, Wu Y, Li L, Zhou J (2022) Gray matter abnormalities and associated familial risk endophenotype in individuals with first-episode bipolar disorder: evidence from whole-brain voxel-wise meta-analysis. Asian J Psychiatr 74:103179. https://doi.org/10.1016/j.ajp.2022.103179
18. Woo Y, Kang W, Kang Y, Kim A, Han KM, Tae WS et al (2021) Cortical thickness and surface area abnormalities in bipolar I and II disorders. Psychiatry Investig 18(9):850–863. https://doi.org/10.30773/pi.2021.0074
19. Jabbi M, Arasappan D, Eickhoff SB, Strakowski SM, Nemeroff CB, Hofmann HA (2020) Neuro-transcriptomic signatures for mood disorder morbidity and suicide mortality. J Psychiatr Res 127:62–74. https://doi.org/10.1016/j.jpsychires.2020.05.013

20. Wise T, Radua J, Via E, Cardoner N, Abe O, Adams TM et al (2017) Common and distinct patterns of grey-matter volume alteration in major depression and bipolar disorder: evidence from voxel-based meta-analysis. Mol Psychiatry 22(10):1455–1463. https://doi.org/10.1038/mp.2016.72
21. Schmaal L, Hibar DP, Samann PG, Hall GB, Baune BT, Jahanshad N et al (2017) Cortical abnormalities in adults and adolescents with major depression based on brain scans from 20 cohorts worldwide in the ENIGMA Major Depressive Disorder Working Group. Mol Psychiatry 22(6):900–909. https://doi.org/10.1038/mp.2016.60
22. Goodkind M, Eickhoff SB, Oathes DJ, Jiang Y, Chang A, Jones-Hagata LB et al (2015) Identification of a common neurobiological substrate for mental illness. JAMA Psychiatry 72(4):305–315. https://doi.org/10.1001/jamapsychiatry.2014.2206
23. Wylie KP, Tregellas JR (2010) The role of the insula in schizophrenia. Schizophr Res 123(2–3):93–104. https://doi.org/10.1016/j.schres.2010.08.027
24. Stubbs B, Thompson T, Acaster S, Vancampfort D, Gaughran F, Correll CU (2015) Decreased pain sensitivity among people with schizophrenia: a meta-analysis of experimental pain induction studies. Pain 156(11):2121–2131. https://doi.org/10.1097/j.pain.0000000000000304
25. de la Fuente-Sandoval C, Favila R, Gómez-Martin D, Pellicer F, Graff-Guerrero A (2010) Functional magnetic resonance imaging response to experimental pain in drug-free patients with schizophrenia. Psychiatry Res 183(2):99–104. https://doi.org/10.1016/j.pscychresns.2010.05.003
26. Linnman C, Coombs G 3rd, Goff DC, Holt DJ (2013) Lack of insula reactivity to aversive stimuli in schizophrenia. Schizophr Res 143(1):150–157. https://doi.org/10.1016/j.schres.2012.10.038
27. Marwick K, Hall J (2008) Social cognition in schizophrenia: a review of face processing. Br Med Bull 88(1):43–58. https://doi.org/10.1093/bmb/ldn035
28. Quintana J, Wong T, Ortiz-Portillo E, Marder SR, Mazziotta JC (2003) Right lateral fusiform gyrus dysfunction during facial information processing in schizophrenia. Biol Psychiatry 53(12):1099–1112. https://doi.org/10.1016/s0006-3223(02)01784-5
29. Walther S, Federspiel A, Horn H, Bianchi P, Wiest R, Wirth M et al (2009) Encoding deficit during face processing within the right fusiform face area in schizophrenia. Psychiatry Res 172(3):184–191. https://doi.org/10.1016/j.pscychresns.2008.07.009
30. Jáni M, Kašpárek T (2018) Emotion recognition and theory of mind in schizophrenia: a meta-analysis of neuroimaging studies. World J Biol Psychiatry 19(sup3):S86–s96. https://doi.org/10.1080/15622975.2017.1324176
31. Sommer IE, Diederen KM, Blom JD, Willems A, Kushan L, Slotema K et al (2008) Auditory verbal hallucinations predominantly activate the right inferior frontal area. Brain 131(Pt 12):3169–3177. https://doi.org/10.1093/brain/awn251
32. Hoffman RE, Anderson AW, Varanko M, Gore JC, Hampson M (2008) Time course of regional brain activation associated with onset of auditory/verbal hallucinations. Br J Psychiatry 193(5):424–425. https://doi.org/10.1192/bjp.bp.107.040501
33. Shergill SS, Cameron LA, Brammer MJ, Williams SC, Murray RM, McGuire PK (2001) Modality specific neural correlates of auditory and somatic hallucinations. J Neurol Neurosurg Psychiatry 71(5):688–690. https://doi.org/10.1136/jnnp.71.5.688
34. Dunn BD, Dalgleish T, Ogilvie AD, Lawrence AD (2007) Heartbeat perception in depression. Behav Res Ther 45(8):1921–1930. https://doi.org/10.1016/j.brat.2006.09.008
35. Wiebking C, Bauer A, de Greck M, Duncan NW, Tempelmann C, Northoff G (2010) Abnormal body perception and neural activity in the insula in depression: an fMRI study of the depressed "material me". World J Biol Psychiatry 11(3):538–549. https://doi.org/10.3109/15622970903563794
36. Pastrnak M, Simkova E, Novak T (2021) Insula activity in resting-state differentiates bipolar from unipolar depression: a systematic review and meta-analysis. Sci Rep 11(1):16930. https://doi.org/10.1038/s41598-021-96319-2
37. Ellard KK, Gosai AK, Felicione JM, Peters AT, Shea CV, Sylvia LG et al (2019) Deficits in frontoparietal activation and anterior insula functional connectivity during regulation of cognitive-affective interference in bipolar disorder. Bipolar Disord 21(3):244–258. https://doi.org/10.1111/bdi.12709
38. Nord CL, Lawson RP, Dalgleish T (2021) Disrupted dorsal mid-insula activation during Interoception across psychiatric disorders. Am J Psychiatr 178(8):761–770. https://doi.org/10.1176/appi.ajp.2020.20091340
39. McTeague LM, Huemer J, Carreon DM, Jiang Y, Eickhoff SB, Etkin A (2017) Identification of common neural circuit disruptions in

cognitive control across psychiatric disorders. Am J Psychiatry 174(7):676–685. https://doi.org/10.1176/appi.ajp.2017.16040400

40. Friston KJ (2011) Functional and effective connectivity: a review. Brain Connect 1(1): 13–36. https://doi.org/10.1089/brain.2011.0008

41. Friston KJ, Kahan J, Biswal B, Razi A (2014) A DCM for resting state fMRI. NeuroImage 94: 396–407. https://doi.org/10.1016/j.neuroimage.2013.12.009

42. Manoliu A, Riedl V, Zherdin A, Mühlau M, Schwerthöffer D, Scherr M et al (2014) Aberrant dependence of default mode/central executive network interactions on anterior insular salience network activity in schizophrenia. Schizophr Bull 40(2):428–437. https://doi.org/10.1093/schbul/sbt037

43. Wotruba D, Michels L, Buechler R, Metzler S, Theodoridou A, Gerstenberg M et al (2014) Aberrant coupling within and across the default mode, task-positive, and salience network in subjects at risk for psychosis. Schizophr Bull 40(5):1095–1104. https://doi.org/10.1093/schbul/sbt161

44. White TP, Joseph V, Francis ST, Liddle PF (2010) Aberrant salience network (bilateral insula and anterior cingulate cortex) connectivity during information processing in schizophrenia. Schizophr Res 123(2–3):105–115. https://doi.org/10.1016/j.schres.2010.07.020

45. Kapur S (2003) Psychosis as a state of aberrant salience: a framework linking biology, phenomenology, and pharmacology in schizophrenia. Am J Psychiatry 160(1):13–23. https://doi.org/10.1176/appi.ajp.160.1.13

46. Wang C, Ji F, Hong Z, Poh JS, Krishnan R, Lee J et al (2016) Disrupted salience network functional connectivity and white-matter microstructure in persons at risk for psychosis: findings from the LYRIKS study. Psychol Med 46(13):2771–2783. https://doi.org/10.1017/s0033291716001410

47. Pu W, Li L, Zhang H, Ouyang X, Liu H, Zhao J et al (2012) Morphological and functional abnormalities of salience network in the early-stage of paranoid schizophrenia. Schizophr Res 141(1):15–21. https://doi.org/10.1016/j.schres.2012.07.017

48. Brandl F, Avram M, Weise B, Shang J, Simões B, Bertram T et al (2019) Specific substantial dysconnectivity in schizophrenia: a transdiagnostic multimodal meta-analysis of resting-state functional and structural magnetic resonance imaging studies. Biol Psychiatry 85(7):573–583. https://doi.org/10.1016/j.biopsych.2018.12.003

49. Aryutova K, Paunova R, Kandilarova S, Maes M, Stoyanova K, Stoyanov D (2021) Differential aberrant connectivity of precuneus and anterior insula may underpin the diagnosis of schizophrenia and mood disorders. World J Psychiatry, in press

50. Stoyanov D, Aryutova K, Kandilarova S, Paunova R, Arabadzhiev Z, Todeva-Radneva A et al (2021) Diagnostic task specific activations in functional MRI and aberrant connectivity of insula with middle frontal gyrus can inform the differential diagnosis of psychosis. Diagnostics (Basel) 11(1). https://doi.org/10.3390/diagnostics11010095

51. Kandilarova S, Stoyanov DS, Paunova R, Todeva-Radneva A, Aryutova K, Maes M (2021) Effective connectivity between major nodes of the limbic system, salience and frontoparietal networks differentiates schizophrenia and mood disorders from healthy controls. J Pers Med 11(11). https://doi.org/10.3390/jpm11111110

52. Liu CH, Ma X, Wu X, Li F, Zhang Y, Zhou FC et al (2012) Resting-state abnormal baseline brain activity in unipolar and bipolar depression. Neurosci Lett 516(2):202–206. https://doi.org/10.1016/j.neulet.2012.03.083

53. Ellard KK, Zimmerman JP, Kaur N, Van Dijk KRA, Roffman JL, Nierenberg AA et al (2018) Functional connectivity between anterior insula and key nodes of Frontoparietal executive control and salience networks distinguish bipolar depression from unipolar depression and healthy control subjects. Biol Psychiatry Cogn Neurosci Neuroimag 3(5):473–484. https://doi.org/10.1016/j.bpsc.2018.01.013

54. Pang Y, Chen H, Wang Y, Long Z, He Z, Zhang H et al (2018) Transdiagnostic and diagnosis-specific dynamic functional connectivity anchored in the right anterior insula in major depressive disorder and bipolar depression. Prog Neuro-Psychopharmacol Biol Psychiatry 85:7–15. https://doi.org/10.1016/j.pnpbp.2018.03.020

55. Han S, Cui Q, Wang X, Li L, Li D, He Z et al (2020) Resting state functional network switching rate is differently altered in bipolar disorder and major depressive disorder. Hum Brain Mapp 41(12):3295–3304. https://doi.org/10.1002/hbm.25017

56. Yu H, Li ML, Li YF, Li XJ, Meng Y, Liang S et al (2020) Anterior cingulate cortex, insula and amygdala seed-based whole brain resting-state functional connectivity differentiates bipolar from unipolar depression. J Affect Disord 274:38–47. https://doi.org/10.1016/j.jad.2020.05.005

57. Manoliu A, Meng C, Brandl F, Doll A, Tahmasian M, Scherr M et al (2013) Insular dysfunction within the salience network is associated with severity of symptoms and aberrant inter-network connectivity in major depressive disorder. Front Hum Neurosci 7. https://doi.org/10.3389/fnhum.2013.00930
58. Kandilarova S, Stoyanov D, Kostianev S, Specht K (2018) Altered resting state effective connectivity of anterior insula in depression. Front Psych 9:83. https://doi.org/10.3389/fpsyt.2018.00083
59. Hwang JW, Egorova N, Yang XQ, Zhang WY, Chen J, Yang XY et al (2015) Subthreshold depression is associated with impaired resting-state functional connectivity of the cognitive control network. Transl Psychiatry 11:e683
60. Mulders PC, van Eijndhoven PF, Schene AH, Beckmann CF, Tendolkar I (2015) Resting-state functional connectivity in major depressive disorder: a review. Neurosci Biobehav Rev 56:330–344. https://doi.org/10.1016/j.neubiorev.2015.07.014
61. Kaiser RH, Andrews-Hanna JR, Wager TD, Pizzagalli DA (2015) Large-scale network dysfunction in major depressive disorder: a meta-analysis of resting-state functional connectivity. JAMA Psychiatry 72(6):603–611. https://doi.org/10.1001/jamapsychiatry.2015.0071

Chapter 12

Magnetic Resonance Spectroscopy

Letizia Squarcina and Paolo Brambilla

Abstract

Magnetic resonance spectroscopy (MRS) is a completely noninvasive technique which allows the detection of brain metabolites' concentrations in vivo. Brain metabolites are involved in energy processes, neuronal and membrane health, and in neurotransmission. The possibility of identifying changes in their concentration may open the way to early diagnosis and treatment evaluation in various pathologies.

In this chapter, the main principles behind brain MRS are briefly discussed, together with an overview of the primary clinical applications. Moreover, some concepts about data acquisition and processing are discussed, with the aim of facilitating MRS spectra analysis and interpretation.

Key words Magnetic resonance spectroscopy, Brain metabolites, Proton spectroscopy, Functional spectroscopy

1 Introduction

1.1 Magnetic Resonance Spectroscopy: Principles

Magnetic resonance spectroscopy (MRS) is the only technique that can noninvasively quantify the concentration of metabolites in the brain in vivo. While MR imaging methods give images as an output, which depict, for example, brain structure, function, integrity, or blood perfusion, with MR spectroscopy the outputs are signals, that is, spectra, which provide chemical information on the tissue of interest.

The principle behind MRS functioning is that different molecules experience different magnetic fields inside an MR scanner, depending on the different distribution of electrons inside atoms. This technique is, in fact, identical to nuclear magnetic resonance (NMR) used in chemistry. Each molecule of the considered tissue generates a slightly different signal, resulting from the different resonant frequencies. The generated signal can be viewed as a spectrum, where each resonance frequency (chemical shift) corresponds to a signal intensity, which is related to the concentration of the specific metabolite. The chemical shift is usually reported in parts per million (ppm), which are independent from the magnetic

Drozdstoy Stoyanov et al. (eds.), *Computational Neuroscience*, Neuromethods, vol. 199,
https://doi.org/10.1007/978-1-0716-3230-7_12,

field of the specific scanner used for the acquisitions. In this way, for example, the N-acetylaspartate (NAA) peak is located at 2.0 ppm, which corresponds to 64 MHz if acquired at 1.5 T and to 128 Mhz at 3 T.

A spectrum can be obtained from any nucleus having a non-zero spin, and the most used in the clinical practice for in vivo brain MRS are protons (1H) and Phosphorus-31 (31P), but there are also applications for fluorine (19F), carbon (13C), and sodium (23NA). 31P-MRS is essentially used in studies investigating the cellular energetic behavior, in particular investigating ATP and energy metabolism. Protons, on the other hand, are abundant in the human brain, and 1H-MRS enables the quantification of brain metabolites. Moreover, 1H-MRS only requires the standard coils also used for MR imaging, while other nuclei need coils specifically tuned to their Larmor frequencies, and are therefore less available in standard clinical settings. Noteworthy, the proton spectrum is altered in many neurologic and psychiatric disorders.

Metabolites with concentrations in the range of micromoles per gram can be quantified with MRS. The 1H metabolites that are most commonly quantified include NAA, a marker of functioning neurons, phosphocreatine plus creatine (PCr + Cr), involved in energetic processes, choline-containing compounds (Ch), glutamine (Gln), glutamate (Glu), and gamma amino-butyric acid (GABA) (Fig. 1). Concentrations of metabolites can be reported as absolute values, whose unit is mM, or in relative values, which means that the values are reported as ratio, with PCr + Cr as denominator. To obtain absolute concentration a reference is needed: the water signal, obtained from the same volume without water suppression, is often used as internal reference. It has to be noted that, despite reporting MRS results in absolute units, results obtained from different scanners and with different post-processing methods are not entirely comparable across platforms.

The MRS spectrum is composed of different peaks, generated by specific metabolites.

The peak of NAA has a resonance of 2.0 ppm, with a concentration of around 8 mM in healthy adults, and is the biggest metabolite peak in healthy MRS spectra. NAA, which is synthesized in neuronal mitochondria, is found only in the nervous system, in gray and white matter, and its concentration has a strong clinical meaning: in fact, it is reduced in processes that damage neurons. A reduced NAA is also found in psychiatric pathologies, as will be discussed in the next paragraphs.

The PCr + Cr peak can be found at 3.0 ppm. Creatine is involved in storage and transfer of energy. Its concentration is relatively constant; thus, it is considered as a stable metabolite. This is the reason why it is used as a standard when reporting ratio results. Importantly, the PCr + Cr signal can be altered in brain tumors and metastasis.

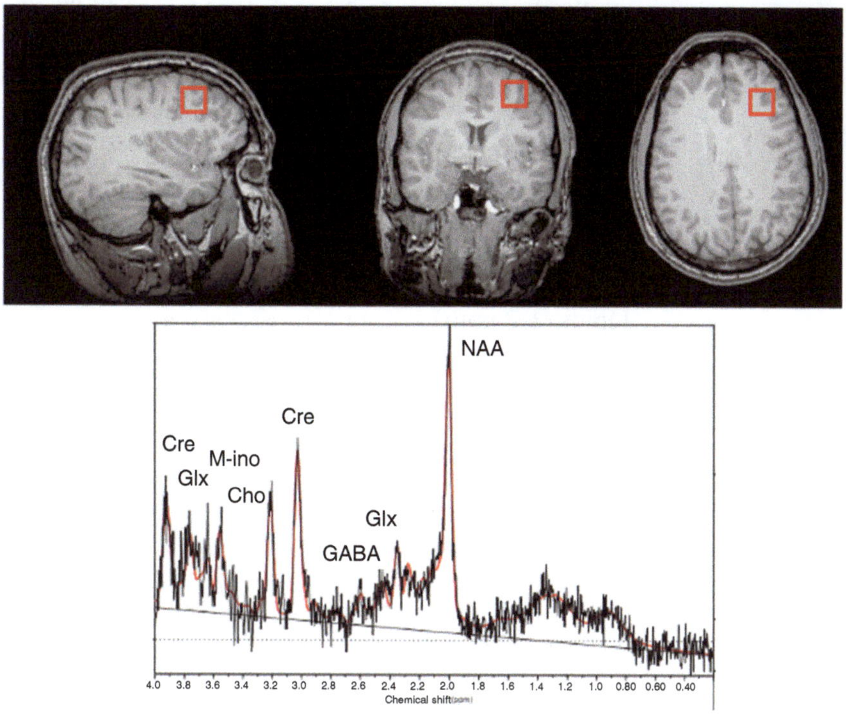

Fig. 1 Magnetic resonance spectroscopy voxel positioned in the left dorsolateral prefrontal cortex (above) and its 1H spectrum (below). The typical metabolites are depicted

The sum of choline-containing compounds (choline, phosphocholine, glycerophosphocholine) resonates at 3.2 ppm: these compounds are metabolites of membrane phospholipids, and are susceptible to various pathologies, including tumors, ischemia, and inflammation. The monitoring of choline metabolism may be important in the study of tumors [1].

Glu and Gln are stored in the glia, and the balance of concentrations of these two neurochemicals is important for the regular functioning of the brain. Moreover, Glu is an important excitatory neurotransmitter, while Gln is a precursor of both excitatory and inhibitory neurotransmitters. They are difficult to separate, and usually the intensity of the glutamate-glutamine (Glx) peak is reported. This peak is located between 2.2 and 2.4 ppm. Anomalies in glutamate signaling have been linked to various pathologies, as attention-deficit hyperactivity disorder (ADHD), mood disorders, schizophrenia, hepatic encephalopathy, brain tumors, and neurodegenerative disorders [2].

GABA is the principal inhibitory neurotransmitter in the human brain, and it resonates at 2.2–2.4 ppm. Since its peak overlaps the Glx and NAA peak, the MRS experiment has to be specifically thought to isolate the GABA peak from the spectrum.

Following is a list of other measurable metabolites:

- Lactate (1.3 ppm)
- Myo-inositol (3.5 ppm)
- Citrate (2.6 ppm)
- Taurine (3.4 ppm)
- Glucose (3.4 and 3.8 ppm)
- Mannitol (3.8 ppm)
- Lipids (1.3 ppm)
- Alanine (1.48 ppm)

The metabolites that can be quantified depend on the chosen echo time (TE) (which will be discussed in the following text) and on the pathological status of the patients: some metabolites (as lactate, mannitol, citrate) are detectable only in presence of specific pathologies and are not detectable in healthy tissue.

1.2 Clinical Applications of MRS

The first clinical applications of brain MRS, in the 1990s, focused on 31P-MRS, with the aim of investigating energetic processes. To date, the majority of MRS studies consist in 1H-MRS studies, because the advancements of the technique and the fact that it does not need additional equipment in respect to conventional imaging, make it definitely more convenient than using other nuclei.

Brain MRS has a variety of clinical applications and is currently applied to diverse pathologies, especially in neurology and psychiatry.

1.3 MRS Applications in Neurology

Early studies found disrupted metabolites' levels in stroke, multiple sclerosis, and brain tumors, when compared with healthy controls [3]. A group of neuroradiologists, neurologists, oncologists, and clinical neuroscientists recently published a consensus article [3] regarding brain MRS, with the aim of documenting the impact of this technique in clinical settings, encouraging the standardization of acquisition and processing protocols, to facilitate a broader use of MRS for the diagnosis and assessment of central nervous system (CNS) pathologies. It is worth noticing that, while MR imaging is a very effective means for identifying brain lesions, thanks to the contrast given by soft tissues, it does not allow the identification of possible changes in chemical composition, which is possible to obtain with MRS.

In particular, spectra of cancer tissues are very different from spectra obtained from normal brain tissue [4, 5], and MRS has been proven useful for cancer diagnosis and for the evaluation of treatment response. In fact, cancerous tissues show reduced levels of NAA, choline, and creatine compounds, and higher levels of choline, glutamine, citric acid, and lactate: these two metabolites, interestingly, are not detectable in healthy controls. Lactate is also found in brain injury and mitochondrial diseases [4, 5].

Reduced levels of NAA, considered as a marker of neuronal dysfunction, are found in stroke and multiple sclerosis. Strokes are also associated with lower concentrations of choline and creatine compounds.

1H-MRS is useful in the evaluation of the newborn brain in case of hypoxia: in fact, one of the earliest signs of perinatal brain injury is the presence of lactate in MRS spectra, which is also associated with long-term outcome [3]. Moreover, specific spectra in the neonatal brain are indicative of particular metabolic and genetic syndromes as, for example, Canavan disease, a genetic leukodystrophy, and may be diriment in the diagnostic process.

Consistent results have been found also in patients with neurodegenerative pathologies, as Alzheimer's, Parkinson, and Huntington diseases, who show reduced NAA and Glu concentrations, and elevated levels of tCho, myo-inositol, and glucose [3]. Myo-inositol, in particular, seems to be a marker of dementia and brain atrophy.

1.4 MRS Applications in Psychiatry

One of the fields where brain MRS has been fruitfully applied is psychiatry. In recent years, there has been a growing interest toward spectroscopy, given that it makes it possible to quantify the concentrations of, in particular, glutamate (the most common excitatory neurotransmitter), and GABA (the most common inhibitory neurotransmitter). In fact, one of the hypotheses on the biological basis of schizophrenia (SCZ) postulates an imbalance of glutamate and GABA homeostasis [6], with an excessive presence of glutamate. Studies have focused on the quantification of GABA in various brain regions, with mixed results. For example, a recent meta-analysis [7] identified lower levels of GABA in SCZ patients in respect to controls in frontal brain areas. On the other hand, another meta-analysis [8] found mixed results, with some studies reporting increased GABA and others reporting reduced GABA in SCZ. It has to be noted that, to date, MRS studies on GABA suffer from some limitations, as a generally poor SNR, due to the properties of the GABA signal itself, as stated in the previous section, and sparse sampling, as the quantification is usually limited to two or three single voxels, which do not cover the entire brain. Thus, it cannot be ruled out that the technical limitations of the technique, at this time, prevent consistent findings of altered GABA levels, which seem to be confirmed in post-mortem studies.

Interestingly, studies on glutamate support glutamatergic abnormalities in SCZ. In particular, studies found increased Glx levels in the basal ganglia, in the hippocampus, in the thalamus, and in frontal regions [9, 10]. The possibility of measuring glutamate levels in vivo has the important implications of making it possible to measure the effects of potential novel glutamatergic medications: this concept is corroborated by the fact that antipsychotics seem to decrease the levels of glutamate [9]. Finally, it is worth noting that two separate meta-analyses [11, 12] identified decreased levels of NAA in SCZ patients in prefrontal and temporal regions, and thalamus, supporting the theory of neuronal involvement in SCZ.

A glutamatergic involvement has been hypothesized also for mood disorders: in particular, major depression disorder (MDD) has been linked to decreased levels of Glu, possibly due to neurotransmission abnormalities, since the 1990s. Similarly, to what has been found in SCZ, also in this case medication seems to normalize glutamate levels. Moreover, this theory is corroborated by the fact that ketamine, a *N*-methyl-d-aspartate receptor antagonist, is effective in reducing depressive symptoms. Despite the limitations intrinsic to the technique, discussed in the case of SCZ, meta-analyses and original studies consistently demonstrated decreased glutamate levels especially in the prefrontal cortex [13] and in the anterior cingulate cortex [14] of MDD patients compared to controls. There also seems to be a role of illness chronicity in the neuronal abnormalities detected with 1H-MRS, with levels of NAA correlating with chronicity and possibly the progression of illness [15].

In bipolar disorder (BD) 1H-MRS studies consistently found decreased levels of NAA and increased concentrations of Cho and Glx [16]. These results confirm, from a biochemical point of view, the neuronal damage happening in BD, characterized by the changes in NAA and Cho in particular, which are a marker of loss of neuronal integrity and membrane damage. The most investigated regions, which show metabolic changes in BD patients compared to controls, are the frontal areas and in particular the anterior cingulate cortex (ACC) and the prefrontal cortex [17].

It has to be noticed that NAA can be considered also a marker of mitochondrial metabolism, because its levels correlate with the concentration of ATP [18]: many studies pinpointed an involvement of a dysfunction of mitochondria in the pathogenesis of BD. A recent meta-analysis [18] found lower levels of NAA in the prefrontal cortex of BD patients during a depressive episode and in euthymic state, and higher levels of Gln levels in the ACC of euthymic patients. An earlier study considering BD regardless of patients' state found decreased NAA in the basal ganglia and in the hippocampus [19].

Notably, elevated Glx levels found in patients with BD support the glutamatergic model of BD pathogenesis [20]. The concentration of Glx has been found higher in several brain areas, in particular in the ACC, and in prefrontal and hippocampal regions.

The study of first episode psychosis (FEP) is of particular interest in the field of psychiatry, because the effects of chronicity and long-term medication can be ruled out. The study of prefrontal integrity is particularly important in FEP, given its involvement in the hypothesis of a frontal lobe biochemical alteration [21]. Results on the ACC metabolic changes investigated with 1H-MRS are partially conflicting, even if there are findings on increased glutamatergic levels in FEP [22]. A recent 7 T study demonstrated a link between ACC glutamate and GABA and functional connectivity in FEP [23]. Also the hippocampus and the striatum showed metabolic changes in FEP, in particular lower Glx levels have been reported in the hippocampus [24] and higher concentrations of choline in the striatum [25], possibly indicating mitochondrial and membrane dysfunctions in FEP.

It has to be taken into account that 1H-MRS results in the clinical settings may be influenced by the heterogeneity intrinsic to the technique (magnetic field intensity, voxel size, SNR) and by treatment. For example, the interaction of medication with lithium and MRS spectra has been investigated, with mixed results and studies finding changes in metabolites' concentration after the administration of lithium [17, 26], while other studies found no significant associations between lithium and metabolites' levels [27].

1.5 Functional MRS

In the latest years the advances in MR technology, in particular related to higher magnetic fields and robust acquisition sequences, are allowing faster acquisitions of 1H-MRS signals, paving the way for the acquisition of neurotransmitters' dynamic signals. In fact, the measures of metabolites' levels in conventional MRS acquisitions depict steady-state concentrations. Considering that understanding the relationships between neurotransmitters would be of paramount importance in the study of human behavior, and that the most relevant excitatory and inhibitory neurotransmitters, Glu and GABA, are both quantifiable with MRS, it is undeniable how the possibility of studying the dynamics between transmitters could provide a more direct measure of behaviorally relevant neural activity [28]. One of the main advantages of functional 1H-MRS (1H fMRS) over functional magnetic resonance imaging (fMRI) is the possibility of measuring directly task-related changes in Glu, without the mediation of neurovascular effects. Nevertheless, to date there are no studies demonstrating the reliability and stability of the task-related Glu measures. This would be necessary for the reliability of the 1H-fMRS studies themselves. Moreover, despite the fact that consistent results are reported on Glu concentration changes during functional tasks, the magnitude of these changes is not stable across studies, ranging from 2% to 3% in visual stimulation studies considering the occipital cortex to 18% in the case of painful stimulations, in respect to the baseline measured in the ACC or insula [29], with an average of an increase of around 7% during neuronal activation. Studies considering cognitive tasks report heterogeneous results in terms of magnitude of changes in Glu concentration.

Even if there are to date still several aspects to be considered for a reliable interpretation of 1H-fMRS studies, this technique may offer a complementary measure of brain activation in respect to fMRI and allow a direct measure of the dynamic relationships between neurotransmitters.

2 Materials

MR scanner, a minimum field strength of 1.5 T is required for reliable spectra, with best results at 3 T.

MR coil, depending on the investigated nucleus. 1H-MRS requires the standard radiofrequency coil also used for MR imaging. Other nuclei require ad-hoc coils tuned to their Larmor frequencies.

Software package, of choice of the researcher, LCModel [30] and jMRUI [31], are among the most widely used and are introduced in the following text.

3 Methods

3.1 Data Acquisition

The MRS spectrum is obtained from specifically selected regions of the brain: this means that MRS is not, in general, a whole-brain technique but it is focused on specific brain areas. There is the possibility of acquiring signal from a single voxel (single voxel spectroscopy, SVS), or from many adjacent voxels (MRS imaging, MRSI, chemical-shift imaging, CSI), covering part of the brain. Whole brain coverage is possible using MRSI.

However, the first step is to acquire an anatomical image, usually a T1-weighted image, which can be used to precisely position the MRS volume of interest (VOI).

The choice on the specific acquisition protocol should be driven by the particular disease investigated. If the affected region of the brain is well defined in advance, SVS may be the most practical method, whereas MRSI may be more indicated if the pathology of interest is not defined by focal lesions, or if the lesions are expected to be diffuse in the brain, or heterogeneous.

3.1.1 Single Voxel Spectroscopy

SVS techniques are the most widely used in the clinical practice, particularly because the results are easy to interpret, the acquisition is fairly simple and fast (a few minutes), and they provide a good signal-to-noise ratio (SNR), which can be used for quantitative analysis of the metabolites. Moreover, since the investigated volume is small (typically around 8 cm^3) a good shim can usually be obtained, which leads to good peaks' separation in the spectra. Pointed-resolved spectroscopy (PRESS) or stimulated echo acquisition mode (STEAM) is used for acquisition of SVS: PRESS technique is usually favored because it gives a better SNR, which is essential for metabolite estimation. However, the SNR is proportional to the voxel size, so larger voxels lead to better SNR.

3.1.2 MRS Imaging

In this case, the output of the MRS acquisition is a matrix (or grid) of voxels, each containing an MRS spectrum. The size of the matrix is chosen in advance and corresponds to the size of the field of view (FOV). With the acquisition of many voxels, there are some obvious advantages, such as the possibility of investigating larger brain areas and higher spatial sensitivity, since the individual size of a CSI voxel is smaller than that of an SVS voxel. Nevertheless, there are some disadvantages of this technique that should not be underestimated. First, the acquisition is slower and the shim is difficult to obtain, because of the greater size of the region of interest. Moreover, the signal is intrinsically noisier than that obtained with SVS, so metabolites' quantification is not as reliable as with SVS. It is also important to notice that, since the shape of the matrix is rectangular and cannot be adjusted to the shape of the head, the signal can be corrupted by tissues outside the brain, as fat. Finally, MRSI spectra are often corrupted by spectral contamination by adjacent voxels, an effect called "voxel bleed," which affects the quantification. Also in this case, PRESS or STEAM techniques are used.

In general, SVS is to be preferred when precise quantification is required, while MRSI may be chosen if the aim of the acquisition is to identify metabolites' spatial distribution.

3.1.3 Water and Lipid Suppression

Water is predominant in all brain tissues, and each water molecule contains two protons: thus, if not suppressed, the water signal, with a resonance around 4.7 ppm, has an intensity of over 10,000 times that of other metabolites (metabolite concentrations are in the millimolar range, while water is usually tens of moles), which overpowers the signals from other components, making their quantifications difficult. Lipid-suppression techniques are aimed at eliminating spurious signals coming from non-brain regions, as subcutaneous fat, coming from the scalp in the case of brain MRS. Their signal is also order of magnitudes greater than that of the metabolites of interest. Both water and lipid signals have a spectral position close to that of metabolites, and if not suppressed lead to baseline distortions. It is worth noticing that some pathologies (as neurodegeneration or certain tumors) may cause the accumulation of lipids (which are usually not present) in the brain, which may be detected with MRS. These lipids should not be affected by lipid suppression techniques, because they may obviously be of clinical interest. Lipid suppression is particularly important in the case of MRS imaging.

Water suppression can be achieved either during the acquisition, with ad-hoc sequences, or during the data analysis step. If the decision is to acquire water-suppressed spectra, a combination of RF pulses is used, which saturate the water spins before the acquisition of the metabolic signals. A variety of techniques have been presented, as CHESS, WET, and VAPOR. Using these sequences, the water peak is usually reduced and its size becomes comparable to the metabolites' peaks. It is to be remembered, though, that the acquisition of a non-suppressed water signal is important as the water peak may be used as a reference for the absolute quantification of metabolites' concentrations.

There are also several methods aimed at removing the water peak during the data processing phase, for example, the Hankel-Lanczos Singular Value Decomposition (HLSVD), which decomposes the signal in order to identify the water components, or methods based on the Fourier transform, filters, or methods based on optimization, as the Advanced Methods for Accurate, Robust, and Efficient Spectral fitting (AMARES). These post-processing methods are heavily dependent on the signal quality, as the shape of the water peak or the SNR affect their performances [32].

Regarding lipid suppression, advanced SVS techniques eliminate the signal coming from non-brain tissues thanks to the excellent localization of the acquisition voxel [32]. In the case of MRSI, it is not easily solved due to the intrinsic shape of the acquisition matrix. In this case, usually saturation bands (outer volume suppression (OVS) bands) are placed over regions containing lipids, which eliminate the signal from the regions where they are placed (Fig. 2).

3.2 Data Processing

The choice of the post-processing technique is a very important one. To date, there are various reliable software packages available, which automatically and reliably quantify MR spectra. Some of the most widely used are LCModel [30], TARQUIN [33], jMRUI [31]: they give reliable and reproducible quantifications of metabolites' levels, with varying degrees of user intervention. This is an important advance in respect to the past, when authors developed in-house software [6] which made it difficult to replicate the results.

With LCModel, frequency domain data are fitted to model spectra obtained in vitro using the same sequence as during the in vivo acquisition. With this model, each metabolite is fitted to the model, which makes it very popular for the quantification of 1H-MRS acquired at short TE. Using jMRUI, data are analyzed in the time domain, using a priori knowledge [34].

When processing the acquired data, it is important to remember that the number of quantifiable metabolites depends on the sequence settings, and that the quality of the results is heavily affected by the signal resolution and the SNR, which in turn

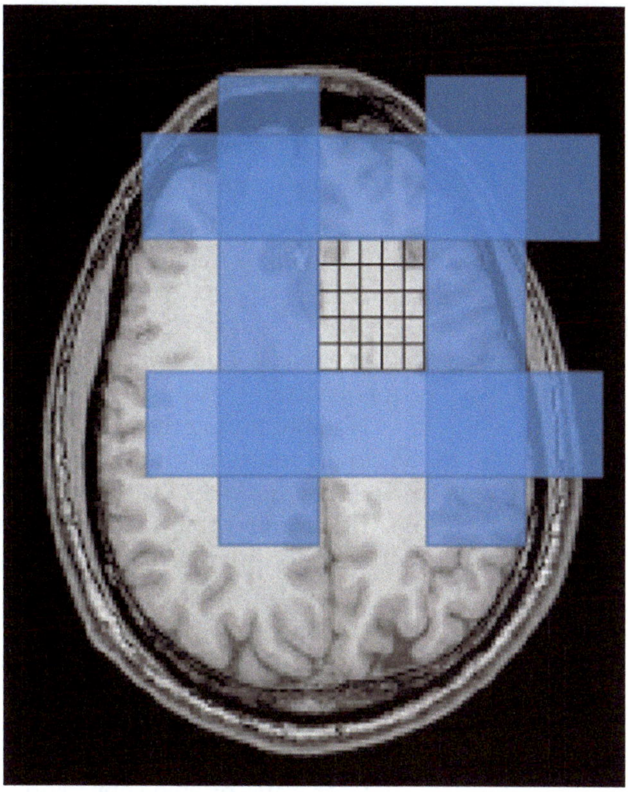

Fig. 2 Example of outer volume saturation bands around an MRSI matrix. The bands are set during the acquisition to avoid spurious signal coming from outside the volume of interest

depend on many variables, including magnetic field strength, B0 inhomogeneities, and type of coil used. Using the standard processing software available, it is also possible to evaluate error estimates on the quantifications, which are a measure of results reliability.

4 Notes

4.1 Understanding Data Quality

One of the most important variables which influence SNR, thus data quality, is the magnitude of the magnetic field. 3 T scanners ensure higher levels of SNR in respect to 1.5 T, but other variables must be taken into account to comprehensively evaluate data quality. It is important to remember that, while both 1.5 and 3 T scanner can be used reliably for MRS quantification, a spectrum acquired at 1.5 T with a protocol aimed at a high spectral quality is more accurate than a poor spectrum acquired at 3 T. In particular, as a general guideline [3], a spectrum should be rejected from the analysis if:

- The full width at half maximum (FWHM) is higher than 0.1 ppm

- The line shape is asymmetric
- There are obvious artifacts
- The water suppression is poor (<98%)
- The SNR is lower than 3
- Residuals contain unexplained features
- The Cramer-Rao lower bound (given as an output, e.g., by LCModel) is higher than 50%

4.2 Artifacts

The MRS signal may be affected by artifacts, which are detrimental in the process of quantification of brain metabolites and may affect data quality.

Notably, during the acquisition the shimming process needs to be closely addressed. Its aim is to make the magnetic field homogeneity better inside the spectroscopy VOI. This process is automatic in MRI scanners, but may need manual adjustments. If the shimming is poor, some artifacts may arise: elevated FWHM, low signal amplitude, with consequent signal lower SNR, and difficulty separating metabolites' peaks. Some regions, located near the bone or close to air–tissue interfaces (e.g., temporal or orbitofrontal regions, or regions close to the sinuses) are more prone to this kind of artifact, and the placement of voxels should be avoided in those areas [35].

Poor water or lipid suppression may lead to artifacts: in particular, if the water is not adequately suppressed, the spurious water signal may be too high and hides the metabolites' spectra.

Spectra acquired at higher magnetic fields are also prone to the chemical shift displacement (CSD) artifact: different molecules have different chemical shift, and it translates to a different frequency-encoded location, meaning that the signals originate in different locations and do not originate in the same volume.

It is also worth mentioning that MRS is sensitive to motion. Spectra corrupted by movement may have broader peaks and phase fluctuations, which lead to incorrect quantification.

Motion during an MRS scan may also induce errors which are not noticed during the quality control phase. For example, if a subject moves inside a scanner after the MRS voxel has been positioned, the acquired signal, although of good quality, may originate from a location which is not the region of interest. There are methods that track subject movements using cameras or navigators [36] which are effective in eliminating the artifacts caused by localization errors due to motion. Head movement during the scan may also lead to magnetic field inhomogeneities, which may be corrected using ad-hoc techniques.

Artifacts may also be caused by eddy currents, which cause distortions to the spectral peaks: eddy current correction is often performed during the postprocessing phase.

4.3 Long and Short TE

The spectra obtained from the acquisition depends on the TE. A short TE (20–40 ms) leads to high SNR and spectra with more quantifiable metabolites than those obtained with long TEs (135–238 ms). Spectra acquired with long TEs have lower SNR, but are simpler. Importantly, TEs between 135–144 ms lead to spectra showing the lactate peak inverted below the baseline: this allows the identification of the lactate peak, which otherwise overlaps with lipids.

At 1.5 T, sequences with long TEs allow the quantification of NAA, Cr, Cho, lactate, and alanine, which are quantifiable also with short TEs, with the addition of lipids, myo-inositol, Glx, and glucose [35].

Acknowledgments

This work was partially funded by grants from the Italian Ministry of Health.

(RF-2019-12371349), Fondazione Cariplo (2019-3416), and ERA-NET NEURON (JTC2018: UNMET).

References

1. Glunde K, Jiang L, Moestue SA, Gribbestad IS (2011) MRS and MRSI guidance in molecular medicine: targeting and monitoring of choline and glucose metabolism in cancer. NMR Biomed 24(6):673–690. https://doi.org/10.1002/nbm.1751. PMID: 21793073; PMCID: PMC3146026
2. Ramadan S, Lin A, Stanwell P (2013) Glutamate and glutamine: a review of in vivo MRS in the human brain. NMR Biomed 26(12):1630–1646. https://doi.org/10.1002/nbm.3045. Epub 2013 Oct 4. PMID: 24123328; PMCID: PMC3849600
3. Oz G, Alger JR, Barker PB, Bartha R, Bizzi A, Boesch C, Bolan PJ, Brindle KM, Cudalbu C, Dinçer A, Dydak U, Emir UE, Frahm J, González RG, Gruber S, Gruetter R, Gupta RK, Heerschap A, Henning A, Hetherington HP, Howe FA, Hüppi PS, Hurd RE, Kantarci K, Klomp DW, Kreis R, Kruiskamp MJ, Leach MO, Lin AP, Luijten PR, Marjańska M, Maudsley AA, Meyerhoff DJ, Mountford CE, Nelson SJ, Pamir MN, Pan JW, Peet AC, Poptani H, Posse S, Pouwels PJ, Ratai EM, Ross BD, Scheenen TW, Schuster C, Smith IC, Soher BJ, Tkáč I, Vigneron DB, Kauppinen RA, MRS Consensus Group (2014) Clinical proton MR spectroscopy in central nervous system disorders. Radiology 270(3):658–679. https://doi.org/10.1148/radiol.13130531. PMID: 24568703; PMCID: PMC4263653
4. Horská A, Barker PB (2010) Imaging of brain tumors: MR spectroscopy and metabolic imaging. Neuroimaging Clin N Am 20(3):293–310. https://doi.org/10.1016/j.nic.2010.04.003. PMID: 20708548; PMCID: PMC2927327
5. Zhu H, Barker PB (2011) MR spectroscopy and spectroscopic imaging of the brain. Methods Mol Biol 711:203–226. https://doi.org/10.1007/978-1-61737-992-5_9. PMID: 21279603; PMCID: PMC3416028
6. Port JD (2020) Magnetic resonance spectroscopy for psychiatry: progress in the last decade. Neuroimaging Clin N Am 30(1):25–33. https://doi.org/10.1016/j.nic.2019.09.002. Epub 2019 Nov 8. PMID: 31759569
7. Kumar V, Vajawat B, Rao NP (2021) Frontal GABA in schizophrenia: a meta-analysis of ^{1}H-MRS studies. World J Biol Psychiatry 22(1):1–13. https://doi.org/10.1080/15622975.2020.1731925. Epub 2020 Mar 4. PMID: 32067569
8. Egerton A, Modinos G, Ferrera D, McGuire P (2017) Neuroimaging studies of GABA in schizophrenia: a systematic review with meta-analysis. Transl Psychiatry 7(6):e1147. https://doi.org/10.1038/tp.2017.124. PMID: 28585933; PMCID: PMC5537645

9. Poels EM, Kegeles LS, Kantrowitz JT, Javitt DC, Lieberman JA, Abi-Dargham A, Girgis RR (2014) Glutamatergic abnormalities in schizophrenia: a review of proton MRS findings. Schizophr Res 152(2–3):325–332. https://doi.org/10.1016/j.schres.2013.12.013. Epub 2014 Jan 11. PMID: 24418122; PMCID: PMC3951718
10. Merritt K, Egerton A, Kempton MJ, Taylor MJ, McGuire PK (2016) Nature of glutamate alterations in schizophrenia: a meta-analysis of proton magnetic resonance spectroscopy studies. JAMA Psychiatry 73(7):665–674. https://doi.org/10.1001/jamapsychiatry.2016.0442. PMID: 27304221
11. Brugger S, Davis JM, Leucht S, Stone JM (2011) Proton magnetic resonance spectroscopy and illness stage in schizophrenia--a systematic review and meta-analysis. Biol Psychiatry 69(5):495–503. https://doi.org/10.1016/j.biopsych.2010.10.004. Epub 2010 Dec 8. PMID: 21145039
12. Iwata Y, Nakajima S, Plitman E, Mihashi Y, Caravaggio F, Chung JK, Kim J, Gerretsen P, Mimura M, Remington G, Graff-Guerrero A (2018) Neurometabolite levels in antipsychotic-naïve/free patients with schizophrenia: a systematic review and meta-analysis of ^{1}H-MRS studies. Prog Neuro-Psychopharmacol Biol Psychiatry 86:340–352. https://doi.org/10.1016/j.pnpbp.2018.03.016. Epub 2018 Mar 23. PMID: 29580804
13. Moriguchi S, Takamiya A, Noda Y, Horita N, Wada M, Tsugawa S, Plitman E, Sano Y, Tarumi R, ElSalhy M, Katayama N, Ogyu K, Miyazaki T, Kishimoto T, Graff-Guerrero A, Meyer JH, Blumberger DM, Daskalakis ZJ, Mimura M, Nakajima S (2019) Glutamatergic neurometabolite levels in major depressive disorder: a systematic review and meta-analysis of proton magnetic resonance spectroscopy studies. Mol Psychiatry 24(7):952–964. https://doi.org/10.1038/s41380-018-0252-9. Epub 2018 Oct 12. PMID: 30315224; PMCID: PMC6755980
14. Luykx JJ, Laban KG, van den Heuvel MP, Boks MP, Mandl RC, Kahn RS, Bakker SC (2012) Region and state specific glutamate downregulation in major depressive disorder: a meta-analysis of (1)H-MRS findings. Neurosci Biobehav Rev 36(1):198–205. https://doi.org/10.1016/j.neubiorev.2011.05.014. Epub 2011 Jun 6. PMID: 21672551
15. Brambilla P, Stanley JA, Nicoletti MA, Sassi RB, Mallinger AG, Frank E, Kupfer DJ, Keshavan MS, Soares JC (2005) 1H Magnetic resonance spectroscopy study of dorsolateral prefrontal cortex in unipolar mood disorder patients. Psychiatry Res 138(2):131–139. https://doi.org/10.1016/j.pscychresns.2004.12.001. PMID: 15766636
16. Sosa-Moscoso B, Ullauri C, Chiriboga JD, Silva P, Haro F, Leon-Rojas JE (2022) Magnetic resonance spectroscopy and bipolar disorder: how feasible is this pairing? Cureus 14(3):e23690. https://doi.org/10.7759/cureus.23690. PMID: 35505758; PMCID: PMC9056012
17. Sassi RB, Stanley JA, Axelson D, Brambilla P, Nicoletti MA, Keshavan MS, Ramos RT, Ryan N, Birmaher B, Soares JC (2005) Reduced NAA levels in the dorsolateral prefrontal cortex of young bipolar patients. Am J Psychiatry 162(11):2109–2115. https://doi.org/10.1176/appi.ajp.162.11.2109. PMID: 16263851
18. Chabert J, Allauze E, Pereira B, Chassain C, De Chazeron I, Rotgé JY, Fossati P, Llorca PM, Samalin L (2022) Glutamatergic and N-acetylaspartate metabolites in bipolar disorder: a systematic review and meta-analysis of proton magnetic resonance spectroscopy studies. Int J Mol Sci 23(16):8974. https://doi.org/10.3390/ijms23168974. PMID: 36012234; PMCID: PMC9409038
19. Kraguljac NV, Reid M, White D, Jones R, den Hollander J, Lowman D, Lahti AC (2012) Neurometabolites in schizophrenia and bipolar disorder – a systematic review and meta-analysis. Psychiatry Res 203(2–3):111–125. https://doi.org/10.1016/j.pscychresns.2012.02.003. Epub 2012 Sep 13. PMID: 22981426; PMCID: PMC3466386
20. Maddock RJ, Buonocore MH (2012) MR spectroscopic studies of the brain in psychiatric disorders. Curr Top Behav Neurosci 11:199–251. https://doi.org/10.1007/7854_2011_197. PMID: 22294088
21. Zabala A, Sánchez-González J, Parellada M, Moreno DM, Reig S, Burdalo MT, Robles O, Desco M, Arango C (2007) Findings of proton magnetic resonance spectometry in the dorsolateral prefrontal cortex in adolescents with first episodes of psychosis. Psychiatry Res 156:33–42
22. Squarcina L, Stanley JA, Bellani M, Altamura CA, Brambilla P (2017) A review of altered biochemistry in the anterior cingulate cortex of first-episode psychosis. Epidemiol Psychiatr Sci 26(2):122–128. https://doi.org/10.1017/S2045796016000895. Epub 2017 Jan 20. PMID: 28103961; PMCID: PMC6998769
23. Overbeek G, Gawne TJ, Reid MA, Kraguljac NV, Lahti AC (2021) A multimodal

neuroimaging study investigating resting-state connectivity, glutamate and GABA at 7 T in first-episode psychosis. J Psychiatry Neurosci 46(6):E702–E710. https://doi.org/10.1503/jpn.210107. PMID: 34933941; PMCID: PMC8695527

24. Shakory S, Watts JJ, Hafizi S, Da Silva T, Khan S, Kiang M, Bagby RM, Chavez S, Mizrahi R (2018) Hippocampal glutamate metabolites and glial activation in clinical high risk and first episode psychosis. Neuropsychopharmacology 43(11):2249–2255. https://doi.org/10.1038/s41386-018-0163-0. Epub 2018 Jul 28. PMID: 30087434; PMCID: PMC6135774

25. Sivaraman S, Kraguljac NV, White DM, Morgan CJ, Gonzales SS, Lahti AC (2018) Neurometabolic abnormalities in the associative striatum in antipsychotic-naïve first episode psychosis patients. Psychiatry Res Neuroimaging 281:101–106. https://doi.org/10.1016/j.pscychresns.2018.06.003. Epub 2018 Jun 9. PMID: 30286325; PMCID: PMC7874514

26. Soeiro-de-Souza MG, Otaduy MCG, Machado-Vieira R, Moreno RA, Nery FG, Leite C, Lafer B (2018) Lithium-associated anterior cingulate neurometabolic profile in euthymic Bipolar I disorder: a ^{1}H-MRS study. J Affect Disord 241:192–199. https://doi.org/10.1016/j.jad.2018.08.039. Epub 2018 Aug 10. PMID: 30130684

27. Brambilla P, Stanley JA, Sassi RB, Nicoletti MA, Mallinger AG, Keshavan MS, Soares JC (2004) 1H MRS study of dorsolateral prefrontal cortex in healthy individuals before and after lithium administration. Neuropsychopharmacology 29(10):1918–1924. https://doi.org/10.1038/sj.npp.1300520. PMID: 15257303

28. Stanley JA, Raz N (2018) Functional magnetic resonance spectroscopy: The "new" MRS for cognitive neuroscience and psychiatry research. Front Psychiatry. 9:76. https://doi.org/10.3389/fpsyt.2018.00076. PMID: 29593585; PMCID: PMC5857528

29. Mullins PG (2018) Towards a theory of functional magnetic resonance spectroscopy (fMRS): a meta-analysis and discussion of using MRS to measure changes in neurotransmitters in real time. Scand J Psychol 59(1):91–103. https://doi.org/10.1111/sjop.12411. PMID: 29356002

30. Provencher SW (2001) Automatic quantitation of localized in vivo 1H spectra with LCModel. NMR Biomed 14(4):260–264. https://doi.org/10.1002/nbm.698. PMID: 11410943

31. Stefan D, Di Cesare F, Andrasescu A, Popa E, Lazariev A, Vescovo E, Strbak O, Williams S, Starcuk Z, Cabanas M, van Ormondt D, Graveron-Demilly D (2009) Quantitation of magnetic resonance spectroscopy signals: the jMRUI software package. Meas Sci Technol 20:104035

32. Tkáč I, Deelchand D, Dreher W, Hetherington H, Kreis R, Kumaragamage C, Považan M, Spielman DM, Strasser B, de Graaf RA (2021) Water and lipid suppression techniques for advanced ^{1}H MRS and MRSI of the human brain: experts' consensus recommendations. NMR Biomed 34(5):e4459. https://doi.org/10.1002/nbm.4459. Epub 2020 Dec 16. PMID: 33327042; PMCID: PMC8569948

33. Wilson M, Reynolds G, Kauppinen RA, Arvanitis TN, Peet AC (2011) A constrained least-squares approach to the automated quantitation of in vivo 1 H magnetic resonance spectroscopy data. Magn Reson Med 65(1):1–12

34. van der Graaf M (2010) In vivo magnetic resonance spectroscopy: basic methodology and clinical applications. Eur Biophys J 39(4):527–540. https://doi.org/10.1007/s00249-009-0517-y. Epub 2009 Aug 13. PMID: 19680645; PMCID: PMC2841275

35. Bertholdo D, Watcharakorn A, Castillo M (2013) Brain proton magnetic resonance spectroscopy: introduction and overview. Neuroimaging Clin N Am 23(3):359–380. https://doi.org/10.1016/j.nic.2012.10.002. Epub 2013 Jan 20. PMID: 23928194

36. Saleh MG, Edden RAE, Chang L, Ernst T (2020) Motion correction in magnetic resonance spectroscopy. Magn Reson Med. 84(5):2312–2326. https://doi.org/10.1002/mrm.28287. Epub 2020 Apr 17. PMID: 32301174; PMCID: PMC8386494

Chapter 13

The Effect of Exogenous and Endogenous Parameters on Group Resting-State Effective Connectivity and BOLD Signal

Liucija Vaisvilaite, Meng-Yun Wang, Micael Andersson, and Karsten Specht

Abstract

fMRI has gained popularity in research and clinical settings. The majority of work relies on group comparisons to identify the resting-state network structure, connectivity, and differences between diverse populations. However, the reliability of fMRI has been doubted together with an increasing number of publications on individual differences and other factors affecting the resting-state fMRI. The current publication addresses the factors that are often overlooked or not accounted for when selecting the "healthy control group" as a baseline for comparison. Our results indicate that additional biological factors, such as blood pressure, level of hemoglobin, waist circumference, time of image acquisition, as well as age and sex introduce variability in the group data. It is therefore likely to conclude that this variability on a group level might in turn affect the differences or changes observed between populations.

Key words fMRI, DCM, Group average, Resting-state, Individual differences

1 Introduction

One overarching goal of neuroscience is to uncover how the brain works [1]. With the recent advances in technology, a number of various neuroimaging methods can monitor brain activity noninvasively. For example, EEG (electroencephalography) and MEG (magnetoencephalography) have been ubiquitously utilized to explore brain oscillations with their high temporal resolution characteristics [2–4]. Meanwhile, fMRI (functional magnetic resonance imaging), with its high spatial resolution, has been predominantly adopted in studies that explore cognitive functions of deep brain regions [5–8].

fMRI has been developed in the early 1990s and records blood oxygen level-dependent (BOLD) signals, which are used as a proxy for neuronal signaling [9, 10]. fMRI allows researchers to localize

Drozdstoy Stoyanov et al. (eds.), *Computational Neuroscience*, Neuromethods, vol. 199,
https://doi.org/10.1007/978-1-0716-3230-7_13,

the active brain region or network involved in particular cognitive processing [7]. The task fMRI paradigm, as suggested by its name, requires an experimental manipulation involving a task that participants engage in [11, 12]. Interestingly, there has been a substantial shift from task fMRI to resting-state fMRI, in which the intrinsic brain activities are recorded without receiving any explicit tasks, for example, engaging in inner mentation [13]. Due to its simple implementation, resting-state fMRI has been increasingly used in the clinical field, in which it is difficult for some vulnerable populations to carry out a particular cognitive task in MRI, such as patients with autism [14].

Undoubtedly, the insights into how the brain works have been unarguably benefited from fMRI results. For instance, it is now established that brain regions coordinate with each other and constellate into different networks even at the resting state, which are known as resting-state networks [15–17]. The noticeable, and most researched resting-state networks include the default mode network (DMN) [18, 19], salience network (SN) [20], and central executive network (CEN) [20, 21].

Although our understanding of how the brain works has been substantially improved based on fMRI studies, the reliability of the fMRI results, however, has been questioned [22–24]. For example, a recent publication, allowed 70 independent research teams to analyze the same database to test 9 hypotheses. The results indicated that as high as 20% of teams reported a result that differs from the majority of teams on average across all the hypotheses [24]. The variability they reported had an impact on the reliability of fMRI, which in turn directly affected the clinical application of fMRI [25, 26]. Contemporary literature in the field suggests that resting-state fMRI is sensitive to individual variability, for instance, it is reflective of personality traits such as extraversion [27] and other personality traits, such as openness to experience [28]. In addition to personality reflective differences, research suggests gender dimorphisms contribute to the resting-state network connectivity variability between the sexes [29]. Furthermore, it is added and supported that not only gender but also age has a significant impact on the resting-state network connectivity that possibly stems from underlying hormonal differences [30]. The amount of literature on differences in network connectivity that is associated with one or the other behavioral trait, socioeconomic status, and marital status is staggering [31–33]. Even though the benefit of research in various aspects and factors influencing their resting-state connectivity should not be overlooked, the consensus and solid ground for building reliable resting-state fMRI protocol and procedures are undoubtedly lacking [34].

Importantly, most publications in the field of functional neuroimaging have relied on group analyses, for example, averaging data [35] to obtain an informed view on the generic aspects of functional connectivity (FC). The group averaging data does provide the benefit of increasing the signal-to-noise ratio (SNR) which in turn allows for in-group comparisons often used in clinical studies. Not limited to but also inclusive of the clinical research, the control group as commonly referred to as the healthy participant group is often described with just a little specificity as "healthy young adults." Some studies do include a more comprehensive description, providing both the age (mean) and gender of the participants. To date, however, there is a lack of a descriptive procedure for participants inclusion/exclusion criteria for control groups, with the majority of studies relying on the convenience sampling procedures, for example, undergraduate university student participants. The concern here is that relatively loose inclusion requirements may hinder potential differences in resting-state networks given a naturally occurring individual variation as stated above.

Given the outlined potential effects of healthy individual variability on fMRI, it is therefore essential to explore the biological factors that might have a significant effect on group comparisons in fMRI. To date, there are several studies published that investigated the effects of endogenous and exogenous factors in healthy participants, for example, representative population [5, 36]. We have previously reported findings that the time of image acquisition has an effect on the resting-state fMRI data and should be noted in the fMRI protocols [5]. Furthermore, a combination of BMI and diastolic blood pressure values has been linked to variability in effective connectivity [36]. The level of exogenous estradiol in females has been shown to affect both task- and resting-state network connectivity [37]. Finally, individual variability and exogenous and endogenous parameters affecting fMRI have been noted as one of the current challenges in the way of achieving high reliability in neuroimaging studies [26, 37].

With the current report, we aim to build on the existing findings and explore additional biological factors, namely, blood pressure, waist circumference, levels of hemoglobin, sex, age, and time of image acquisition, that we believe have an effect on resting-state fMRI.

For the purposes of exploring the impact that the named factors may invoke on resting-state fMRI, we have chosen to use spectral Dynamic Causal Modeling (DCM) for resting-state for analyzing the data [38]. It has been proposed that DCM, which models the effective (dynamic) connectivity and parameters that define the BOLD signal separately, has an advantage over the mainstream fMRI data analysis techniques [39]. Based on the parameters available, we believe that the effects might differ for the parameters defining the BOLD signal and effective connectivity (EC).

Thus, boldly, we presently propose that the reliability crisis can be mitigated by controlling additional exogenous and endogenous factors, especially in group comparison studies. Specifically, we aim to suggest that the definition of "healthy controls" should be taken with caution and controlled for additional variables that currently are rarely or not considered at all. Therefore, we compare two groups of "healthy" participants, where one group is controlled for several biological factors, while the "random" group consists of randomly selected subjects out of the same data pool.

2 Methods

2.1 Participants

For the purposes of the current study, readily collected data from the BETULA project on memory, health, and aging were used [40]. It is an ongoing longitudinal study conducted in Umeå, Sweden. There has been a total of seven data collection occasions (time point of collection (T)) in the BETULA cohort study from 1988 until now. The data from data collection occasion T5 which involved both biological and neuroimaging data collection was used.

Structural and functional MRI scans were collected during T5 for 375 individuals in total. Due to technical error and/or severe motion within the scanner, 37 participants were excluded from the total sample. The remaining sample of 339 participants from T5 were ranging from 25 to 80 years of age (mean age = 61.5, *SD* = 13.3), 163 males and 176 females. Eighty-four participants had to be excluded from the follow-up data collection because of depression, Parkinson's disease, and dementia [41]. The remaining 292 participant data were quality checked for mean framewise displacement (FD) in the scanner. A conservative threshold of FD < 0.2 was chosen [42] resulting in the removal of additional 56 participants. The sample of neuropsychologically healthy and quality-checked participants was 236.

For the purposes of the current study, the following additional selection of participants has been executed:

For the healthy participant sample the following thresholds for biological parameters were used: blood pressure (lying down) systolic <120, diastolic <80, hemoglobin levels >111 (hemoglobin levels were additionally screened for gender, 2 participants included in the healthy sample had hemoglobin levels of 111 and 119, deeming them eligible for the healthy sample), waist circumference < 94 cm for males and < 80 cm for females. The final sample for the analysis was 41 participants (17 males, 24 females), age range from 26 to 76 years of age at the fMRI image acquisition.

For the random sample of participants, firstly the already selected pool of healthy participants has been removed from the selection. A random number generator was used to determine random participant IDs of 40 participants (14 males, 26 females), age range from 36 to 81 years of age. For the random sample, no additional thresholds were used, aside from the original restrictions of neuropsychological health and FD value.

2.2 BETULA Data Collection

MRI data were acquired using a 3-T General Electric scanner equipped with a 32-channel head coil. For the resting-state fMRI gradient echo-planar imaging sequence was used with the following specifications: 37 transaxial slices, thickness; 3.4 mm, gap; 0.5 mm, repetition time (TR); 2000 ms, echo time (TE); 30 ms, flip angle; 80°, field of view; 25 × 25 cm, 170 volumes. Ten dummy scans were collected and then discarded before the experimental procedure. A 3D fast spoiled gradient-echo sequence was used to collect high-resolution T1-weighted structural images, with the following specifications: 180 slices; thickness, 1 mm; TR, 8.2 ms; TE, 3.2 ms; flip angle, 12°; field of view, 25 × 25 cm. The pCASL sequence was set using a stack of spiral fast spin-echo read-out, with 512 sampling points on 8 spirals, FOV, 240 × 240 mm^2; slice thickness 4 mm; arterial labeling at the cerebellar base; 1500 ms labeling duration, 1525 ms post labeling delay, 30 control/label pairs. Cerebral blood flow (CBF) maps were computed showing tissue CBF in ml/min/100 g [41].

The participants of the BETULA study at each data collection occasion visited the lab at Umeå University, Umeå, Sweden, two times. The first visit primarily assessed health conditions, whereas the second focused on cognitive function assessment. A trained nurse was responsible for collecting blood, urine, and saliva samples.

The health visit lasted from 1.5 to 2 h. 40 mL of blood was taken during the first visit and 25 mL during the second, where visits were scheduled 3 months apart. The total amount of 55 mL was analyzed for complete blood chemistry, where 10 mL were frozen for perspectives analyses. Blood pressure, height, weight, and waist circumference measurements were taken during the first participant visit. For more information, please see [40].

During the MRI session, the participants were instructed to lay as still as possible with their eyes open, looking at a white fixation cross on the dark screen.

2.3 Data Analysis

2.3.1 Pre-processing

The brain imaging data were pre-processed using the Statistical Parametric Mapping software (SPM12; Welcome Department of Cognitive Neurology, University College London, London, United Kingdom). The pre-processing consisted of the following steps and were in the order as stated; slice-timing correction, realign and un-warp, co-registration of the structural and the functional scans, segmentation of the structural scans, normalization of the functional scan, and, finally, smoothing of the data. For the slice-timing correction, the temporal middle slice (number 2 in an interleaved slice acquisition) was used as the reference image. The data were then realigned and un-warped. For the time series co-registration with the structural T1 image, the estimated mean images across the time series were used. The co-registered structural images were segmented for gray matter, white matter, and cerebrospinal fluid. Thereafter, all functional volumes were normalized into a standard anatomical reference space, provided by the Montreal Neurological Institute (MNI) space, using the segmented deformation fields. Finally, the data were smoothed with a Gaussian kernel of 8 mm. Additionally, the data were further processed for the DCM analysis. The head movement effects in total 12 movement parameters and the signal from white matter and cerebrospinal fluid areas were regressed out from the fMRI time series using linear general models, as implemented in SPM12. In detail, the time course was extracted from a spherical volume with a radius of 6 mm at MNI coordinate [0–24 −33] and [0–40 −5], respectively. The time courses and the movement parameters were then incorporated in a general linear model, using SPM12, which again formed the basis for extracting the time courses for the subsequent DCM analyses. This procedure has been performed separately for each individual.

2.3.2 ROI Selection

In total eight regions of interest (ROIs) have been selected due to computational limitations for a spectral DCM analysis. The DMN, CEN, and SN networks were selected based on their prominent use in the previous literature. Eight pre-selected regions of interest, namely, medial prefrontal cortex, posterior cingulate cortex, right inferior parietal cortex, left inferior parietal cortex, dorsal anterior cingulate cortex, left dorsolateral prefrontal cortex, left intraparietal sulcus/superior parietal lobule, and right intraparietal sulcus/superior parietal lobule, were selected to represent DMN, CEN and SN networks. The MNI coordinates of the mentioned networks can be found in Table 1.

2.3.3 DCM Analysis

Spectral dynamic causal modeling (DCM) was used to estimate the effective connectivity, and parameters defining the blood oxygen level-dependent (BOLD) signal, *epsilon*, *decay*, and *transit*. The parametric empirical Bayes (PEB) framework was used to estimate the joint effective connectivity for each group, in this case, healthy and random groups separately [38]. A second-level PEB analysis was performed to assess the model fit to the group data. In total there were 21 models specified and listed in Table 2.

Bayesian model comparison (BMC) was used to estimate the model fit to the group data. The models with the best fit were explored at a cut-off of the posterior probability of $pp > 0.95$. The

Table 1
MNI coordinates for selected ROIs

	x	y	z	
SN	10	34	24	dACC
SN	−32	44	16	LDLPFC
CEN	−38	−78	36	LIPS
CEN	36	−80	26	RIPS
DMN	3	54	−2	mPFC
DMN	0	−52	26	PCC
DMN	48	−69	35	RIPC
DMN	−50	−63	32	LIPC

Table 2
The list of models specified for PEB analysis

1. Sex
2. Age
3. Waist circumference
4. Systolic blood pressure
5. Diastolic blood pressure
6. Hemoglobin levels
7. Time of image acquisition
8. Sex and age
9. Systolic blood pressure and diastolic blood pressure
10. Systolic blood pressure, diastolic blood pressure, and hemoglobin levels
11. Waist circumference, systolic blood pressure, and diastolic blood pressure
12. Waist circumference, systolic blood pressure, diastolic blood pressure, and hemoglobin levels
13. Sex, age, waist circumference, systolic blood pressure, diastolic blood pressure, and hemoglobin levels
14. Sex, age, waist circumference, systolic blood pressure, diastolic blood pressure, hemoglobin levels, and time of image acquisition
15. Sex, age, and waist circumference
16. Sex, age, and diastolic blood pressure
17. Sex, age, and hemoglobin levels
18. Sex, age, systolic blood pressure, and diastolic blood pressure
19. Sex, age, systolic blood pressure, diastolic blood pressure, and time of image acquisition
20. Sex, age, and time of image acquisition
21. Null model: expectation that none of the models explain the variability in the data

BMC and posterior probability estimation were conducted for the effective connectivity matrix (A-matrix, 8 × 8 parameter). The same procedure was repeated for the hemodynamic parameters of the Balloon model, namely, *transit time, epsilon, decay*, and the parameters α (reflecting amplitude) and β (reflecting the spectral density of the neural fluctuations). The parameter transit time was estimated for each of the eight ROIs, where *decay, epsilon*, α, and β were global parameters.

3 Results

3.1 Effective Connectivity

The conditional model probabilities under the flat priors for the healthy group effective resting-state connectivity indicated that the highest model evidence (model accuracy minus model complexity) was model no. 17: sex, age, and hemoglobin levels. In contrast, the model probabilities under the flat priors for the random group effective resting-state connectivity indicated that the highest model evidence (model accuracy minus model complexity) was model no. 12: waist circumference, systolic and diastolic blood pressure, and hemoglobin levels. In other words, the Bayesian model comparison suggests that effective connectivity variability was best explained by the aforementioned parameters.

Figure 1 illustrates the effective connectivity matrix for both samples. Applying a posterior probability threshold of $pp > 0.95$, the A matrix for both samples show a similar pattern, namely, a strong inhibitory connection from all nodes of the SN to those of the DMN and CEN, while the nodes within the SN showed strong excitatory effective connectivity. Further, there was an inhibitory connection from the CEN to the DMN (RIPS to LIPC), and an excitatory connection from the mPFC of the DMN to the LIPS of the CEN. The main difference between the groups is that the random sample demonstrated more excitatory connections within the DMN, an additional inhibitory connection from the CEN to the DMN, and an excitatory connection from the CEN to SN.

3.2 Parameters Defining the BOLD Signal

The Bayesian model comparison performed on the parameters defining the BOLD signal, namely, transit time, epsilon, decal, amplitude (α), and spectral density (β), indicated that for the healthy group there were two competing models that best fit the data, model no. 13: sex, age, waist circumference, systolic and diastolic blood pressure, and hemoglobin levels, and model no. 15: sex, age, and waist circumference. For the randomly selected participant group the Bayesian model comparison for the parameters that define the BOLD signal indicated that model 14: sex, age, systolic and diastolic blood pressure, hemoglobin levels, and time of image acquisition was the model that explained data best (Fig. 2).

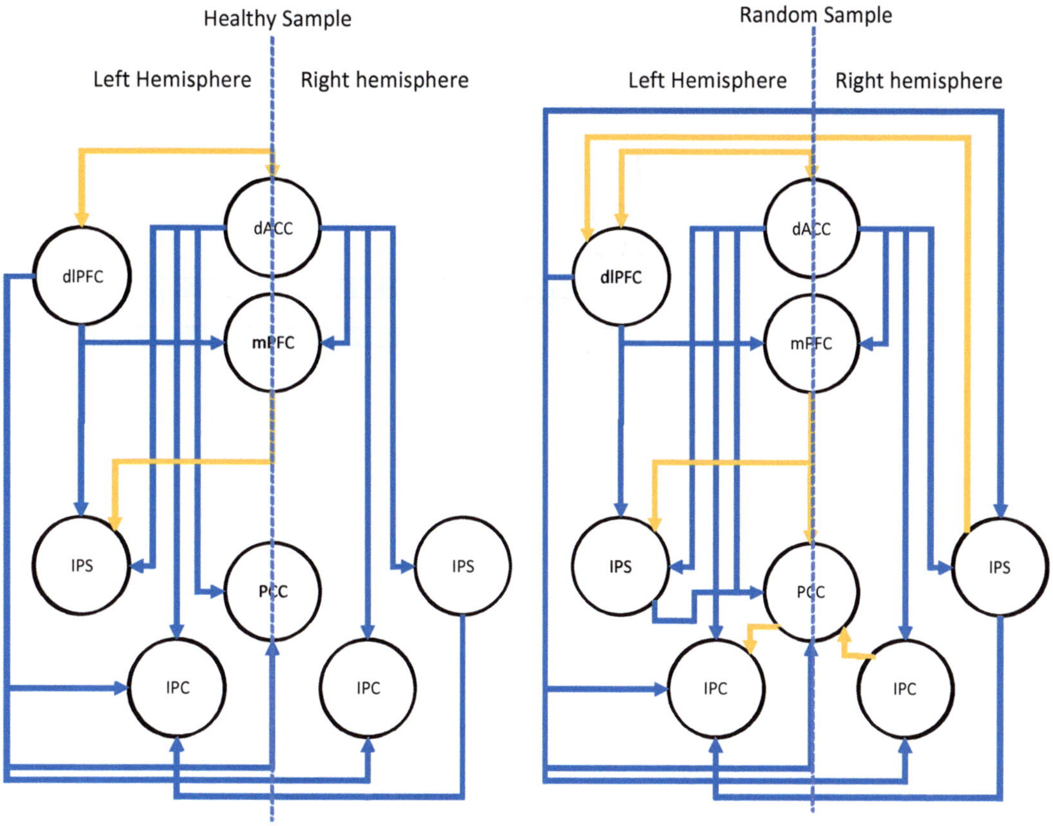

Fig. 1 The figure summarizes results for effective connectivity for both samples, where inhibitory connections are displayed in blue and excitatory in orange. Self-inhibitory connections are not displayed in the figure

4 Discussions

The aims of the current study were twofold. Firstly, to explore the endogenous and exogenous factors that may affect the individual variability in resting-state fMRI results. Secondly, to revisit the current working definition of "healthy control group" often used in clinical and research settings.

The overall results indicate that, despite a similar connectivity pattern (see Fig. 1), effective connectivity for large-scale resting-state networks, for example, DMN, SN, and CEN, in both healthy and randomly selected participant groups did exhibit variability based on the selected parameters, given that the null model yielded no evidence. Furthermore, the level of hemoglobin has notably appeared in all the models that yielded significant posterior probabilities, meaning, carried explanatory power to the data. The BOLD signal forms the basis of the fMRI technique, given it serves as a proxy for neuronal activity [43]. The changes in blood oxygenation, in which hemoglobin is directly involved, are captured by

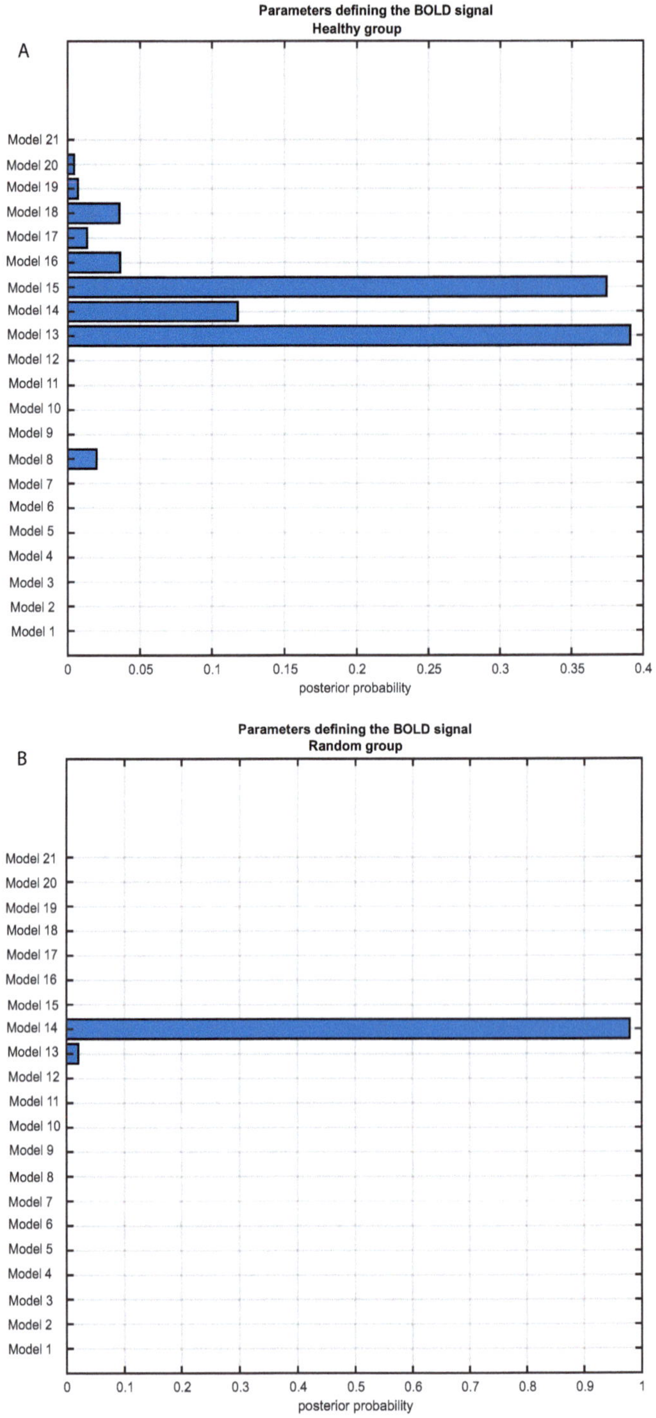

Fig. 2 Bayesian Model Comparison for the hemodynamic parameters for both samples, healthy sample (**a**) and random sample (**b**)

the MRI, oxygen molecule binding to hemoglobin as transporter forming oxy-hemoglobin, where the rate of oxy-hemoglobin and deoxy-hemoglobin in areas of presumed neuronal activation differs [43]. Hence, it is not surprising that the levels of hemoglobin appear to be a vital component when accounting for data variability.

More specifically, for the healthy group EC analyses, the model with the highest fit to the data was model 17: sex, age, and hemoglobin levels. For the random group, model no. 12: waist circumference, systolic and diastolic blood pressure, and hemoglobin levels exhibited the highest model evidence. These results are in line with previously reported results on age and gender differences in functional connectivity [28, 30], as well as findings on BMI and blood pressure effects in resting-state fMRI [5, 36, 44]. It is, however, important to note that the latter is only true for the random group. Even though the previous publication used BMI not the waist circumference measures, it has been suggested that the correlation between the two measures is high [45]. A recent publication from Ward and colleagues on the elderly population suggested that even in the absence of cognitive differences between the low and high oxy-hemoglobin groups, the connectivity measures exhibited differences [46]. This stresses that despite cognitive symptomatology, the hemoglobin levels are essential for reliable between-group comparison in functional and effective connectivity studies. The likely reason for the healthy group variability explained by the model with hemoglobin is that even though a recommended cut-off was used (120 and above, note that two individuals in the sample had lower hemoglobin levels: 111 and 119), there was substantial variability of hemoglobin levels between the healthy participants. For the random group, we observe that all of the selected health-related measures explain the data. These findings are in agreement with a recent publication where the noise influence on the resting-state data introduced by respiration, blood pressure, and arterial CO_2 concentration [47] was explored. The authors suggest that blood pressure rather than respiration may introduce greater variability in data, [47]. For randomly selected groups, even though inclusive of only "healthy individuals" as per the common definition in fMRI studies, "no preexisting neuropsychological disorders or cardiovascular diseases" indicates that blood pressure, waist circumference, and hemoglobin levels contribute to explaining the generated effective connectivity. Given the outlined findings, it is plausible to suggest that controlling for measures such as blood pressure measurements and waist circumference should come in addition to controlling for gender and age in healthy participant groups in resting-state fMRI studies.

The conditional model probabilities under flat priors for the BOLD signal for the selected healthy participant group indicate that both model 13 and model 15 fit the generated data best.

Specifically, model 13: sex, age, waist circumference, systolic and diastolic blood pressure, and level of hemoglobin and model 15: sex, age, and waist circumference. These findings indicate that the additional health variables are essential to consider when analyzing fMRI data. It has been previously urged that both the neurovascular and neuronal signals are contributing to the BOLD signal captured by fMRI [48]. Tsvetanov and colleagues [48] suggest that the plausible misinterpretation of fMRI results might stem from the lack of understanding of underlying mechanisms in neurovascular coupling and neuronal firing. In addition to the latter, the current results do support previously reported findings on gender [29], age [30], blood pressure, and BMI [36] contribution to the variability in resting-state fMRI findings.

The model that represented the data best for the parameters defining the BOLD signal for the randomly selected groups was model 14: sex, age, waist circumference, diastolic and systolic, levels of hemoglobin, and time of image acquisition. An interesting difference between the healthy and randomly selected participant groups for the BOLD signal parameters was that for the random group time of image acquisition seemed to be the contributing factor in terms of conditional model probabilities under flat priors. The time of image acquisition variability that affected only the parameters that define the BOLD signal, but not the effective connectivity, has been previously shown in Vaisvilaite and colleagues [5]. The fact that the time of image acquisition has been part of the best-fit model for the random group, but not healthy, might reflect the skewed distribution of the randomly selected group in terms of the time of image acquisition. For participants of the healthy group the image acquisition time was relatively equally distributed throughout the day, from 7:30 to 18:00, specifically, 7:30–9.30 $N = 14$; 12:00–13:30 $N = 13$; 14:00–18:00 $N = 12$. For the participants of the random group the image acquisition time was skewed to the first part of the day, from 7:30 to 18:00, specifically, 7:30–9.30 $N = 23$; 11:15 $N = 2$; 12:00–13:30 $N = 7$; 14:00–18:00 $N = 7$.

There are several limitations to the current work that should be accounted for in future studies. The current results indicated that for the healthy group parameters that define the BOLD signal there were two competing models that fit the data. Therefore, it is likely that it might indicate a factor that is not part of the generated models. Furthermore, the number of participant groups, for example, "healthy" and "random" was limited by the overall participant pool. We suggest that future studies could investigate several distinct factor groups to gain a better understanding of how the various components affect the effective connectivity and parameters that define the BOLD signal, for example, comparing overweight, high blood pressure, and anemic groups with healthy participants. Given that data was provided by the BETULA study on aging and cognition, the age of the current sample was broadly distributed. Ideally, the participants should be age and other variable-matched between groups in order to narrow down single or several factor/factors that may introduce the most variability in the data.

5 Conclusions

In summary, the findings suggest that the levels of hemoglobin overall were the factor that all the winning models had in common. With this in mind, we urge that it is possibly the factor that may contribute to fMRI data variability most significantly and should be accounted for, especially when conducting test-retest or clinical comparison studies. Furthermore, in terms of effective connectivity, the results suggest that controlling for blood pressure, waist circumference, time of data acquisition, gender, and age is beneficial when collecting resting-state fMRI data for group comparisons.

Acknowledgments

The study was financed by the Research Council of Norway (Project no. 276044: When Default is not Default: Solutions to the replication crisis and beyond). The data were provided by the Betula prospective cohort study, Umeå University, Sweden. The Betula Project is supported by Knut and Alice Wallenberg Foundation (KAW) and the Swedish Research Council (K2010-61X-21446-01).

References

1. Adolphs R (2015) The unsolved problems of neuroscience. Trends Cogn Sci 19:173–175
2. Wang M-Y, Yuan Z (2021) EEG decoding of dynamic facial expressions of emotion: evidence from SSVEP and causal cortical network dynamics. Neuroscience 459:50–58
3. Biasiucci A, Franceschiello B, Murray MM (2019) Electroencephalography. Curr Biol 29: R80–R85
4. da Silva FL (2013) EEG and MEG: relevance to neuroscience. Neuron 80:1112–1128
5. Vaisvilaite L, Hushagen V, Grønli J, Specht K (2021) Time-of-day effects in resting-state functional magnetic resonance imaging: changes in effective connectivity and blood oxygenation level dependent signal. Brain Connect 12:515. https://doi.org/10.1089/brain.2021.0129
6. Kazimierczak K, Craven AR, Ersland L et al (2021) Combined fMRI region- and network-analysis reveal new insights of top-down modulation of bottom-up processes in auditory laterality. Front Behav Neurosci 15: 802319
7. Laeng B, Flaaten CB, Walle KM et al (2021) "Mickey Mousing" in the brain: motion-sound synesthesia and the subcortical substrate of audio-visual integration. Front Hum Neurosci 15:605166
8. Poldrack RA, Farah MJ (2015) Progress and challenges in probing the human brain. Nature 526:371–379
9. Ogawa S, Tank DW, Menon R et al (1992) Intrinsic signal changes accompanying sensory stimulation: functional brain mapping with magnetic resonance imaging. Proc Natl Acad Sci U S A 89:5951–5955
10. Kwong KK, Belliveau JW, Chesler DA et al (1992) Dynamic magnetic resonance imaging of human brain activity during primary sensory

stimulation. Proc Natl Acad Sci U S A 89: 5675–5679
11. Clark VP (2012) A history of randomized task designs in fMRI. NeuroImage 62:1190–1194
12. Liu TT (2012) The development of event-related fMRI designs. NeuroImage 62:1157–1162
13. Biswal B, Yetkin FZ, Haughton VM, Hyde JS (1995) Functional connectivity in the motor cortex of resting human brain using echo-planar MRI. Magn Reson Med 34:537–541
14. Sharda M, Silani G, Specht K et al (2019) Music therapy for children with autism: investigating social behaviour through music. Lancet Child Adolesc Health 3:759–761
15. Smith SM, Fox PT, Miller KL et al (2009) Correspondence of the brain's functional architecture during activation and rest. Proc Natl Acad Sci U S A 106:13040–13045
16. Fox MD, Raichle ME (2007) Spontaneous fluctuations in brain activity observed with functional magnetic resonance imaging. Nat Rev Neurosci 8:700–711
17. Damoiseaux JS, Rombouts SARB, Barkhof F et al (2006) Consistent resting-state networks across healthy subjects. Proc Natl Acad Sci U S A 103:13848–13853
18. Raichle ME (2015) The brain's default mode network. Annu Rev Neurosci 38:433–447
19. Raichle ME, MacLeod AM, Snyder AZ et al (2001) A default mode of brain function. Proc Natl Acad Sci U S A 98:676–682
20. Uddin LQ (2015) Salience processing and insular cortical function and dysfunction. Nat Rev Neurosci 16:55–61
21. Seeley WW, Menon V, Schatzberg AF et al (2007) Dissociable intrinsic connectivity networks for salience processing and executive control. J Neurosci 27:2349–2356
22. Noble S, Scheinost D, Constable RT (2019) A decade of test-retest reliability of functional connectivity: a systematic review and meta-analysis. NeuroImage 203:116157
23. Bennett CM, Miller MB (2010) How reliable are the results from functional magnetic resonance imaging? Ann N Y Acad Sci 1191: 133–155
24. Botvinik-Nezer R, Holzmeister F, Camerer CF et al (2020) Variability in the analysis of a single neuroimaging dataset by many teams. Nature 582:84–88
25. Rosen BR, Savoy RL (2012) fMRI at 20: has it changed the world? NeuroImage 62:1316–1324
26. Specht K (2019) Current challenges in translational and clinical fMRI and future directions. Front Psych 10:924
27. Tian F, Wang J, Xu C et al (2018) Corrigendum: focusing on the differences of resting-state brain networks, using a data-driven approach to explore the functional neuroimaging characteristics of extraversion trait. Front Neurosci 12:380
28. Dubois J, Galdi P, Han Y et al (2018) Resting-state functional brain connectivity best predicts the personality dimension of openness to experience. Personal Neurosci 1:E6
29. Hjelmervik H, Hausmann M, Osnes B et al (2014) Resting states are resting traits--an FMRI study of sex differences and menstrual cycle effects in resting state cognitive control networks. PLoS One 9:e103492
30. Sie J-H, Chen Y-H, Shiau Y-H, Chu W-C (2019) Gender- and age-specific differences in resting-state functional connectivity of the central autonomic network in adulthood. Front Hum Neurosci 13:369
31. Baird A (2019) Sex in the brain: how seizures, strokes, dementia, tumors, and trauma can change your sex life. Columbia University Press, New York Chichester
32. Tooley UA, Mackey AP, Ciric R et al (2021) Erratum to: associations between neighborhood SES and functional brain network development. Cereb Cortex 31:2307
33. Bellucci G, Hahn T, Deshpande G, Krueger F (2019) Functional connectivity of specific resting-state networks predicts trust and reciprocity in the trust game. Cogn Affect Behav Neurosci 19:165–176
34. Cole DM, Smith SM, Beckmann CF (2010) Advances and pitfalls in the analysis and interpretation of resting-state FMRI data. Front Syst Neurosci 4:8
35. Dubois J, Adolphs R (2016) Building a science of individual differences from fMRI. Trends Cogn Sci 20:425–443
36. Sjuls GS, Specht K (2021) Why blood pressure and body mass should be controlled for in resting-state functional magnetic resonance imaging studies. bioRxiv. https://doi.org/10.1101/2021.06.02.446721
37. Wen Z, Hammoud MZ, Scott JC et al (2021) Impact of exogenous estradiol on task-based and resting-state neural signature during and after fear extinction in healthy women. Neuropsychopharmacology 46:2278–2287
38. Zeidman P, Jafarian A, Corbin N et al (2019) A guide to group effective connectivity analysis,

part 1: first level analysis with DCM for fMRI. NeuroImage 200:174–190

39. Friston KJ (2011) Functional and effective connectivity: a review. Brain Connect 1:13–36
40. Nilsson L-G, Bäckman L, Erngrund K et al (1997) The Betula prospective cohort study: memory, health, and aging. Aging Neuropsychol Cognit 4:1–32
41. Salami A, Wåhlin A, Kaboodvand N et al (2016) Longitudinal evidence for dissociation of anterior and posterior MTL resting-state connectivity in aging: links to perfusion and memory. Cereb Cortex 26:3953–3963
42. Lanka P, Deshpande G (2019) Combining Prospective Acquisition CorrEction (PACE) with retrospective correction to reduce motion artifacts in resting state fMRI data. Brain Behav 9:e01341
43. Buxton RB (2013) The physics of functional magnetic resonance imaging (fMRI). Rep Prog Phys 76:096601
44. Park B-Y, Park H, Morys F et al (2021) Interindividual body mass variations relate to fractionated functional brain hierarchies. Commun Biol 4:735
45. Wilmet G, Verlinde R, Vandevoorde J et al (2017) Correlation between Body Mass Index and abdominal circumference in Belgian adults: a cross-sectional study. Rom J Intern Med 55: 28–35
46. Ward PGD, Orchard ER, Oldham S et al (2020) Individual differences in haemoglobin concentration influence bold fMRI functional connectivity and its correlation with cognition. NeuroImage 221:117196
47. Murphy K, Birn RM, Bandettini PA (2013) Resting-state fMRI confounds and cleanup. NeuroImage 80:349–359
48. Tsvetanov KA, Henson RNA, Rowe JB (2021) Separating vascular and neuronal effects of age on fMRI BOLD signals. Philos Trans R Soc Lond Ser B Biol Sci 376:20190631

Chapter 14

Utility of Computational Approaches for Precision Psychiatry: Applications to Substance Use Disorders

Jasmin Vassileva, Jeung-Hyun Lee, Elena Psederska, and Woo-Young Ahn

Abstract

Revolutionary advances in neuroscience and genetics over the past two decades have provided unprecedented opportunities for increasing our understanding of the etiology and pathogenesis of psychiatric disorders. Despite these advances, the translation of this knowledge into clinical practice has been hindered by the significant heterogeneity within disorders and the neurobiologically imprecise categorization of patients. Traditional diagnostic categories do not capture the underlying neurobiological processes and etiological mechanisms of psychiatric disorders and cannot adequately inform prognosis and treatment. To address this gap, the National Institute on Mental Health (NIMH) has proposed an alternative research framework based on Research Domain Criteria (RDoC), which yields psychiatric classification grounded on discoveries from neuroscience and genomics. Recently, the RDoC approach has been adapted for the study of addictions with the Addictions Neuroclinical Assessment (ANA) framework (Kwako et al., Biol Psychiatry 80:179–189, 2016), which proposes that the assessment of addictions should cover multiple systems and focus on three key neurofunctional domains: *Executive Function, Incentive Salience,* and *Negative Emotionality.* Building upon the aims of the RDoC framework, a new field of "precision psychiatry" has emerged, considered to be a paradigm shift in psychiatry. Both the RDoC framework and precision psychiatry are predicated on precise measurement, which has necessitated the development of novel analytic approaches for classification and novel dimensional tools for phenotyping that can identify the unique mechanisms of psychiatric disorders at an individual level.

In this chapter, we make the case that theory-driven and data-driven computational approaches have enormous potential for increasing the precision and reliability of measurement, and the accuracy of diagnosis and prognosis in psychiatry. We review three types of computational approaches and their utility for precision psychiatry: (1) *Theory-driven approaches,* such as *computational modeling*; (2) *Data-driven approaches,* using various *machine learning* methods; and (3) *Hybrid approaches*, such as *joint modeling* and *adaptive design optimization.* We focus more narrowly on the application of these approaches to substance use disorders (SUD), where we attempt to map them on the current RDoC framework for addictions.

Key words Computational modelling, Machine learning, Precision psychiatry, Addictions

Drozdstoy Stoyanov et al. (eds.), *Computational Neuroscience*, Neuromethods, vol. 199,
https://doi.org/10.1007/978-1-0716-3230-7_14,

1 Introduction

Revolutionary advances in neuroscience and genetics over the past two decades have provided unprecedented opportunities for increasing our understanding of the etiology and pathogenesis of psychiatric disorders. Despite these advances, the translation of this knowledge into clinical practice has been hindered by the significant heterogeneity within disorders and the neurobiologically imprecise categorization of patients. It is now well known that just as multiple etiological pathways may lead to the same clinical presentation, clinically distinct diagnostic categories may also be related to the same underlying transdiagnostic mechanisms. It has become increasingly apparent that traditional diagnostic categories do not capture the underlying neurobiological processes and etiological mechanisms of psychiatric disorders and cannot adequately inform prognosis and treatment. To address the gap between etiology and nosology and to improve treatment outcomes, the National Institute on Mental Health (NIMH) has proposed an alternative research framework based on Research Domain Criteria (RDoC), which yields psychiatric classification grounded on discoveries from neuroscience and genomics as a complement to the existing classification system [2]. According to the RDoC framework, psychiatric disorders should be investigated at several interacting levels of analysis, from genes, to molecules, cells, brain circuits, cognition, behavior, and environment, which has led to a progressive transition from categorical to dimensional approaches to measurement and classification.

More recently, the RDoC approach has been adapted for the study of addictions with the Addictions Neuroclinical Assessment framework (ANA; [1]) which maps on three recurrent stages of addiction: *binge/intoxication, withdrawal/negative affect, and preoccupation/anticipation*, each associated with different neurocircuitry and functional domains [3, 4]. The ANA proposes that the assessment of addictions should cover multiple systems and focus on three key neurofunctional domains: (1) *executive function (EF)*, associated with reduced prefrontal cortex (PFC)-mediated top-down impulse control characterizing the *preoccupation/anticipation* stage of the addiction cycle, commonly associated with craving and relapse; (2) *incentive salience (IS)*, associated with phasic reward-based dopaminergic activation in the basal ganglia and the *binge/intoxication* stage of addiction; and (3) *negative emotionality (NE)*, associated with engagement of brain stress systems and the *withdrawal/negative affect* stage of addiction, characterizing periods of abstinence. Though this framework has been applied primarily to alcohol use disorder [5], more recently it has been extended to other types of substance use disorders (SUD), including opioid use disorder, cocaine use disorder, and cannabis use disorder [6, 7]. It has been employed predominantly in the context of assessment and treatment of SUD [8], but has recently been proposed for SUD prevention [9]. A nontrivial practical limitation of the RDoC approach is that the multi-dimensional assessment that it requires entails administering lengthy assessment batteries, which may take up to 10 h of testing [1]. This is a significant rate- and cost-limiting factor, which prevents the wider implementation of the RDoC approach in clinical research and practice. Another practical limitation is that current methods for biomarker discovery such as neuroimaging and various-omics approaches are costly, invasive, and not suitable for clinical practice. Neurobehavioral assessments can overcome some of these limitations as they are noninvasive and relatively inexpensive; however, they seem to have stagnated, with many clinical neuropsychological tests developed decades ago still in use [10]. They also rely on crude summary statistics that are not particularly informative about the underlying neurocognitive processes and are minimally sensitive to individual differences, a prerequisite if a test is to have good diagnostic and predictive utility. The behavioral metrics of these tasks are most often atheoretical with respect to the underlying mechanisms and their neurocircuit signatures are poorly understood because the tasks are designed to measure broad cognitive functions (e.g., "executive functions") rather than specific neurocognitive processes. Of particular concern is the low replicability of neuropsychological and neuroimaging findings [11–13] and the surprisingly low test-retest reliability of even the most well-established neurocognitive tasks [14], where effects are reliable when measuring the behavior of groups of individuals but not when examining how an individual performs across repeated assessments [15]. This has led to a "crisis of confidence" in psychological science [13], which has made some question the existence of key psychological constructs central to addiction, such as impulsivity [16]. Critically, the "reliability paradox" [14] suggests that many fMRI-based biomarkers that use such neurocognitive tasks may also be unreliable [17–19] and that well-established approaches in cognitive psychology and neuropsychology may not directly translate to the study of individual differences in brain structure and function (but see [20, 21]). The lack of reliable, precise, and efficient neurobehavioral measures is therefore one of the most formidable challenges in measuring RDoC constructs. Clearly, novel alternate metrics are needed to measure complex behavior and neurofunctional domains.

Building upon the aims of the RDoC framework to map clinical observations on neurobiological mechanisms, a new field of "precision psychiatry" has emerged over the past few years, considered to be a paradigm shift in the field of psychiatry [22, 23]. It integrates advances in neuroscience and technology into a computational framework, with the goal to develop personalized therapeutic approaches tailored to the specific characteristics of each individual [23]. The RDoC framework and precision psychiatry are both predicated on precise measurement, which has necessitated the development of new analytic approaches for classification and novel dimensional tools for phenotyping that can identify the unique mechanisms of psychiatric disease at an individual level. Dimensional approaches have helped identify key trans-disease processes such as delay discounting, considered key diagnostic and prognostic biomarkers of addiction [24, 25], and other types of reinforcement pathology [26]. Such approaches have helped address a critical challenge in the treatment of addiction: the significant heterogeneity within addiction to a specific drug, and the similarity of underlying processes across addictions to different drugs.

In this chapter, we make the case that theory-driven and data-driven computational approaches have enormous potential for increasing the precision and reliability of measurement, and the accuracy of diagnosis and prognosis in psychiatry. We review three types of computational approaches and their utility for precision psychiatry: (1) *theory-driven approaches,* such as *computational modeling*, which have shown utility as novel phenotyping tools that may increase the precision of neurocognitive phenotyping, refine diagnosis, and aid intervention selection; (2) *hybrid approaches*, such as *joint modeling* and *adaptive design optimization*, which increase the efficiency and reliability of neurocognitive phenotyping; and (3) *data-driven approaches* using various *machine learning* methods that are particularly useful for predictive modeling, stratification, classification, and biotyping in psychiatry. We focus more narrowly on the application of these approaches to substance use disorders (SUD), where we attempt to map them on current RDoC frameworks for addictions. We argue that increasing the precision of clinical phenotyping by integrating genetically informed personality, neurocognitive, and neuroimaging approaches within a computational framework will be critical for identifying etiological markers of different biotypes of addiction that could be targeted by modular combinations of behavioral, neurostimulatory, and pharmacological interventions, personalized to individual multivariate computational profiles.

2 Theory-Driven Approaches: Computational Modeling and Computational Phenotyping

The human brain has long been considered the archetype of computation [27], as its key function is to compute by storing and summarizing information and using that information to make predictions about the future [28, 29]. The past 10 years have witnessed the emergence of the discipline of computational psychiatry [30], which seeks to characterize mental dysfunction in terms of aberrant computations [31]. Theory-driven computational approaches such as computational modeling have been proposed to provide a new paradigm for understanding psychopathology [28], which can help address the "explanatory gap" and lack of suitable levels of description that link findings at the molecular level to clinical entities, such as addictions and other psychiatric disorders [31, 32].

In the field of substance use disorders, computational modeling has been applied primarily to the study of decision-making, as impulsive and maladaptive decision-making is considered one of the core neurocognitive deficits of individuals with SUD and other addictive disorders [33, 34]. Indeed, many of the diagnostic criteria for SUD could be considered directly or indirectly related to abnormalities in decision-making (e.g., persistent drug use despite negative consequences, consuming larger amounts and for longer period of time than intended, persistent desire, and sense of compulsion to take the substance). In real life, individuals with SUD show profound impairments in judgment and decision-making, characterized by a tendency to choose immediate rewards, at the expense of often devastating negative consequences in the future. Such real-life impairments are typically measured with neurocognitive tasks that mimic major life contingencies in a realistic manner, such as gambling and discounting tasks that involve different reward and punishment contingencies [34, 35]. Abnormally steep delay discounting rates, indicating preference for immediate but smaller rewards, have been reliably associated with both quantity-frequency of use and with severity of SUD [35], including alcohol [36], nicotine [37], heroin [38–40], and cocaine use disorders [38, 41]. The neural systems probed by delay discounting paradigms are also well-known; therefore, delay discounting is a promising candidate for the RDoC approach [42, 43].

It is important to study decision-making with different decision tasks, to obtain converging evidence about the cognitive and affective mechanisms underlying decision-making deficits [44]. Of the various decision tasks used in the literature, the Iowa Gambling Task (IGT) [33, 45] is one of the oldest, originally developed in the early 1990s as an attempt to capture the prominent difficulties in day-to-day functioning displayed by patients with lesions of the ventromedial prefrontal cortex (vmPFC), who otherwise showed no demonstrable deficits on standard intellectual and neuropsychological tasks. It was designed to simulate real-life decision-making, defined as the ability to select the most advantageous course of action from a set of possible alternative behaviors, where decision makers learn by trial and error to choose among four decks of cards that produce both wins and losses. Impaired performance on the task is taken as an indicator of insensitivity to future consequences or "myopia for the future" [45]. The task has become one of the most widely used decision tasks in the addiction literature and is one of the earliest for which computational cognitive models were developed. The first cognitive model for the IGT was the Expectancy Valence Learning (EVL) model, developed by Busemeyer and Stout [46]. Since then, a number of additional models have been developed to better capture the behavioral patterns of the task, such as the Prospect Valence Learning (PVL) model [47], the Value-Plus-Perseverance (VPP) model [48], and most recently the Outcome-Representation Learning (ORL) model [49]. Though individuals with SUD are consistently impaired on the task [33, 50–58], it has been difficult to discern differences in decision-making between people with different types of SUD or with different comorbid disorders based solely on the standard performance indices, because even though the task has high sensitivity to decision-making impairments in individuals with SUD, it has proven equally sensitive to capturing decision-making impairments in patients with other externalizing disorders such as antisocial personality disorder, psychopathy, and ADHD [57, 59–61].

Though gambling and discounting tasks are ecologically valid and capture important aspects of real-life functioning, they are designed to be complex and involve numerous motivational, learning, and choice processes [44, 46]. Similar to the heterogeneity of substance use and other psychiatric disorders, there is an inherent heterogeneity in the computational processes involved in such tasks, reflecting different mechanisms underlying decision-making, such as sensitivity to reward, sensitivity to loss, risk aversion, risk tolerance, ambiguity tolerance, and exploration/exploitation, among others. Consequently, impaired performance on the tasks may have many different causes [44]. Similar to how the current diagnostic classification system does not capture the numerous etiological mechanisms underlying psychiatric disorders, traditional neurobehavioral performance indices on these tasks do not capture the different underlying causes of impaired performance. Computational modeling of such cognitively complex tasks have proven much more informative in this regard, as they deconstruct neurobehavioral performance into underlying latent processes, and use the parameter estimates of these processes to understand the specific mechanisms underlying the neurocognitive deficits manifested by different clinical populations [46, 50, 62].

Findings in the addiction literature consistently reveal that computational model parameter estimates of different psychological processes involved in decision-making are more sensitive to dissociating substance-specific and disorder-specific neurocognitive profiles than standard neurobehavioral performance indices across a variety of decision tasks [49, 50, 62–67]. For example, computational model parameters of the IGT robustly discriminate between opiate and stimulant-dependent individuals even in protracted abstinence [49, 50]. Despite no group differences on the traditional performance index on the task (net score), computational modeling has uncovered notable differences in the underlying processes driving the decision-making performance of different types of substance users: reduced sensitivity to loss in opiate users [49, 50] vs increased sensitivity to reward [50] and preference for switching selections (exploration) in stimulant users [49].

Unlike the IGT, which measures decision-making under uncertainty and ambiguity and involves learning by trial and error, the Cambridge Gambling Task (CGT) is a probabilistic task measuring decision-making and risk-taking outside of a learning context, where no uncertainty is involved [68]. Though not as extensively studied in the addiction field as the IGT, findings are consistent and reveal decision-making impairments in different types of substance users [60, 66, 68–71]. Romeu et al. [66] recently developed the first computational model for the task, which, similar to findings with the IGT, revealed differences in decision-making between healthy controls and individuals with different types of SUD ("pure" heroin, "pure" amphetamine, polysubstance) that were not observable with traditional metrics. All three types of substance users were characterized by lower sensitivity to loss and higher delay aversion than controls, though mono-substance dependent (i.e., "pure") heroin and amphetamine users were more sensitive to loss than polysubstance users. In addition, pure amphetamine users showed lower probability distortion than pure heroin users and controls, reflecting greater willingness to make less optimal choices. These findings were recently replicated by Todesco et al. [72] who revealed lower sensitivity to loss and lower probability distortion in polysubstance users relative to controls. Computational modeling of another decision task, the Balloon Analogue Risk Task (BART; [73]) measuring risky decision-making, revealed similarly reduced loss aversion and increased risk preference in heroin users [65]. The application of computational models may therefore elucidate the unique effects of various features of addictive disorders, such as the influence of specific drug classes or comorbid psychopathology. This may help identify neurocomputational biotypes of individuals with SUD, characterized by unique and computationally distinct decision-making deficits and accompanying neural circuit abnormalities. Such parameters of gain- and loss-related sensitivity show significant potential as novel "computational signatures" for different types of SUD and other forms of psychopathology, which could help refine neurocognitive addiction phenotypes, identify computational biomarkers, and develop more rigorous models of addiction.

Computational phenotypes have already led to many new insights into the neurobehavioral mechanisms underlying substance use and addictive disorders, but their clinical utility has only recently started to be considered [30]. To date, theory-driven computational approaches have demonstrated clinical utility primarily via back translation from "clinic-to-computation," by demonstrating how specific disorders map onto specific computational processes [74]. In contrast, forward translation, from computation to clinic, is still rare [74] and is the next frontier in computational research. Clinically, theory-driven computational phenotypes hold promise for improving treatment success by providing novel actionable targets for prevention and intervention and increasing the precision and efficacy of treatment interventions for addictive disorders [8]. For example, delay discounting has been successfully targeted by novel interventions such as episodic future thinking, which have resulted in reductions not only in delay discounting but also in substance use [75–77]. The utility of computational modeling to precision psychiatry has recently been shown in a treatment context, where computational methods were able to capture treatment-sensitive aspects of decision-making, such as changes in loss sensitivity that were not accessible via traditional methods [72]. Further, dynamic changes in specific computational parameters of decision-making over time, such as daily fluctuations in ambiguity tolerance and risk preference, have been shown to predict imminent relapse in abstinent opioid-dependent individuals [63]. This suggests that computational parameters may have prognostic and diagnostic utility to inform not only with whom to intervene, but also when to intervene. Such computational signatures and within-person fluctuations in computational parameters could provide a dynamic characterization of different addiction trajectories and transitions between different stages of addiction [78].

2.1 Joint Modeling

Novel computational modeling approaches such as "joint modeling" aim to link behavior across different tasks and measurement modalities [64, 79]. In many cases, multiple neurocognitive tasks and cognitive models purport to describe similar processes, but it is difficult to evaluate whether they measure the same latent processes or traits. To address this question, recent studies have modeled behavior across decision tasks by connecting cognitive model parameters from different tasks to common latent constructs, such as impulsivity. For example, a recent joint modeling study of the CGT and the Monetary Choice Questionnaire (MCQ) of delay discounting [64] revealed that the tasks appear to index separate neurocognitive dimensions of impulsivity, with the MCQ indexing choice impulsivity [80], whereas the CGT tapped on action impulsivity [81]. Further, the temporal discounting parameter on the delay-discounting task (MCQ) was more closely related to trait measures of externalizing psychopathology and aggression, whereas temporal discounting on the CGT was related to neurobehavioral response inhibition failures [64]. A key feature of the joint modeling approach is that it allows linking neurobehavioral data directly to neural activity [79, 82–84]; therefore, joint modeling holds promise for linking behaviors not only across different tasks, but also across different domains of functioning (e.g., neurocognitive, neurocircuitry) [79]. Computational models have also been used to guide model-based neuroimaging approaches and by providing a framework to study neural mechanisms of various cognitive processes, show distinct advantages, and offer insights into *how* a particular process is implemented in the brain as opposed to merely identifying *where* the process is located [85, 86]. In general, joint models exhibit greater predictive validity than neural or behavioral data alone, allowing more precise and mechanistically informative characterization of neurocognitive profiles and neural function associated with different types of SUD and psychiatric disorders [79].

Within the framework of the three stages of addiction [3], computational modeling parameters that parse different neurocognitive functions such as decision-making, could increase understanding of the role of different decision-making processes at different stages of the addiction cycle and mapping these processes to the ANA domains. For instance, *increased sensitivity to reward* may most closely characterize the *binge/intoxication* stage of addiction and the *incentive salience* ANA domain, driven by positive reinforcement mechanisms involving preferentially the dopaminergic and opioid systems. In contrast, *reduced sensitivity to loss (or reduced loss aversion)* may better characterize the *withdrawal/negative affect stage* of addiction and the *negative emotionality* ANA domain, driven by negative reinforcement mechanisms that preferentially engage the extended amygdala and its projections. The *executive function* ANA domain and the *preoccupation/anticipation stage* of addiction may be characterized by an imbalance between model-based and model-free decision-making systems [87] and a conflict between Pavlovian and instrumental systems [88]. Using computational approaches to characterize and map the ANA domains promises to identify novel actionable targets for prevention and treatment of SUD that may supplement existing programs and inform the development of new programs [8, 89–91]. The application of computational models may also elucidate the specific effects of various features of addictive disorders, such as the influence of specific drug classes or comorbid externalizing and internalizing psychopathology, thereby leading to identification of subtypes of individuals with SUD, characterized by specific types of cognitive and affective deficits. This could further increase the validity of the tasks as sensitive and specific measures of distinct "computational signatures," which could be targeted by interventions tailored to the specific type of neurocomputational risk profile.

3 Hybrid Approaches/Adaptive Design Optimization

While computational tools have increased the knowledge extracted from neurocognitive tasks, there are surprisingly few high-quality assays for monitoring and characterizing neurocognitive domains. One of the major problems in identifying reliable biomarkers of addiction and other psychiatric disorders is the low reproducibility of neurocognitive findings and the surprisingly low test-retest reliability of well-established and widely used neuropsychological tasks. Computational modeling holds promise for addressing the "Reliability Paradox," or the failure of robust cognitive paradigms to produce reliable individual differences [14]. For example, a recent study that compared traditional neurobehavioral indices from working memory, priming, associative interference, and impulsivity tasks against computational models of these tasks revealed that computational model parameters show substantially better test-retest reliability than the standard behavioral indices, increasing reliability by as much as 0.8 on a −1 to 1 scale [15]. Advances in Bayesian statistics and machine learning offer algorithm-based ways to generate optimal and efficient experimental designs so as to minimize uninformative and wasted experimental trials [92]. Bayesian computational approaches can improve not only the reliability but also the efficiency of neurocognitive assessment and help develop RDoC measures that provide more rapid and precise behavioral markers of different types of SUDs. One such approach is *adaptive design optimization (ADO)* [93], which aims to find the most informative design for estimating model parameters on the fly during an experiment. ADO is a "smart" search machine learning algorithm, whose search is guided by one or more computational models, depending on the objectives of the research. When comparing competing models (e.g., of decision-making), it searches for the stimulus that is most likely to discriminate the models. When its goal is to estimate model parameters, it presents the stimulus that is expected to generate the most informative response for parameter estimation. The model(s), combined with participants' responses, are updated at each trial to optimize stimulus selection with the goal of achieving efficiency, precision, and reliability. This approach has a promising track record for improving the efficiency and precision of psychiatric assessment [94–96]. ADO is a model-based machine-learning approach to optimization in the sense that it requires a quantitative model that predicts experimental outcomes based on the model's parameters and design variables. ADO has been successfully applied for identifying best-fitting models in gambling tasks [97] and delay discounting tasks [98], as well as to optimally assess visual acuity [96] in neurotypical individuals. All of these studies demonstrate that ADO substantially reduced the number of trials required to do model comparisons or parameter estimation. This indicates that ADO may significantly increase task efficiency, reduce the length and burden of administration of current RDoC assessment batteries, and facilitate their wider implementation in clinical practice.

ADO has also been shown to dramatically increase the test-retest reliability of common tasks compared to non-ADO methods. For example, a recent ADO study [99] revealed 0.95 and higher test-retest reliability of the discounting rate within only 10–20 trials (under 1–2 min of testing), which captured approximately 10% more variance in test-retest reliability, was 3–5 times more precise, and 3–8 times more efficient than the staircase method. Of note, ADO shows excellent reliability in different populations, including college students, patients with SUDs, and online Amazon MTurk workers [99]. Critically, ADO task parameters demonstrate linkages with real-world substance use outcomes akin to computational parameters from longer and more burdensome tasks. For example, preliminary findings suggest that model parameters from ADO-based tasks could predict future cigarette use, which further suggests that ADO is a promising approach for assessing and predicting addictive and other psychiatric conditions [100].

4 Data-Driven Approaches/Machine Learning

The scientific community is always looking for well-powered and unbiased methods for identifying features of interest. Combining neurocomputational signatures with clinical, behavioral, neuroimaging, genetic, and other types of data in large multivariate datasets promises to increase the interpretability of neurocognitive phenotyping and increase the precision of prediction and classification in psychiatry [101]. Rapid improvements in computational resources and the quality of big data nowadays allows combining multiple sources of data in large datasets and freely sharing the data and tools with the scientific community. This has led to rapid development of data-driven computational methods using various machine learning (ML) approaches that have increased our understanding of the multi-dimensional features and associated neural substrates and genetic underpinnings of different forms of psychopathology. Machine-learning has found steadily increasing applications in the addiction literature. Supervised ML methods have been used to predict adolescent alcohol use [102] and misuse [103], distinguish between smokers and non-smokers [104–106], between people with and without cocaine use disorder [107, 108] or cannabis use disorder [109–111], and between people with different types of

SUD [107, 112–116]. These ML studies have identified multivariate neurobiological, neurocognitive, psychiatric, and personality profiles that differentiate addictions to different classes of drugs. Some studies have identified common features for multiple SUD, emphasizing the trans-diagnostic utility of certain neurocognitive and personality characteristics that may increase vulnerability to addiction in general, regardless of drug class. For example, higher delay discounting [113] and impulsive/antisocial features of psychopathy [112] have emerged as significant trans-diagnosing markers classifying alcohol-, opiate-, and stimulant use disorders. In addition, delay discounting has been identified as the most prominent predictor of successful smoking cessation [117], underscoring its significant role in the recovery stage of the addiction cycle. On the other hand, substance-specific markers classifying addictions to different classes of drugs have also been identified. For example, Ahn and Vassileva [112] identified unique multivariate personality, psychiatric, and neurocognitive features that classified opiate and stimulant addictions with a high degree of accuracy. Amphetamine users were (uniquely) characterized by higher sensation-seeking, hostility, response deliberation time, and delay discounting. Heroin users were uniquely characterized by attention deficits, impaired decision-making, lower risk-taking, callous/unemotional features of psychopathy, impulsivity under negative emotional states ("negative urgency"), depression, anxiety, and aggression. Out of 54 features, the impulsive/antisocial factor of the Hare Psychopathy Checklist (PCL:SV) was the strongest and only common classification marker of both heroin and amphetamine dependence [112]. Others have used connectome-based modeling to identify substance-specific neural networks involved in abstinence from opiates and cocaine [115], and brain morphology to differentially predict alcohol, tobacco, and cannabis use initiation in adolescents [114], highlighting the importance of studying both common and unique markers of different types of SUD.

A few large multidisciplinary collaborations have produced large phenotypically rich datasets that have served as key drivers and accelerators of data-driven computational research. The IMAGEN consortium [118] is a multidisciplinary European collaboration in imaging genomics aiming to detect longitudinal associations between genotype and brain structure and function and disentangle gene–environment interactions. The Human Connectome Project [119, 120] is another large imaging genetics study that uses multimodal imaging technology to understand the network of human brain functions and map its neuroanatomical connectivity patterns. The Enhancing Neuroimaging Genetics through Meta-Analysis (ENIGMA) consortium [121] brings together researchers in imaging genomics, neurology, and psychiatry, and involves 30 working groups spanning 185 institutions in 35 countries worldwide to understand brain structure and function in different psychiatric and neurological disorders. The Adolescent Brain Cognitive Development[SM] (ABCD) Study [122–125] is the largest ($N = 11{,}875$) longitudinal study of brain development in the United States, examining risk and resilience factors associated with substance use and other psychiatric and physical outcomes from middle childhood to early adulthood. These research initiatives have been supplemented by government-sponsored big genomics projects such as the UK Biobank in the United Kingdom with over 500,000 participants [126] and All of Us in the United States created by President Obama's Precision Medicine Initiative [127], which aims to recruit over 1 million Americans. These large citizen-science projects have shed light on numerous psychological facets of cognitive, social, emotional, and physical development and have generated vast amounts of high-dimensional data that requires increasingly sophisticated methods for processing and analysis.

Unsupervised ML methods such as clustering have proven to be especially useful for data-driven disease subtyping and identification of biotypes. Biotypes are subtypes of a broader syndrome or disorder defined by distinct aggregations of behavioral, mood, and genetic markers with specific dysfunctions in the functional and structural connectivity of large-scale neural circuits that govern mood, behavior control, and self-reflective functions [128, 129]. Biotypes are increasingly proposed as an alternative to clinical phenomenology in the classification of disease because their organizing features center on neurobiological mechanisms rather than differentiations based on broad symptomatology. Though applied to other psychiatric disorders, such as anxiety, depression [23, 129, 130], schizophrenia, and bipolar disorder [128, 131–133], the biotype approach is still relatively unexplored in the study of addictions. Few studies to date have applied this approach to SUD and other addictive disorders, which is a significant gap in the literature. Zhu et al. [134] used connectivity features from resting state fMRI to identify three biotypes of alcohol misuse—mild, comorbid, and moderate—which demonstrated significant differences in alcohol use frequency and connectivity involving the frontal, parietal, subcortical, and default mode networks. To understand comorbidities between addictions and the transition/replacement of one addiction form with another, Zarate et al. [135] conducted a network analysis of 10 forms of addictive behaviors (alcohol, drugs, tobacco, sex, online gambling, Internet use, Internet gaming, social media use, shopping, and exercise). Findings suggest that most forms of addictive behaviors are uniquely different and that there are clusters of addiction symptoms (e.g., drug and alcohol misuse, gambling) that have particularly strong influence on the entire network of addictive behaviors. Developing a taxonomy of neurocomputational addiction biotypes based on the ANA framework that maps onto neural circuits involved in addictions may facilitate the translation of empirical data into clinical practice by informing the development of novel, innovative treatment alternatives that are individually tailored to specific subgroups of individuals who share common addiction vulnerabilities. The biotype approach has used neuroimaging, genetic, physiological, clinical, cognitive, environmental, and other sources of data for classification and prediction. This approach could benefit substantially from the inclusion of theoretically derived "computational signatures" of decision-making and other cognitive and affective functions implicated in addiction into the machine learning models along with other relevant sources of data to identify neurocomputational biotypes, which, in turn, could be targeted by modular interventions tailored to individual neurocomputational risk profiles.

5 Summary and Conclusion

Computational approaches have vast potential for optimizing precision psychiatry at a few different levels (Fig. 1). Theory-based approaches such as computational modeling and joint modeling could be integrated into multimodal assessments of RDoC constructs of mechanistic significance for addictions and other psychiatric disorders. Novel interventions could be developed that target the computational signatures (e.g., reward sensitivity, loss aversion, ambiguity tolerance, etc.) identified by theory-based approaches. Hybrid computational approaches like ADO could help develop "smart" assessment batteries, comprised of efficient, reliable, and precise neurocognitive tasks with superior psychometric properties. These batteries could be easily adapted to web-based platforms and mobile apps, as well as in longitudinal designs using ecological momentary assessment to track daily within-subject variability. Data-driven computational approaches could help develop novel mechanistic taxonomies and identify neurocomputational biotypes of specific psychiatric disorders, which could help guide the choice of intervention(s). Translating these computational approaches to clinical practice can be facilitated by open-source and user-friendly software packages, such as *hBayesDM* (hierarchical Bayesian modeling of decision-making tasks) [136] that offers computational models of an array of decision tasks, *easyML* (easy Machine Learning) [137] for machine learning approaches, and *ADOpy* [138] for adaptive design optimization.

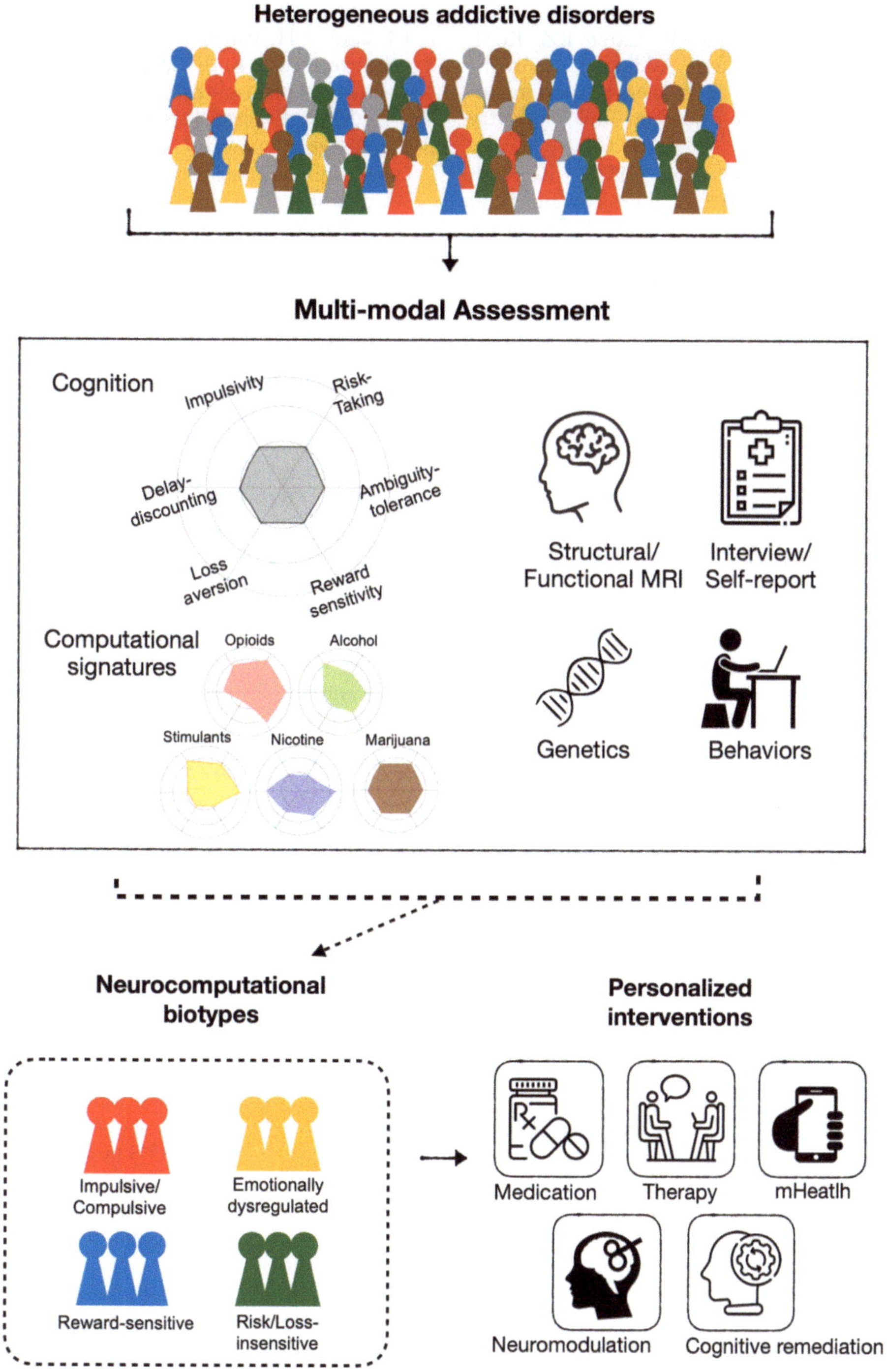

Fig. 1 A theoretical example of potential applications of theory-driven, data-driven, and hybrid computational approaches for assessment and treatment of substance use disorders

References

1. Kwako LE, Momenan R, Litten RZ, Koob GF, Goldman D (2016) Addictions Neuroclinical assessment: a neuroscience-based framework for addictive disorders. Biol Psychiatry 80(3):179–189
2. Insel TR (2014) The NIMH Research Domain Criteria (RDoC) Project: precision medicine for psychiatry. Am J Psychiatry 171(4):395–397
3. Koob GF, Volkow ND (2010) Neurocircuitry of addiction. Neuropsychopharmacology 35(1):217–238
4. Volkow ND, Koob GF, McLellan AT (2016) Neurobiologic advances from the brain disease model of addiction. N Engl J Med 374(4):363–371
5. Kwako LE, Schwandt ML, Ramchandani VA, Diazgranados N, Koob GF, Volkow ND et al (2019) Neurofunctional domains derived from deep behavioral phenotyping in alcohol use disorder. Am J Psychiatry: appiajp201818030357
6. Keyser-Marcus LA, Ramey T, Bjork J, Adams A, Moeller FG (2021) Development and feasibility study of an addiction-focused phenotyping assessment battery. Am J Addict 30(4):398–405
7. Ramey T, Regier PS (2018) Cognitive impairment in substance use disorders. CNS Spectr 24:1–12
8. Vassileva J, Conrod PJ (2019) Impulsivities and addictions: a multidimensional integrative framework informing assessment and interventions for substance use disorders. Philos Trans R Soc Lond Ser B Biol Sci 374(1766):20180137
9. Rezapour T, Rafei P, Baldacchino A, Conrod PJ, Dom G, Fishbein DH et al (2022) Neuroscience-informed classification of prevention interventions in substance use disorders: an RDoC-based approach. MedRxiv: 2022.09.28.22280342
10. Kessels RPC (2019) Improving precision in neuropsychological assessment: bridging the gap between classic paper-and-pencil tests and paradigms from cognitive neuroscience. Clin Neuropsychol 33(2):357–368
11. Kelly RE, Jr., Hoptman MJ (2022) Replicability in Brain Imaging. Brain Sci 12(3)
12. Nosek BA, Alter G, Banks GC, Borsboom D, Bowman SD, Breckler SJ et al (2015) SCIENTIFIC STANDARDS. Promoting an open research culture. Science 348(6242): 1422–1425
13. Pashler H, Wagenmakers EJ (2012) Editors' introduction to the special section on replicability in psychological science: a crisis of confidence? Perspect Psychol Sci 7(6):528–530
14. Hedge C, Powell G, Sumner P (2018) The reliability paradox: why robust cognitive tasks do not produce reliable individual differences. Behav Res Methods 50(3):1166–1186
15. Haines N, Kvam PD, Irving LH, Smith C, Beauchaine TP, Pitt MA, Ahn WY, Turner B (2021) Theoretically informed generative models can advance the psychological and brain sciences: lessons from the reliability Paradox. PsyArXiv
16. Strickland JC, Johnson MW (2021) Rejecting impulsivity as a psychological construct: A theoretical, empirical, and sociocultural argument. Psychol Rev 128(2):336–361
17. Elliott ML, Knodt AR, Ireland D, Morris ML, Poulton R, Ramrakha S et al (2020) What is the test-retest reliability of common task-functional MRI measures? New empirical evidence and a meta-analysis. Psychol Sci 31(7): 792–806
18. Kennedy JT, Harms MP, Korucuoglu O, Astafiev SV, Barch DM, Thompson WK et al (2022) Reliability and stability challenges in ABCD task fMRI data. NeuroImage 252: 119046
19. Whelan R, Garavan H (2014) When optimism hurts: inflated predictions in psychiatric neuroimaging. Biol Psychiatry 75(9):746–748
20. Han X, Ashar YK, Kragel P, Petre B, Schelkun V, Atlas LY et al (2022) Effect sizes and test-retest reliability of the fMRI-based neurologic pain signature. NeuroImage 247: 118844
21. Kragel PA, Han X, Kraynak TE, Gianaros PJ, Wager TD (2021) Functional MRI can be highly reliable, but it depends on what you measure: a commentary on Elliott et al. (2020). Psychol Sci 32(4):622–626
22. Fernandes BS, Williams LM, Steiner J, Leboyer M, Carvalho AF, Berk M (2017) The new field of 'precision psychiatry'. BMC Med 15(1):80
23. Williams LM (2016) Precision psychiatry: a neural circuit taxonomy for depression and anxiety. Lancet Psychiatry 3(5):472–480
24. Bickel WK, Jarmolowicz DP, Mueller ET, Koffarnus MN, Gatchalian KM (2012) Excessive discounting of delayed reinforcers as a trans-disease process contributing to addiction and other disease-related vulnerabilities:

emerging evidence. Pharmacol Ther 134(3): 287–297

25. Kwako LE, Bickel WK, Goldman D (2018) Addiction biomarkers: dimensional approaches to understanding addiction. Trends Mol Med 24(2):121–128
26. Bickel WK, Johnson MW, Koffarnus MN, MacKillop J, Murphy JG (2014) The behavioral economics of substance use disorders: reinforcement pathologies and their repair. Annu Rev Clin Psychol 10:641–677
27. Turing AMhdops (1937) On computable numbers, with an application to the Entscheidungsproblem Proc Lond Math Soc s2-42: 230–65
28. Huys QJM, Maia TV, Frank MJ (2016) Computational psychiatry as a bridge from neuroscience to clinical applications. Nat Neurosci 19(3):404–413
29. Huys QJM, Browning M, Paulus MP, Frank MJ (2021) Advances in the computational understanding of mental illness. Neuropsychopharmacology 46(1):3–19
30. Paulus MP, Huys QJ, Maia TV (2016) A roadmap for the development of applied computational psychiatry. Biol Psychiatry Cogn Neurosci Neuroimaging 1(5):386–392
31. Montague PR, Dolan RJ, Friston KJ, Dayan P (2012) Computational psychiatry. Trends Cogn Sci 16(1):72–80
32. Friston KJ, Stephan KE, Montague R, Dolan RJ (2014) Computational psychiatry: the brain as a phantastic organ. Lancet Psychiatry 1(2):148–158
33. Bechara A, Dolan S, Denburg N, Hindes A, Anderson SW, Nathan PE (2001) Decision-making deficits, linked to a dysfunctional ventromedial prefrontal cortex, revealed in alcohol and stimulant abusers. Neuropsychologia 39(4):376–389
34. Chen S, Yang P, Chen T, Su H, Jiang H, Zhao M (2020) Risky decision-making in individuals with substance use disorder: a meta-analysis and meta-regression review. Psychopharmacology 237(7):1893–1908
35. Amlung M, Vedelago L, Acker J, Balodis I, MacKillop J (2017) Steep delay discounting and addictive behavior: a meta-analysis of continuous associations. Addiction 112(1): 51–62
36. Petry NM (2001) Delay discounting of money and alcohol in actively using alcoholics, currently abstinent alcoholics, and controls. Psychopharmacology 154(3): 243–250
37. Bickel WK, Odum AL, Madden GJ (1999) Impulsivity and cigarette smoking: delay discounting in current, never, and ex-smokers. Psychopharmacology 146(4): 447–454
38. Kirby KN, Petry NM (2004) Heroin and cocaine abusers have higher discount rates for delayed rewards than alcoholics or non-drug-using controls. Addiction 99(4): 461–471
39. Kirby KN, Petry NM, Bickel WK (1999) Heroin addicts have higher discount rates for delayed rewards than non-drug-using controls. J Exp Psychol Gen 128(1):78–87
40. Petry NM, Bickel WK, Arnett M (1998) Shortened time horizons and insensitivity to future consequences in heroin addicts. Addiction 93(5):729–738
41. Cox DJ, Dolan SB, Johnson P, Johnson MW (2020) Delay and probability discounting in cocaine use disorder: comprehensive examination of money, cocaine, and health outcomes using gains and losses at multiple magnitudes. Exp Clin Psychopharmacol 28(6):724–738
42. Anokhin AP, Grant JD, Mulligan RC, Heath AC (2015) The genetics of impulsivity: evidence for the heritability of delay discounting. Biol Psychiatry 77(10):887–894
43. Lempert KM, Steinglass JE, Pinto A, Kable JW, Simpson HB (2019) Can delay discounting deliver on the promise of RDoC? Psychol Med 49(2):190–199
44. Ahn WY, Dai J, Vassileva J, Busemeyer JR, Stout JC (2016) Computational modeling for addiction medicine: from cognitive models to clinical applications. Prog Brain Res 224:53–65
45. Bechara A, Damasio AR, Damasio H, Anderson SW (1994) Insensitivity to future consequences following damage to human prefrontal cortex. Cognition 50(1–3):7–15
46. Busemeyer JR, Stout JC (2002) A contribution of cognitive decision models to clinical assessment: decomposing performance on the Bechara gambling task. Psychol Assess 14(3): 253–262
47. Ahn WY, Busemeyer JR, Wagenmakers EJ, Stout JC (2008) Comparison of decision learning models using the generalization criterion method. Cogn Sci 32(8):1376–1402
48. Worthy DA, Pang B, Byrne KA (2013) Decomposing the roles of perseveration and expected value representation in models of the Iowa gambling task. Front Psychol 4:640
49. Haines N, Vassileva J, Ahn WY (2018) The outcome-representation learning model: a novel reinforcement learning model of the Iowa gambling task. Cogn Sci 42(8): 2534–2561

50. Ahn WY, Vasilev G, Lee SH, Busemeyer JR, Kruschke JK, Bechara A et al (2014) Decision-making in stimulant and opiate addicts in protracted abstinence: evidence from computational modeling with pure users. Front Psychol 5:849
51. Bechara A, Martin EM (2004) Impaired decision making related to working memory deficits in individuals with substance addictions. Neuropsychology 18(1):152–162
52. Gonzalez R, Bechara A, Martin EM (2007) Executive functions among individuals with methamphetamine or alcohol as drugs of choice: preliminary observations. J Clin Exp Neuropsychol 29(2):155–159
53. Gonzalez R, Schuster RM, Mermelstein RJ, Vassileva J, Martin EM, Diviak KR (2012) Performance of young adult cannabis users on neurocognitive measures of impulsive behavior and their relationship to symptoms of cannabis use disorders. J Clin Exp Neuropsychol 34(9):962–976
54. Grant S, Contoreggi C, London ED (2000) Drug abusers show impaired performance in a laboratory test of decision making. Neuropsychologia 38(8):1180–1187
55. Martin EM, DeHaan S, Vassileva J, Gonzalez R, Weller J, Bechara A (2013) Decision making among HIV+ drug using men who have sex with men: a preliminary report from the Chicago Multicenter AIDS Cohort Study. J Clin Exp Neuropsychol 35(6): 573–583
56. Vassileva J, Gonzalez R, Bechara A, Martin EM (2007) Are all drug addicts impulsive? Effects of antisociality and extent of multidrug use on cognitive and motor impulsivity. Addict Behav 32(12):3071–3076
57. Vassileva J, Petkova P, Georgiev S, Martin EM, Tersiyski R, Raycheva M et al (2007) Impaired decision-making in psychopathic heroin addicts. Drug Alcohol Depend 86(2–3):287–289
58. Verdejo-Garcia AJ, Perales JC, Perez-Garcia M (2007) Cognitive impulsivity in cocaine and heroin polysubstance abusers. Addict Behav 32(5):950–966
59. Mellentin AI, Skot L, Teasdale TW, Habekost T (2013) Conscious knowledge influences decision-making differently in substance abusers with and without co-morbid antisocial personality disorder. Scand J Psychol 54(4): 292–299
60. Psederska E, Thomson ND, Bozgunov K, Nedelchev D, Vasilev G, Vassileva J (2021) Effects of psychopathy on neurocognitive domains of impulsivity in abstinent opiate and stimulant users. Front Psych 12:660810
61. Segala L, Vasilev G, Raynov I, Gonzalez R, Vassileva J (2015) Childhood symptoms of ADHD and impulsivity in abstinent heroin users. J Dual Diagn 11(3–4):174–178
62. Yechiam E, Busemeyer JR, Stout JC, Bechara A (2005) Using cognitive models to map relations between neuropsychological disorders and human decision-making deficits. Psychol Sci 16(12):973–978
63. Konova AB, Lopez-Guzman S, Urmanche A, Ross S, Louie K, Rotrosen J et al (2020) Computational markers of risky decision-making for identification of temporal windows of vulnerability to opioid use in a real-world clinical setting. JAMA Psychiatry 77(4): 368–377
64. Kvam PD, Romeu RJ, Turner BM, Vassileva J, Busemeyer JR (2020) Testing the factor structure underlying behavior using joint cognitive models: impulsivity in delay discounting and Cambridge gambling tasks. Psychol Methods
65. Park H, Yang J, Vassileva J, Ahn WY (2021) Development of a novel computational model for the balloon analogue risk task: the exponential-weight mean-variance model. J Math Psychol 102:102532
66. Romeu RJ, Haines N, Ahn WY, Busemeyer JR, Vassileva J (2020) A computational model of the Cambridge gambling task with applications to substance use disorders. Drug Alcohol Depend 206:107711
67. Vassileva J, Ahn WY, Weber KM, Busemeyer JR, Stout JC, Gonzalez R et al (2013) Computational modeling reveals distinct effects of HIV and history of drug use on decision-making processes in women. PLoS One 8(8):e68962
68. Rogers RD, Everitt BJ, Baldacchino A, Blackshaw AJ, Swainson R, Wynne K et al (1999) Dissociable deficits in the decision-making cognition of chronic amphetamine abusers, opiate abusers, patients with focal damage to prefrontal cortex, and tryptophan-depleted normal volunteers: evidence for monoaminergic mechanisms. Neuropsychopharmacology 20(4):322–339
69. Fishbein D, Hyde C, Eldreth D, London ED, Matochik J, Ernst M et al (2005) Cognitive performance and autonomic reactivity in abstinent drug abusers and nonusers. Exp Clin Psychopharmacol 13(1):25–40
70. Tolomeo S, Davey F, Steele JD, Baldacchino A (2021) Compulsivity and impulsivity in opioid dependence. Drug Alcohol Depend 229 (Pt A):109018
71. Wilson MJ, Vassileva J (2018) Decision-making under risk, but not under ambiguity,

predicts pathological Gambling in discrete types of abstinent substance users. Front Psychiatry 9:239

72. Todesco S, Chao T, Schmid L, Thiessen KA, Schutz CG (2021) Changes in loss sensitivity during treatment in concurrent disorders inpatients: a computational model approach to assessing risky decision-making. Front Psych 12:794014
73. Lejuez CW, Read JP, Kahler CW, Richards JB, Ramsey SE, Stuart GL et al (2002) Evaluation of a behavioral measure of risk taking: the Balloon Analogue Risk Task (BART). J Exp Psychol Appl 8(2):75–84
74. Yip SWB DM, Chase HW, Flagel S, Huys QJM, Konova AB, Montague R, Paulus M (2022) From computation to clinic. Biol Psychiatry Global Open Sci
75. Athamneh LN, Brown J, Stein JS, Gatchalian KM, LaConte SM, Bickel WK (2022) Future thinking to decrease real-world drinking in alcohol use disorder: repairing reinforcer pathology in a randomized proof-of-concept trial. Exp Clin Psychopharmacol 30(3): 326–337
76. Ruhi-Williams P, King MJ, Stein JS, Bickel WK (2022) Episodic future thinking about smoking-related illness: a preliminary investigation of effects on delay discounting, cigarette craving, and cigarette demand. Int J Environ Res Public Health 19(12)
77. Stein JS, Wilson AG, Koffarnus MN, Daniel TO, Epstein LH, Bickel WK (2016) Unstuck in time: episodic future thinking reduces delay discounting and cigarette smoking. Psychopharmacology 233(21–22):3771–3778
78. Gueguen MCM, Schweitzer EM, Konova AB (2021) Computational theory-driven studies of reinforcement learning and decision-making in addiction: what have we learned? Curr Opin Behav Sci 38:40–48
79. Turner BM, Rodriguez CA, Norcia TM, McClure SM, Steyvers M (2016) Why more is better: simultaneous modeling of EEG, fMRI, and behavioral data. NeuroImage 128:96–115
80. Hamilton KR, Mitchell MR, Wing VC, Balodis IM, Bickel WK, Fillmore M et al (2015) Choice impulsivity: definitions, measurement issues, and clinical implications. Personal Disord 6(2):182–198
81. Hamilton KR, Littlefield AK, Anastasio NC, Cunningham KA, Fink LH, Wing VC et al (2015) Rapid-response impulsivity: definitions, measurement issues, and clinical implications. Personal Disord 6(2):168–181
82. Palmeri TJ, Love BC, Turner BM (2017) Model-based cognitive neuroscience. J Math Psychol 76(Pt B):59–64
83. Turner BM, Forstmann BU, Wagenmakers EJ, Brown SD, Sederberg PB, Steyvers M (2013) A Bayesian framework for simultaneously modeling neural and behavioral data. NeuroImage 72:193–206
84. Turner BM, Wang T, Merkle EC (2017) Factor analysis linking functions for simultaneously modeling neural and behavioral data. NeuroImage 153:28–48
85. O'Doherty JP, Hampton A, Kim H (2007) Model-based fMRI and its application to reward learning and decision making. Ann N Y Acad Sci 1104:35–53
86. Yu JC, Fiore VG, Briggs RW, Braud J, Rubia K, Adinoff B et al (2020) An insula-driven network computes decision uncertainty and promotes abstinence in chronic cocaine users. Eur J Neurosci 52(12): 4923–4936
87. Daw ND, Gershman SJ, Seymour B, Dayan P, Dolan RJ (2011) Model-based influences on humans' choices and striatal prediction errors. Neuron 69(6):1204–1215
88. Guitart-Masip M, Huys QJ, Fuentemilla L, Dayan P, Duzel E, Dolan RJ (2012) Go and no-go learning in reward and punishment: interactions between affect and effect. NeuroImage 62(1):154–166
89. Moutoussis M, Shahar N, Hauser TU, Dolan RJ (2018) Computation in psychotherapy, or how computational psychiatry can aid learning-based psychological therapies. Comput Psychiatr 2:50–73
90. Nair A, Rutledge RB, Mason L (2020) Under the Hood: using computational psychiatry to make psychological therapies more mechanism-focused. Front Psych 11:140
91. Paulus MP, Thompson WK (2021) Computational approaches and machine learning for individual-level treatment predictions. Psychopharmacology 238(5):1231–1239
92. Lesmes LA, Lu ZL, Baek J, Albright TD (2010) Bayesian adaptive estimation of the contrast sensitivity function: the quick CSF method. J Vis 10(3):17 1–21
93. Myung JI, Cavagnaro DR, Pitt MA (2013) A tutorial on adaptive design optimization. J Math Psychol 57(3–4):53–67
94. Aranovich GJ, Cavagnaro DR, Pitt MA, Myung JI, Mathews CA (2017) A model-based analysis of decision making under risk in obsessive-compulsive and hoarding disorders. J Psychiatr Res 90:126–132
95. Gu H, Kim W, Hou F, Lesmes LA, Pitt MA, Lu ZL et al (2016) A hierarchical Bayesian approach to adaptive vision testing: a case study with the contrast sensitivity function. J Vis 16(6):15

96. Hou F, Lesmes LA, Kim W, Gu H, Pitt MA, Myung JI et al (2016) Evaluating the performance of the quick CSF method in detecting contrast sensitivity function changes. J Vis 16(6):18
97. Cavagnaro DR, Gonzalez R, Myung JI, Pitt MA (2013) Optimal decision stimuli for risky choice experiments: an adaptive approach. Manag Sci 59(2):358–375
98. Cavagnaro DR, Aranovich GJ, McClure SM, Pitt MA, Myung JI (2016) On the functional form of temporal discounting: an optimized adaptive test. J Risk Uncertain 52(3): 233–254
99. Ahn WY, Gu H, Shen Y, Haines N, Hahn HA, Teater JE et al (2020) Rapid, precise, and reliable measurement of delay discounting using a Bayesian learning algorithm. Sci Rep 10(1):12091
100. Kwon M, Lee SH, Ahn WY (2022) Adaptive design optimization as a promising tool for reliable and efficient computational fingerprinting. Biological Psychiatry: Cognitive Neuroscience and Neuroimaging
101. Wiecki TV, Poland J, Frank MJ (2015) Model-based cognitive neuroscience approaches to computational psychiatry: clustering and classification. Clin Psychol Sci 3(3):378–399
102. Afzali MH, Sunderland M, Stewart S, Masse B, Seguin J, Newton N et al (2019) Machine-learning prediction of adolescent alcohol use: a cross-study, cross-cultural validation. Addiction 114(4):662–671
103. Whelan R, Watts R, Orr CA, Althoff RR, Artiges E, Banaschewski T et al (2014) Neuropsychosocial profiles of current and future adolescent alcohol misusers. Nature 512(7513):185–189
104. Choi J, Jung HT, Ferrell A, Woo S, Haddad L (2021) Machine learning-based nicotine addiction prediction models for youth E-Cigarette and Waterpipe (Hookah) users. J Clin Med 10(5)
105. Ding X, Yang Y, Stein EA, Ross TJ (2017) Combining multiple resting-state fMRI features during classification: optimized frameworks and their application to nicotine addiction. Front Hum Neurosci 11:362
106. Pariyadath V, Stein EA, Ross TJ (2014) Machine learning classification of resting state functional connectivity predicts smoking status. Front Hum Neurosci 8:425
107. Ahn WY, Ramesh D, Moeller FG, Vassileva J (2016) Utility of machine-learning approaches to identify behavioral markers for substance use disorders: impulsivity dimensions as predictors of current cocaine dependence. Front Psych 7:34
108. Jimenez S, Angeles-Valdez D, Villicana V, Reyes-Zamorano E, Alcala-Lozano R, Gonzalez-Olvera JJ et al (2019) Identifying cognitive deficits in cocaine dependence using standard tests and machine learning. Prog Neuro-Psychopharmacol Biol Psychiatry 95: 109709
109. Niklason GR, Rawls E, Ma S, Kummerfeld E, Maxwell AM, Brucar LR et al (2022) Explainable machine learning analysis reveals sex and gender differences in the phenotypic and neurobiological markers of Cannabis Use Disorder. Sci Rep 12(1):15624
110. Rajapaksha R, Hammonds R, Filbey F, Choudhary PK, Biswas S (2020) A preliminary risk prediction model for cannabis use disorder. Prev Med Rep 20:101228
111. Schwebel FJ, Richards DK, Pfund RA, Joseph VW, Pearson MR, Marijuana Outcomes Study T (2022) Using decision trees to identify salient predictors of Cannabis-related outcomes. J Psychoactive Drugs: 1–10
112. Ahn WY, Vassileva J (2016) Machine-learning identifies substance-specific behavioral markers for opiate and stimulant dependence. Drug Alcohol Depend 161:247–257
113. Bickel WK, Moody LN, Eddy CR, Franck CT (2017) Neurocognitive dysfunction in addiction: testing hypotheses of diffuse versus selective phenotypic dysfunction with a classification-based approach. Exp Clin Psychopharmacol 25(4):322–332
114. Boer OD, El Marroun H, Franken IHA (1795) Brain morphology predictors of alcohol, tobacco, and cannabis use in adolescence: a systematic review. Brain Res 2022:148020
115. Lichenstein SD, Scheinost D, Potenza MN, Carroll KM, Yip SW (2021) Dissociable neural substrates of opioid and cocaine use identified via connectome-based modelling. Mol Psychiatry 26(8):4383–4393
116. Ruberu TLM, Kenyon EA, Hudson KA, Filbey F, Ewing SWF, Biswas S et al (2022) Joint risk prediction for hazardous use of alcohol, cannabis, and tobacco among adolescents: a preliminary study using statistical and machine learning. Prev Med Rep 25: 101674
117. Coughlin LN, Tegge AN, Sheffer CE, Bickel WK (2020) A machine-learning approach to predicting smoking cessation treatment outcomes. Nicotine Tob Res 22(3):415–422
118. Schumann G, Loth E, Banaschewski T, Barbot A, Barker G, Buchel C et al (2010) The IMAGEN study: reinforcement-related

behaviour in normal brain function and psychopathology. Mol Psychiatry 15(12): 1128–1139
119. Elam JS, Glasser MF, Harms MP, Sotiropoulos SN, Andersson JLR, Burgess GC et al (2021) The Human Connectome Project: a retrospective. NeuroImage 244:118543
120. Sporns O, Tononi G, Kotter R (2005) The human connectome: a structural description of the human brain. PLoS Comput Biol 1(4): e42
121. Thompson PM, Stein JL, Medland SE, Hibar DP, Vasquez AA, Renteria ME et al (2014) The ENIGMA Consortium: large-scale collaborative analyses of neuroimaging and genetic data. Brain Imaging Behav 8(2): 153–182
122. Barch DM, Albaugh MD, Avenevoli S, Chang L, Clark DB, Glantz MD et al (2018) Demographic, physical and mental health assessments in the adolescent brain and cognitive development study: rationale and description. Dev Cogn Neurosci 32:55–66
123. Luciana M, Bjork JM, Nagel BJ, Barch DM, Gonzalez R, Nixon SJ et al (2018) Adolescent neurocognitive development and impacts of substance use: overview of the adolescent brain cognitive development (ABCD) baseline neurocognition battery. Dev Cogn Neurosci 32:67–79
124. Morris AS, Squeglia LM, Jacobus J, Silk JS (2018) Adolescent brain development: implications for understanding risk and resilience processes through neuroimaging research. J Res Adolesc 28(1):4–9
125. Volkow ND, Koob GF, Croyle RT, Bianchi DW, Gordon JA, Koroshetz WJ et al (2018) The conception of the ABCD study: from substance use to a broad NIH collaboration. Dev Cogn Neurosci 32:4–7
126. Sudlow C, Gallacher J, Allen N, Beral V, Burton P, Danesh J et al (2015) UK biobank: an open access resource for identifying the causes of a wide range of complex diseases of middle and old age. PLoS Med 12(3): e1001779
127. Collins FS, Varmus H (2015) A new initiative on precision medicine. N Engl J Med 372(9): 793–795
128. Clementz BA, Sweeney JA, Hamm JP, Ivleva EI, Ethridge LE, Pearlson GD et al (2016) Identification of distinct psychosis biotypes using brain-based biomarkers. Am J Psychiatry 173(4):373–384
129. Williams LM (2017) Defining biotypes for depression and anxiety based on large-scale circuit dysfunction: a theoretical review of the evidence and future directions for clinical translation. Depress Anxiety 34(1):9–24
130. Drysdale AT, Grosenick L, Downar J, Dunlop K, Mansouri F, Meng Y et al (2017) Resting-state connectivity biomarkers define neurophysiological subtypes of depression. Nat Med 23(1):28–38
131. Clementz BA, Parker DA, Trotti RL, McDowell JE, Keedy SK, Keshavan MS et al (2022) Psychosis biotypes: replication and validation from the B-SNIP consortium. Schizophr Bull 48(1):56–68
132. Ivleva EI, Clementz BA, Dutcher AM, Arnold SJM, Jeon-Slaughter H, Aslan S et al (2017) Brain structure biomarkers in the psychosis biotypes: findings from the bipolar-schizophrenia network for intermediate phenotypes. Biol Psychiatry 82(1):26–39
133. Tamminga CA, Pearlson G, Gershon E, Keedy S, Hudgens-Haney ME, Ivleva EI et al (2022) Using psychosis biotypes and the Framingham model for parsing psychosis biology. Schizophr Res 242:132–134
134. Zhu T, Becquey C, Chen Y, Lejuez CW, Li CR, Bi J (2022) Identifying alcohol misuse biotypes from neural connectivity markers and concurrent genetic associations. Transl Psychiatry 12(1):253
135. Zarate D, Ball M, Montag C, Prokofieva M, Stavropoulos V (2022) Unravelling the web of addictions: A network analysis approach. Addict Behav Rep 15:100406
136. Ahn WY, Haines N, Zhang L (2017) Revealing neurocomputational mechanisms of reinforcement learning and decision-making with the hBayesDM package. Comput Psychiatr 1: 24–57
137. Ahn WY, Hendricks P, Haines N (2017) Easyml: easily build and evaluate machine learning models bioRxiv: 137240
138. Yang J, Pitt MA, Ahn WY, Myung JI (2021) ADOpy: a python package for adaptive design optimization. Behav Res Methods 53(2): 874–897

Part V

Integrative Computational Neuroscience

Chapter 15

Multimodal Integration in Psychiatry: Clinical Potential and Challenges

Eleonora Maggioni, Maria Chiara Piani, Elena Bondi, Anna M. Bianchi, and Paolo Brambilla

Abstract

The integration of complementary neuroimaging and neurosignaling modalities allows us to obtain a comprehensive view of the structural and functional brain properties and to extend the spatiotemporal domain investigated. In this chapter will explore the domains of multimodal integration that seem to be more relevant in the study of the cerebral bases of psychiatric disorders as an essential contribution to the construction of a personalized and broad-spectrum model of brain anatomy and function, useful in the early identification of risk or diagnostic profiles and in the planning of effective treatment strategies.

Key words Neuroimaging, Electrophysiology, MRI, EEG, Psychiatry

1 Introduction

In the last decades, many advances in the techniques of acquisition and preprocessing of brain images and signals have led to significant progress in the knowledge of the structure and function of the human brain. Basic neuroscientific research, diagnosis, and treatment of psychiatric illnesses can all benefit from the use of single neuroimaging and electrophysiological techniques, but even more from their integration. From the available evidence in neuroscientific literature, which is often very fragmented, it is getting clearer and clearer how both the study of cerebral structure and function and the investigation of the complex neurobiological bases of psychiatric disorders require a multimodal approach.

As a matter of fact, each investigation technique is sensitive to very specific brain tissue properties and only provides a partial measure of the structural and functional characteristics of interest, with spatial and temporal resolutions specific to each technique (as shown in Fig. 1). The physical principle of interaction and the acquisition parameters determine the type of process or structure

Drozdstoy Stoyanov et al. (eds.), *Computational Neuroscience*, Neuromethods, vol. 199,
https://doi.org/10.1007/978-1-0716-3230-7_15,

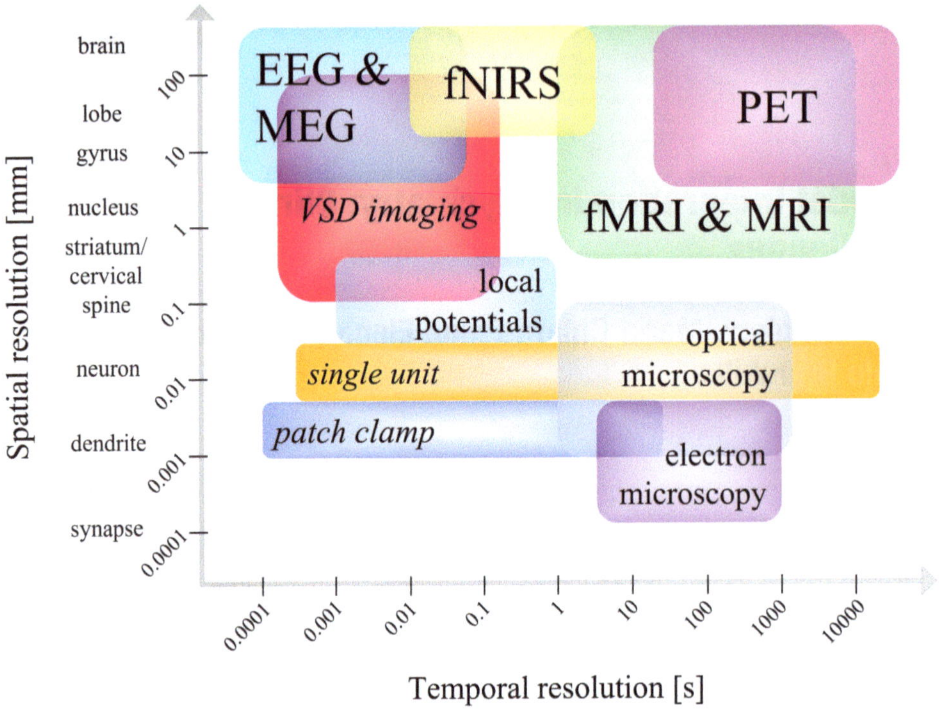

Fig. 1 Spatiotemporal resolution of the main neuroimaging techniques. Each colored region represents an estimation of the spatial (vertical axis) and of the temporal (horizontal axis) investigation domains of each measurement technique. EEG electroencephalography, MEG magnetoencephalography, fNIRS functional near infrared spectroscopy, PET positron emission tomography, VSD voltage-sensitive dye, (f)MRI (functional) magnetic resonance imaging, single unit: registration of the activity of single neurons; patch clamp: measurement of the electrophysiological properties of single ion channels

that is investigated and the level of spatial and/or temporal resolution achieved. Only the integration of complementary neuroimaging and neurosignaling modalities allows us to obtain a comprehensive view of the structural and functional brain properties and to extend the spatiotemporal domain investigated, by operating on different scales. Figure 2 depicts some examples of information that can be obtained with the main techniques discussed in this chapter, taken either individually or in combination. The complex multimodal integration paradigm requires not only expertise in the acquisition and processing of each data type, but also a deep knowledge of their informative content, which is essential to produce a synergic interpretation and to infer new useful information on the phenomenon under investigation.

The term "multimodal integration" usually refers to the combination of datasets acquired through different acquisition instruments, but in a broader sense may indicate the combination of different kinds of information, or of information of the same kind but partially complementary, obtained using a single instrument.

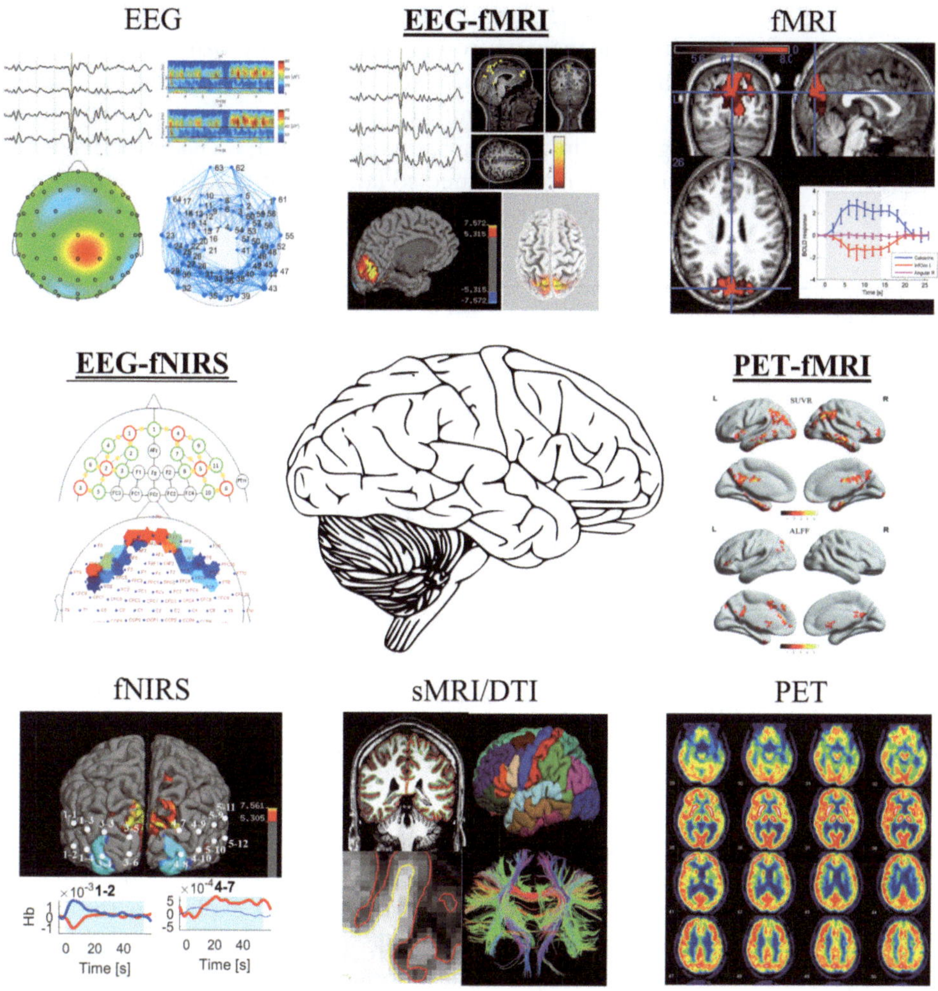

Fig. 2 Types of information obtainable with unimodal and multimodal neuroimaging techniques. Examples of structural and functional brain characteristics investigated by the main neuroimaging techniques, employed either alone or in combination. EEG electroencephalography, fNIRS functional near infrared spectroscopy, PET positron emission tomography, sMRI/fMRI structural/functional magnetic resonance imaging, DTI diffusion tensor imaging

In the following sections we will explore the domains of multimodal integration that seem to be more relevant in the study of the cerebral bases of psychiatric disorders. The research of quantitative brain markers associated with specific clinical dimensions is in line with the recent diagnostic classification scheme based on the Research Domain Criteria (RDoC) proposed by the US National Institute of Mental Health (NIMH), whose objective is to classify psychiatric disorders based on observable behavioral and neurobiological dimensions [1].

We will explain the principles, advantages, and criticalities of each of the main neuroimaging and neurosignaling techniques, and of the relative multimodal integration approaches. Moreover, we will also describe some examples of application of multimodal integration in the neuropsychiatric field, focusing on its potentials, on the evidence obtained so far, and on the possible future applications in psychiatric clinical practice and research.

2 Materials

In this chapter we will explain the principles, advantages, and criticalities of each of the main neuroimaging and neurosignaling techniques, and of the relative multimodal integration approaches. Moreover, we will also describe some examples of application of multimodal integration in the neuropsychiatric field, focusing on its potentials, on the evidence obtained so far, and on the possible future applications in psychiatric clinical practice and research. We will focus on magnetic resonance imaging and spectroscopy, optical imaging, tomography, and electroencephalography techniques and on psychotic disorders as bipolar disorder and schizophrenia.

3 Methods

3.1 Multimodality of Magnetic Resonance Techniques

Magnetic resonance imaging (MRI) represents the most powerful noninvasive technique for the investigation of both structural and functional brain characteristics [2]. By exploiting the interaction between an external magnetic field and the atomic nuclei—mostly hydrogen (^{1}H), but also sodium (^{23}Na), phosphorus (^{31}P), and others—MRI can discriminate the contributions of different cerebral tissues based on multiple factors, such as longitudinal and transverse relaxation times, proton density (or of the atom in question), magnetic susceptibility, magnetization transfer, or local inhomogeneities of the magnetic field. Through appropriate configurations of radiofrequency pulses and magnetic field gradients, which characterize the MRI sequences, it is possible to obtain different image contrasts, which in turn reflect a wide spectrum of physical and chemical properties of the cerebral tissues.

Thanks to its intrinsic multimodality, MRI is able to provide both neuroanatomical and neurofunctional information, as well as information bridging between the two [3]. The cerebral structure is mostly investigated with T1- or T2-weighted sequences, from which it is possible to extract measures of volumetry of the principal brain tissues (volume-based analyses) and of cortical morphology such as thickness, superficial area, and gyrification (surface-based analyses), at the level of the single voxel/vertex or of regions of interest (ROIs). The information on brain function that can be obtained with MRI techniques is relative to the metabolic and hemodynamic correlates of neuronal activity, specifically to variations of cerebral blood flow or volume, through perfusion sequences, or changes in the blood oxygenation state, through blood oxygenation level dependent (BOLD) techniques. Anatomical and functional information can be obtained with diffusion-weighted imaging (DWI) techniques, sensitive to microscopic movements of water molecules in the cerebral tissues. The DWI provides different measures of water diffusion, such as fractional anisotropy and radial diffusion, which in turn reflect the tissue microscopic configuration. Water diffusion properties are also exploited by the modern tractography techniques (diffusion tensor imaging, DTI), which aim at reconstructing the organization of white matter fibers based on the diffusion isotropy/anisotropy, that is, the presence of preferential directions of movement of the water molecules. Furthermore, it is worth mentioning that the physical principle of nuclear magnetic resonance is also exploited in the study of cerebral biochemical processes through magnetic resonance spectroscopy (MRS), which allows us to discriminate between the relative concentrations of specific metabolites in different brain areas.

The enormous potentialities and noninvasiveness of the magnetic resonance techniques are responsible for their increasing relevance in the identification of quantitative biomarkers of psychiatric disorders. In particular, the combined elaboration of MRI and MRS information of different nature enables the extraction of complex anatomo-functional relationships that cannot be derived from the implementation of single techniques independently but are required for the identification of the neurobiological bases of complex mental disorders.

3.2 Multimodal Magnetic Resonance in the Study of Major Psychoses

Integrated MRI analyses of brain morphology and anatomical and functional connectivity can provide key information on the mechanisms at the origin of complex psychiatric disorders like major psychoses, including schizophrenia and bipolar disorder, for which a dysconnectivity hypothesis has been proposed [4]. According to this hypothesis, at the origin of psychotic disorders there are not simply alterations in specific cerebral regions, rather anomalies in the modulation of synaptic efficacy within the brain circuitries.

In the last decades, the ease of obtaining high resolution brain images and the diffusion of sophisticated connectomic investigations have allowed us to deepen the dysconnectivity paradigm in terms of anatomy and functionality of the brain networks in psychotic disorders. Nonetheless, the evidence obtained to date with magnetic resonance techniques, either used singularly or combined, is not enough to draw a complete picture of the cerebral correlates of psychoses.

The most relevant results of magnetic resonance studies on major psychoses are briefly described and discussed in the following paragraphs, highlighting the research topics that deserve further investigations. From the available evidence, it emerges how the use of MRI/MRS fusion techniques is still at the beginning, partly due to the increased complexity of acquisition, analytical, and interpretation steps. Despite being limited, this evidence suggests that data fusion can play a decisive role in the understanding of brain architecture in subjects affected by major psychoses.

The dysconnectivity hypothesis is largely credited in schizophrenia, whose diagnostic formulation includes the concept of disorganization [5]. Numerous structural MRI studies on schizophrenia have reported diffuse anomalies, including (i) gray matter volumetric deficits in cortical, especially temporal, and subcortical regions, including thalamus and amygdala [6], (ii) alterations in fronto-temporal and thalamic-insular volumetric correlations [7], and (iii) thalamic [8] and fronto-temporo-occipital [9] white matter microstructural alterations. Parallelly, functional MRI (fMRI) studies based on the BOLD contrast have reported anomalies in thalamic-cortical functional connectivity [10] and interaction deficits in prefrontal and temporal regions [11], thus converging with the anatomical connectivity results in supporting the notion of schizophrenia as a disconnection syndrome. This hypothesis is further consolidated by MRS studies, which suggest glutamatergic alterations, possibly associated with synaptic plasticity alterations, in the thalamic-cortical-striatal connections [12].

Even if it is plausible to assume a close relationship between brain structure and function, the few multimodal MRI studies available up to now leave many open points on the nature of the relationships between anatomy and function of brain networks in subjects affected by schizophrenia [13]. The results obtained by the integration of DTI and BOLD fMRI data, even if preliminary, suggest that functional connectivity alterations, such as those in the fronto-temporal language network, are associated with alterations in the integrity of the white matter tracts that support those connections [14]. On the same line, integrated structural-functional MRI studies have found both direct and inverse relationships between gray matter volume deficits and functional activations (or deactivations) in different brain regions and networks, including the default mode network, also suggesting alterations in structural-functional integrity during the execution of sensorimotor and working memory exercises [15, 16]. The results of integrated cortical morphology and diffusion investigations are on the same line, suggesting the presence of a close association between white matter and gray matter deficits—in particular in terms of cortical thickness—in schizophrenic subjects [17].

In conclusion, the multimodal MRI studies conducted so far confirm the existence of a link between brain network anatomy and function and schizophrenic symptoms, supporting the dysconnectivity hypothesis, but they do not provide substantial evidence on the primary etiologic mechanisms of the disease. The most recognized theory to date, which requires further investigation, postulates that the alterations in structural connectivity are not the cause, rather a consequence of the schizophrenic pathology, attributing the origin of the symptomatology to functional alterations of synaptic efficacy.

Similar to what has been described for schizophrenia, the multimodal potential of MRI techniques is more and more exploited also in the study of the neural bases of bipolar disorder. Until today, the dysconnectivity hypothesis in bipolar disorder is less explored than in schizophrenia. Indeed, the conceptualization of the disorder in terms of neural circuits has been proposed only recently and requires further neuroscientific confirmations [18].

The most robust evidence obtained so far from neuroimaging investigations suggests the presence of dysfunctions involving cerebral circuits devoted to emotional processing and regulation and to reward, which could determine the behavioral characteristics of bipolar disorder, such as mood swings, emotional lability and dysregulation, and excessive sensitivity to material gains. In this context as well, the use of multimodal MRI techniques turns out to be essential in the comprehension of the cerebral substrates of the disorder. The integration of DTI and BOLD fMRI results has allowed to hypothesize a connection between the impairments in the prefrontal-limbic interactions and the integrity of the subsiding white matter tracts [19], but the causal nature of the connection between structure and function in the cerebral networks involved in bipolar disorder remains largely unknown.

Indeed, most MRI studies conducted so far on individuals affected by bipolar disorder not only have provided a partial vision of the brain correlates of the disorder, but have often produced inconsistent results, thus leaving unclarified the neurobiological mechanisms at the origin of the bipolar symptomatology. This uncertainty reflects at least partially the complexity of bipolar disorder, characterized by heterogeneous psychopathological manifestations and by different levels of cognitive deficits.

By recognizing the non-unitary nature of bipolar disorder, in line with the recent classification scheme based on the RDoC, new neuroimaging analyses targeted at stratifying bipolar disorder based on neurobiological markers associated with specific clinical dimensions are progressively spreading [20]. In this respect, recent literature studies support the hypothesis of a continuum between bipolar disorder and schizophrenia, overcoming the Kraepelinian theory that classifies them as two distinct diagnostic categories. The overlap between the two disorders regards the psychotic dimension, which characterizes a significant proportion of bipolar disorder patients (in particular of type I) and impacts on the disease trajectory, being associated with earlier age of onset, more frequent hospitalizations, more severe symptomatology, and worse prognosis. The few MRI studies that compared psychotic and non-psychotic bipolar disorder patients suggest more pronounced structural deficits in the former, particularly at the level of the right parietal lobe [21, 22].

The presence of psychotic symptoms could thus reflect cerebral alterations common among bipolar disorder, schizoaffective disorder, and schizophrenia. The findings from MRI studies that have compared subjects affected by bipolar disorder and schizophrenia and control subjects partially support the theory of neurobiological continuity between the two disorders. The neuroanatomical comparison between bipolar disorder and schizophrenia showed in the latter more diffused and marked alterations. Nevertheless, shared characteristics have also emerged, including volumetric gray matter reductions, mostly in the cingulate gyrus and in the temporal lobe [6, 23], and reduced structural complexity in the frontal cortex [24]. The morphologic and volumetric cortical alterations discovered in both disorders could reflect shared microstructural anomalies in fronto-temporal and callosal white matter tracts [25]. Parallelly, BOLD fMRI investigations have suggested functional connectivity alterations being shared between bipolar disorder and schizophrenia in many nodes of the default mode, the visual, the sensorimotor, and the cognitive control networks [26].

Even in this field, the integrated anatomical and functional evidence is very scarce. An interesting trimodal study suggested common alterations of gray matter, white matter, and functional connectivity of the hippocampus, being more pronounced in schizophrenia. An integrated DTI and BOLD fMRI study has evidenced both distinct and common characteristics in the two patient groups, showing (1) differences in terms of functionality and structural integrity in the white matter tracts of the frontal and occipital lobes, and (2) similar alterations in the functional connectivity between thalamus and prefrontal cortex and in the structural connectivity in the associated white matter tracts, mostly in the anterior thalamic radiation and in the uncinate fasciculus.

In conclusion, the results from unimodal and multimodal MRI studies support the existence of a transdiagnostic and neurobiologically determined psychotic dimension, showing at the same time brain characteristics specific to each diagnosis. In the next future, the extraction of "cross-information" from structural, functional, and diffusion MRI data will bring an important added value to the study of the specific and shared neurobiological correlates of psychotic disorders.

3.3 Functional Neuroimaging and Neurophysiologic Techniques

Functional brain imaging is a research field characterized by continuous technical advancements, whose objective is to obtain a deep knowledge of the brain activity and functional connectivity both in physiological and in pathological conditions.

As mentioned in the introduction section, the modern neuroimaging and neurophysiological techniques rely on signal "sources" that vary in relation to changes in neuronal activity with characteristic spatial and temporal resolutions. The different modalities are sensitive to electrophysiologic, hemodynamic, and metabolic phenomena, or to a combination of them. Due to the specific sensitivity of each methodology, no single technique can provide a complete picture of brain function, which is obtainable only by integrating multiple complementary modalities. The main functional neuroimaging and neurosignaling methods are described below and cross-referenced based on their characteristic properties, including the type and the scale of the functional information provided, and the ability to give key information in the study of mental illnesses. Despite the still limited use of multimodal functional techniques both in the clinical and in the research fields, we will describe some examples of application of these techniques in the study of psychiatric disorders.

3.3.1 Magnetic Resonance Imaging of the BOLD Effect

Among the different functional neuroimaging modalities, functional magnetic resonance imaging (fMRI) has the unique ability to deliver information on both cortical and subcortical brain function with the highest spatial resolution obtainable in vivo. The fMRI studies are generally based on the BOLD contrast (blood oxygenation level dependent), which originates from local distortions of the magnetic field homogeneity, in turn generated by variations of the cerebral oxygenation status. The BOLD contrast is obtained by applying Echo Planar Imaging techniques, which enable the acquisition of multiple cerebral volumes at a few seconds interval through the application of radiofrequency pulses and magnetic field gradients in very rapid sequence. In the modern magnetic resonance scanners, the acquisition time of a single brain volume generally varies between 1 and 3 s, in a tradeoff between temporal and spatial resolution. Therefore, during a standard fMRI session, hundreds of brain volumes are acquired. The BOLD signal of each voxel—which represents the basic unit of the image—represents the variation in the gray shade of the voxel in consecutive fMRI volumes and reflects the oxygenation changes in the corresponding brain area.

More in detail, the BOLD signal is sensitive to changes in the balance between oxyhemoglobin and deoxyhemoglobin that, due to their different magnetic properties, generate magnetic field distortions with the variation of their reciprocal concentrations. The changes in concentration of oxygenated and deoxygenated hemoglobin are the result of a complex interplay between local cerebral blood flow, local cerebral blood volume, and metabolic rate of oxygen consumption. Since all those factors are indirectly related to the neuronal activity, the BOLD signal itself is often used as a marker of the underlying neuronal activity.

A great merit of the BOLD fMRI technique resides in its ability to provide a measure of brain function extended to the whole brain and at a very high spatial resolution, in the order of millimeters. BOLD fMRI studies allow to map the regional activations in response to various paradigms and to infer about the functional connections, both instantaneous and causal, between spatially remote regions of the brain. Given its suitability as an instrument for broad-spectrum mapping of the brain organization, the BOLD fMRI technique has been used to further explore the brain mechanisms underlying functional segregation and integration in a wide range of researches, both in physiology and in pathology, at resting state and during various sensory, cognitive, social, and emotional experiments. Regarding research in the neuropsychiatric field, as already emerged in the previous section ("Multimodality of Magnetic Resonance Techniques"), BOLD fMRI studies are turning out to be crucial in the understanding of the functional brain mechanisms associated with clinical symptomatology in various psychiatric disorders.

Despite this great potential, the fMRI technique based on BOLD contrast has two main disadvantages, (i) to have a low temporal resolution, in the order of seconds (equal to the repetition time, i.e., between the acquisition of two consecutive brain volumes), and (ii) to be exclusively sensitive to hemodynamic and metabolic processes, which are only indirectly associated with neuronal activity and have a low intrinsic temporal resolution. Moreover, it is worth mentioning that the BOLD signal does not provide an absolute measure of brain oxygenation, rather it represents a relative measure of its variations during the acquisition session. To summarize, the fMRI methodology based on the BOLD contrast is not able to provide direct information on the electrical activity of the brain and suffers from an "ill posed temporal problem," that is, it does not allow to infer about the exact moment of occurrence of the neuronal events underlying the measured hemodynamic and metabolic modifications.

Indeed, the variations of blood flow and volume occur a few seconds after the variations in neuronal activity that cause them, and rapidly reach the peak of response before decaying. It derives that the hemodynamic and metabolic variations measured with the BOLD contrast, other than being sampled at intervals of seconds, reflect changes in neuronal activity that occurred several seconds before. Commonly, fMRI data analyses employ model approaches based on the characteristic temporal properties of the neurovascular coupling, which are often summarized in the canonical hemodynamic response model, and on convolution and deconvolution algorithms for the reciprocal estimation of correlated BOLD signal and neuronal activity variations.

The interpretation of BOLD fMRI data in terms of neuronal function is further complicated by the sensitivity of this technique to oxygenation variations in different vascular compartments—including veins, arteries, and capillaries—which represent differences in metabolism and in coupling with the neuronal activity. Different fMRI sequences have specific metabolic-hemodynamic sensitivity, determined by factors such as the temporal configuration of pulses, the voxel dimension, and the magnitude of the magnetic field. For example, it has been found that some BOLD sequences are mostly sensitive to the oxygenation status in the venous compartment. In those cases, fMRI activation analyses can identify brain areas that do not overlap to the site of neuronal activation.

Furthermore, the technological advancements in the MRI field have enabled the development of sequences with a sub-millimetric spatial resolution. Those high-resolution techniques allow to discriminate between the contributions of different cortical strata and columns and different vascular elements, but at the same time they need the development of more accurate mesoscopic models of BOLD response compared to commonly employed models. In conclusion, given that the nature of neurovascular coupling and the links between metabolism and blood flow/volume vary as a function of the sequence employed, of the cerebral region considered, and also of the pathophysiological mechanisms, the results of BOLD fMRI analyses must be cautiously interpreted.

3.4 MRI Techniques Sensitive to Perfusion and Oxidative Metabolism

It should be stressed that there are other functional neuroimaging modalities sensitive to metabolic or hemodynamic variations, which can be employed in conjunction with BOLD fMRI to discriminate the single mechanisms contributing to the generation of the BOLD signal.

In confirmation of the intrinsic multimodality of fMRI, measures of cerebral blood flow and volume can be computed through arterial spin labelling (ASL) and vascular-space occupancy (VASO) MRI techniques. Blood flow and volume can be computed also from dynamic magnetic susceptibility MRI techniques (dynamic susceptibility contrast, DSC-MRI) based on the injection of a paramagnetic contrast medium, generally gadolinium, and on the measurement of the signal variations dependent on the local concentration of the medium in T2-weighted images. It is worth mentioning that the perfusion parameters that can be computed from DSC-MRI reflect the brain hemodynamics at the capillary level. Regarding the metabolic consumption of oxygen, it is often estimated indirectly from BOLD and ASL/VASO data via the application of a physiological model based on the Fick's principle (through the techniques defined as "calibrated fMRI"). Moreover, ad hoc MRI sequences for the measurement of the cerebral oxidative metabolism have been recently developed, which are able to provide absolute measures of the metabolic rate of consumption of the cerebral oxygen. Among those are included techniques that involve the inhalation of a gas containing the ^{17}O isotope of stable oxygen—present in nature and visible in magnetic resonance—and measure the increase/decrease in metabolism with a direct—through spectroscopy—or an indirect—through protonic imaging—detection of the isotope [27].

In the field of neuropsychiatric research, the aforementioned MRI methods have been less commonly used than BOLD fMRI techniques, and the examples of multimodal integration are limited. Among them, ASL techniques are having more diffusion compared to DSC-MRI, VASO, and ^{17}O-measurement sequences, partly due to their greater ease of use and the absence of contrast media [28]. The obtained evidence, even if limited, has allowed us to confirm and deepen the knowledge of the functional brain mechanisms underlying psychiatric disorders obtained through BOLD fMRI investigations.

The results of ASL studies on schizophrenia suggest perfusion deficits in the bilateral prefrontal cortex, and particularly the middle frontal gyrus. Parallelly, DSC-MRI investigations on schizophrenia have allowed to confirm perfusion alterations in the frontal lobe and to further investigate their clinical correlates, suggesting an impact of the clinical state, age, disease duration, and pharmacological treatment on the deficits of blood flow and volume in the frontal cortex [29, 30]. Overall, these results support the functional connectivity results emerged from BOLD fMRI analyses (described in the section "Multimodality of Magnetic Resonance Techniques"), confirming the role of hemodynamic processes in the deficits of functional interactions at the level of the prefrontal cortex that have been observed in schizophrenia. On the contrary, despite the hypothesized role of cerebral oxidative stress in schizophrenic psychopathology, in vivo evidence on the oxygen metabolism in the brain is very scarce. The literature is devoid of MRI/MRS studies based on ^{17}O measurement, but recent pioneering spectroscopic studies have measured glutathione levels—a tripeptide with antioxidant action—in different brain regions [31]. The obtained results are inconsistent, leaving many open points on the possible relation between oxidative stress, cellular homeostasis, and structural and functional connectivity alterations in schizophrenia. It should be noted that the few multimodal fMRI analyses performed on schizophrenic subjects confirm the importance of further exploring the nature of the discovered BOLD alterations. As an example, the integration of ASL and BOLD results has allowed to detect anomalies in the relationship between the resting cerebral blood flow and the magnitude of variation of the BOLD signal during emotional processing, particularly in the prefrontal and temporo-occipital regions [32].

Analogously to what has been evidenced in schizophrenia, the recent employment of DSC-MRI and ASL techniques in bipolar disorder has allowed us to further investigate the dysfunctional cerebral mechanisms responsible for the disease. Alterations in blood volume and flow have emerged in different areas of the cerebral cortex, at rest and during task execution, and in different mood states, for which it is difficult to provide a conjoint interpretation [33]. Moreover, a recent multimodal study on adolescents with bipolar disorder suggests that perfusion alterations are not associated with alterations in the oxidative metabolism [34]. Overall, these results suggest the presence of hemodynamic alterations dependent on specific clinical dimensions and worthy of further investigations. To confirm the importance of adopting a dimensional approach, a recent ASL study has displayed an association between positive symptoms and altered connectivity between the anterior cingulate cortex and the orbitofrontal cortex, suggesting the presence of frontal hemodynamic alterations shared among the psychotic spectrum disorders [35]. The results of a recent machine learning study on first episode psychosis confirm this hypothesis, highlighting how the DSC-MRI perfusion indexes relative to different brain lobes—especially the right frontal and the left parietal ones—allow to classify patients at their first psychotic episodes in comparison to controls with an accuracy up to 83% [36].

3.5 Optical Neuroimaging Techniques

Further information on the brain hemodynamic correlates of psychiatric disorders can be obtained through optical methodologies, which employ electromagnetic radiations in the near infrared spectrum and measure their absorption dynamics in the cerebral tissues.

In the clinical and neuroscientific research fields, the most frequently used optical methodology is the functional near-infrared spectroscopy (fNIRS), which measures the variations in oxy- and deoxy-hemoglobin concentrations in different cortical regions. fNIRS exploits differences in the absorption spectrum of infrared radiation between the two hemoglobin species to provide separate measures of their concentration changes. The sum of the oxy- and deoxy-hemoglobin variations, proportional to the variation in total hemoglobin, can be used as an estimation of cerebral blood volume changes.

Thanks to its lower cost, its noninvasiveness, and its ease of use, fNIRS is being extensively employed in the clinical and research fields, especially in the study of neonatal or pediatric disorders, such as the attention deficit hyperactivity disorder (ADHD), in which fNIRS could be preferable than fMRI, due to its lower sensitivity to movement and the imposition of less constraints in the experimental setting [37].

Compared to other functional neuroimaging techniques, fNIRS has the unique ability to discriminate between oxy- and deoxy-hemoglobin contributions. fNIRS has the additional advantage to measure the cerebral hemodynamic variations with a good temporal resolution, which can reach hundredths of seconds, at the expense of a series of limitations, including (1) low spatial resolution, in the order of centimeters, (2) sensitivity to the hemodynamic variations in the most superficial layer of the cortex, and (3) low signal to noise ratio due to physiological components—like scalp, skull, and cerebrospinal fluid—that hinder the penetration of the optical radiations in the cerebral cortex. Thanks to the characteristics we just listed, the integration of fNIRS and fMRI results is particularly promising, as it allows further investigations of the hemodynamic mechanisms at the origin of the BOLD effect [38].

Although the use of fNIRS in the study of psychiatric pathologies is still in its early days, the available evidence confirms the utility of this technique in the study of the brain hemodynamic activity during cognitive processes and at rest. The largest study performed up to now has evidenced how frontal fNIRS responses to a verbal fluency task allow to differentiate subjects affected by major depression from those affected by bipolar and schizophrenic depression and from control subjects [39]. An independent fNIRS study has confirmed a potential role of frontal fNIRS activation patterns as differential markers of unipolar and bipolar depression [40]. In parallel, fNIRS investigations in schizophrenia have confirmed the presence of hemodynamic alterations in the frontal lobe during the execution of different working memory tasks [41–43] and at rest [44]. The integration of fNIRS and BOLD fMRI data recorded during the same visuo-spatial task has shown an agreement between the results of the two techniques, confirming the hemodynamic nature of the frontal BOLD alterations emerged in schizophrenic subjects [45].

3.6 Positron Emission Tomography

Most functional neuroimaging techniques described in the previous paragraphs are sensitive to hemodynamic phenomena or to the interaction between hemodynamics and oxidative metabolism. Measures of cerebral oxygen consumption are more difficult to obtain, therefore usually they are indirectly estimated from BOLD, cerebral blood flow or volume, or oxygen saturation data.

The most powerful and versatile neuroimaging technique for the study of cerebral metabolism is undoubtedly the positron emission tomography (PET). PET employs radioisotope-labeled tracers that emit positrons, whose annihilation generates couples of gamma rays at 180° that in turn are recorded by specific detectors. From gamma rays' projections obtained at different angles, the tracer's spatial distribution in the brain is reconstructed, which in turn reflects specific functional characteristics. Generally, PET provides information on the cerebral metabolism of glucose through the injection of ^{18}F-deoxyglucose (^{18}F-FDG), but can also provide information on the oxygen metabolic consumption through the injection of water marked with ^{15}O. Furthermore, dopaminergic, GABAergic, and glutamatergic radiotracers, which allow the investigation of specific biochemical processes associated with cerebral activity, have been recently developed. This unique potentiality is accompanied by a greater invasiveness—due to the use of ionizing radiations—and by a lower spatial resolution in comparison to fMRI techniques.

PET applications in the psychiatric field, despite being less common than in the oncologic and neurologic fields, are proving to be fundamental for the comprehension of the molecular bases of complex psychiatric disorders. In the study of major psychoses, PET has been employed for the identification of biochemical correlates of specific clinical dimensions [46]. A ^{18}F-FDG PET study has evidenced glucose metabolism deficits in fronto-temporal and cerebellar white matter tracts in schizophrenia compared to bipolar disorder [47]. Recent ^{18}F-DOPA PET studies have discovered the existence of two dopaminergic phenotypes in schizophrenia, besides suggesting the presence of a relationship between dopaminergic dysfunction in the striatum and response to antipsychotic therapy in first episode psychosis patients [48]. In the same disorder, PET investigations based on selective radiotracers for type 1 cannabinoid receptors (CBR1) have found deficits in this receptor, being proportional to symptom severity and cognitive deficits in male subjects [49]. The employment of appropriate PET tracers has enabled the monitoring of the cerebral density of dopaminergic receptors in subjects affected by bipolar disorder. The results, although heterogeneous, suggest the presence of dopaminergic variations dependent on the mood of the subjects [50].

The recent development of integrated PET and magnetic resonance scanners has provided simultaneous, broad-spectrum information on the brain function both in physiology and in pathology. Since multimodal scanners are still poorly diffused, most of the PET-MRI applications in psychiatry have regarded a post-hoc integration of the results obtained from the single modalities. A PET-MRS study in psychiatry has highlighted an association between GABA+ levels and the expression of a microglial activation marker in the medial prefrontal cortex in control subjects and in subjects at risk for developing psychosis [51]. A BOLD fMRI-PET study in first episode psychosis has suggested a relationship, in male participants, between basal striatal availability of CBR1 and striatal BOLD response to a working memory exercise [52]. The utility of the PET-MRI integration has been confirmed also in the study of the neurobiological correlates shared and non-shared between bipolar disorder and substance-induced psychosis. A recent study has compared the PET-based evidence on ^{18}F-FDG and structural MRI, highlighting complex relationships between hyper- or hypo-metabolism of the glucose and gray matter deficits being shared and distinct in these forms of psychosis [21]. From the PET-MRI comparison between bipolar disorder in advanced age and fronto-temporal dementia with behavioral variants, specific characteristics have emerged in terms of morphology and glucose metabolism in frontal, temporal, and limbic regions, which can be useful in the differential diagnosis of the two disorders [53].

To summarize, this evidence confirms the validity of PET as a diagnostic and informative instrument in planning and monitoring personalized therapeutic strategies [46] and remarks the importance of obtaining multimodal information, ideally simultaneously, in order to identify robust neurobiological markers of psychiatric disorders.

3.7 Electroencephalography

The techniques presented in the previous paragraphs provide information on the cerebral function through metabolic and hemodynamic phenomena, which are only indirectly associated with the neuronal activity. Direct and temporally precise information on neuronal activity can be obtained through electrophysiologic registrations. The electroencephalographic technique (EEG) measures instantaneously and noninvasively the synchronized electrical activity of large populations of neurons. Its excellent temporal resolution, in the order of milliseconds, makes it optimal for the study of neuronal activity at the same temporal scale of occurrence of the neuronal events. Indeed, the EEG technique is employed to investigate rapid neuronal responses evoked by external stimuli, as well as neuronal synchronization at rest, and to perform analyses in the frequency and time-frequency domains.

On the other hand, since neuronal activity is registered through electrodes placed on the scalp, the EEG presents the drawback of being sensitive to massive neuronal activities and to be characterized by low specificity and low spatial resolution. As opposed to fMRI, the EEG suffers from the "spatial inverse problem," linked to the difficulty of establishing the precise localization of the neuronal sources within the brain from the potentials registered at the scalp level.

The exact complementarity between the EEG and fMRI techniques, especially with BOLD contrast, makes their integration particularly promising. The two techniques provide complementary views of the cerebral function, which, if properly combined, could provide a substantial added value in neuroscientific research and clinical applications.

In the last decade, there has been a progressive spread of the simultaneous EEG-fMRI technique, which offers the unique opportunity of obtaining noninvasively a comprehensive picture of functional brain mechanisms with high spatial and temporal resolutions. The simultaneous acquisition enables the registration of EEG and fMRI data in the same physiologic conditions, allowing to study the neurovascular coupling and the real-time associations between EEG rhythms/event-related potentials (ERPs) and metabolic and blood flow and volume changes in the whole brain [54]. In particular, simultaneous EEG-fMRI analyses can provide a reliable localization of the deep neuronal sources with high spatial and temporal resolution. Recently, simultaneous EEG-fNIRS acquisition systems have also been developed, allowing the spatial co-registration of information on neuronal activity and of the related changes in oxy- and deoxy-hemoglobin. The study of neurovascular coupling during sensory task execution, at rest, or during sleep can provide key insights on the nature of the brain functional dysregulation mechanisms associated with neurologic and psychiatric pathologies.

In the last decades, numerous EEG studies on neuropsychiatric disorders have evidenced highly replicable neuronal dysfunctions, with potential diagnostic and prognostic validity. One of the most robust results regards the presence of altered ERP components associated with cognitive control and response inhibition. Specifically, reduced-amplitude P300 potentials have been observed in different behavioral and psychiatric disorders, including alcohol use disorder, ADHD, and schizophrenia [55]. As described in a recent review of the EEG literature on schizophrenia [56], different studies have reported deficits in the mismatch negativity (MMN) component of auditory ERPs in affected subjects (regardless of the status), in non-affected relatives, and in subjects with subthreshold symptoms. Moreover, the finding of minor MMN alterations in bipolar disorder in comparison to schizophrenia suggests that those deficits could represent an endophenotype for the psychotic dimension. The study of spontaneous EEG rhythms as well has suggested neuronal dysfunctions in major psychoses, including long-term temporal correlations deficits (up to 50 s) in alpha and beta bands, and power increases in theta-alpha and gamma bands. Machine learning studies have confirmed the possibility to employ EEG features in the prediction of diagnosis and treatment response, suggesting that the theta-alpha band coherence between frontal and temporo-parietal electrodes could predict the response to clozapine with high accuracy (85%) [56]. In bipolar disorder, EEG connectivity studies have shown alterations in the fronto-cortical neuronal interactions in different frequency bands and across different mood states [57]. The investigation of EEG rhythms during sleep also turned out to be very useful in the understanding of the pathophysiology of the disorder, evidencing anomalies in sleep patterns in all the phases of the disease [58]. These examples of application confirm the great potential of the EEG technique and suggest how the simultaneous EEG-fMRI investigations can provide added value to the identification of the neurobiological markers of psychiatric disorders.

The available EEG-fMRI evidence on psychiatric disorders, despite being limited, confirms the great potential of this methodology. Preliminary EEG-fMRI studies on ADHD in adulthood have allowed to further investigate the spatiotemporal organization of executive deficits characterizing the disorder [59, 60]. The fusion of ERP and fMRI data based on joint Independent Component Analysis has identified altered neurovascular responses during the hearing of tones in schizophrenic subjects [61]. Furthermore, EEG-driven fMRI analyses on subjects with high psychotic risk have evidenced BOLD response deficits in the brain network responsible for the gamma component of auditory ERPs. The emerged results confirm the potential utility of neurovascular markers not only in the phase of diagnostic characterization but also in the prediction of disease risk.

It should be noticed that the limited diffusion of simultaneous EEG-fMRI studies is largely due to the technical complexity of the EEG-fMRI methodology across multiple stages, from data acquisition to data pre-processing and fusion. Firstly, simultaneous EEG-fMRI acquisitions require EEG systems equipped with amplifiers and headcaps compatible with the magnetic resonance environment. The acquisition has to be performed by expert technicians for the preparation of the equipment and of the EEG cap, and the quality check of the acquired signal. Moreover, the processing of EEG-fMRI data, from the removal of EEG artifacts induced by the MR environment to the integration of the multimodal information itself, is complex and must be performed by engineers or technicians expert in signal and image processing [62]. For all these reasons, the extraction of integrated neuronal-metabolic-hemodynamic information is highly innovative. In this regard, the future of neuroscientific research resides in the integration of two or more modalities simultaneously, made possible by the development of more and more sophisticated acquisition systems, like the PET-MRI-EEG trimodal systems [63], and processing tools.

4 Conclusions

This chapter was dedicated to the description of some of the most promising brain structural and functional imaging modalities. Emphasis was put on the utility of each technique and even more on the potential of the integration of more techniques in the identification of the neurobiological markers of mental illnesses. The available evidence from neuroimaging and electrophysiological studies in psychiatric disorders, despite confirming the utility of the single methodologies, remarks the necessity of adopting a multimodal approach—which should include the integration of information of different kinds and/or relative to multiple observational scales—in order to obtain a more comprehensive picture on the cerebral mechanisms underlying the disease symptomatology. Up to now, data fusion approaches have been employed in a limited way and have regarded various combinations of techniques mainly applied to small samples, producing results that need further confirmations. In the big data era, the synergic integration of neuroimaging and neurosignaling data will make an essential contribution to the construction of a personalized and broad-spectrum model of brain anatomy and function, useful in the early identification of risk or diagnostic profiles and in the planning of effective treatment strategies.

Acknowledgments

Financial support was provided by the Italian Ministry of Health (GR-2019-12370616 to EM), the BIAL Foundation (Grant n° 288/20 to EM), and the Fondazione Cariplo (Award number 2019–3416 to PB, Award number 2020-1366 to EB).

References

1. Insel T, Cuthbert B, Garvey M, Heinssen R, Pine DS, Quinn K et al (2010) Research domain criteria (RDoC): toward a new classification framework for research on mental disorders. Am J Psychiatry 167(7):748–751. https://doi.org/10.1176/appi.ajp.2010.09091379
2. Atlas SW (2009) Magnetic resonance imaging of the brain and spine, vol Vol. 1. Lippincott Williams & Wilkins
3. Maggioni E (2014) Multimodal magnetic resonance imaging at 3 T and challenges for application at ultra-high field
4. O'Donoghue S, Holleran L, Cannon DM, McDonald C (2017) Anatomical dysconnectivity in bipolar disorder compared with schizophrenia: a selective review of structural network analyses using diffusion MRI. J Affect Disord 209:217–228. https://doi.org/10.1016/j.jad.2016.11.015
5. Friston K, Brown HR, Siemerkus J, Stephan KE (2016) The dysconnection hypothesis (2016). Schizophr Res 176(2–3):83–94. https://doi.org/10.1016/j.schres.2016.07.014
6. Maggioni E, Crespo-Facorro B, Nenadic I, Benedetti F, Gaser C, Sauer H et al (2017) Common and distinct structural features of schizophrenia and bipolar disorder: the European Network on Psychosis Affective disorders and Cognitive Trajectory (ENPACT) study. PLoS One 12(11):e0188000. https://doi.org/10.1371/journal.pone.0188000
7. Corradi-Dell'Acqua C, Tomelleri L, Bellani M, Rambaldelli G, Cerini R, Pozzi-Mucelli R et al (2012) Thalamic-insular dysconnectivity in schizophrenia: evidence from structural equation modeling. Hum Brain Mapp 33(3):740–752. https://doi.org/10.1002/hbm.21246
8. Agarwal N, Rambaldelli G, Perlini C, Dusi N, Kitis O, Bellani M et al (2008) Microstructural thalamic changes in schizophrenia: a combined anatomic and diffusion weighted magnetic resonance imaging study. J Psychiatry Neurosci 33(5):440–448. Retrieved from https://www.ncbi.nlm.nih.gov/pubmed/18787663
9. Andreone N, Tansella M, Cerini R, Versace A, Rambaldelli G, Perlini C et al (2007) Cortical white-matter microstructure in schizophrenia. Diffusion imaging study. Br J Psychiatry 191:113–119. https://doi.org/10.1192/bjp.bp.105.020990
10. Anticevic A, Cole MW, Repovs G, Murray JD, Brumbaugh MS, Winkler AM et al (2014) Characterizing thalamo-cortical disturbances in schizophrenia and bipolar illness. Cereb Cortex 24(12):3116–3130. https://doi.org/10.1093/cercor/bht165
11. Pettersson-Yeo W, Allen P, Benetti S, McGuire P, Mechelli A (2011) Dysconnectivity in schizophrenia: where are we now? Neurosci

Biobehav Rev 35(5):1110–1124. https://doi.org/10.1016/j.neubiorev.2010.11.004

12. Brambilla P, Riva MA, Melcangi RC, Diwadkar VA (2007) The role of glutamatergic pathways in schizophrenia: from animal models to human imaging studies. Clinical neuropsychiatry: journal of treatment. Evaluation
13. Sui J, Yu Q, He H, Pearlson G, Calhoun V (2012) A selective review of multimodal fusion methods in schizophrenia. Front Hum Neurosci 6(27). https://doi.org/10.3389/fnhum.2012.00027
14. Leroux E, Delcroix N, Dollfus S (2014) Left fronto-temporal dysconnectivity within the language network in schizophrenia: an fMRI and DTI study. Psychiatry Res 223(3): 261–267. https://doi.org/10.1016/j.pscychresns.2014.06.002
15. Michael AM, Baum SA, White T, Demirci O, Andreasen NC, Segall JM et al (2010) Does function follow form?: methods to fuse structural and functional brain images show decreased linkage in schizophrenia. NeuroImage 49(3):2626–2637. https://doi.org/10.1016/j.neuroimage.2009.08.056
16. Michael AM, King MD, Ehrlich S, Pearlson G, White T, Holt DJ et al (2011) A data-driven investigation of gray matter-function correlations in schizophrenia during a working memory task. Front Hum Neurosci 5:71. https://doi.org/10.3389/fnhum.2011.00071
17. Sasamoto A, Miyata J, Kubota M, Hirao K, Kawada R, Fujimoto S et al (2014) Global association between cortical thinning and white matter integrity reduction in schizophrenia. Schizophr Bull 40(2):420–427
18. Phillips ML, Swartz HA (2014) A critical appraisal of neuroimaging studies of bipolar disorder: toward a new conceptualization of underlying neural circuitry and a road map for future research. Am J Psychiatry 171(8): 829–843. https://doi.org/10.1176/appi.ajp.2014.13081008
19. Wang F, Kalmar JH, He Y, Jackowski M, Chepenik LG, Edmiston EE et al (2009) Functional and structural connectivity between the perigenual anterior cingulate and amygdala in bipolar disorder. Biol Psychiatry 66(5): 516–521. https://doi.org/10.1016/j.biopsych.2009.03.023
20. Delvecchio G, Maggioni E, Squarcina L, Arighi A, Galimberti D, Scarpini E et al (2020) A critical review on structural neuroimaging studies in BD: a Transdiagnostic perspective from psychosis to Fronto-Temporal Dementia. Curr Behav Neurosci Rep 7(2): 86–95
21. Altamura AC, Delvecchio G, Marotta G, Oldani L, Pigoni A, Ciappolino V et al (2017) Structural and metabolic differentiation between bipolar disorder with psychosis and substance-induced psychosis: an integrated MRI/PET study. Eur Psychiatry 41:85–94. https://doi.org/10.1016/j.eurpsy.2016.09.009
22. Maggioni E, Altamura AC, Brambilla P (2017) Exploring the neuroanatomical bases of psychotic features in bipolar disorder. Epidemiol Psychiatr Sci 26(4):358–363. https://doi.org/10.1017/S2045796017000087
23. Ivleva EI, Clementz BA, Dutcher AM, Arnold SJM, Jeon-Slaughter H, Aslan S et al (2017) Brain structure biomarkers in the psychosis biotypes: findings from the bipolar-schizophrenia network for intermediate phenotypes. Biol Psychiatry 82(1):26–39. https://doi.org/10.1016/j.biopsych.2016.08.030
24. Squarcina L, De Luca A, Bellani M, Brambilla P, Turkheimer FE, Bertoldo A (2015) Fractal analysis of MRI data for the characterization of patients with schizophrenia and bipolar disorder. Phys Med Biol 60(4): 1697–1716. https://doi.org/10.1088/0031-9155/60/4/1697
25. Squarcina L, Bellani M, Rossetti MG, Perlini C, Delvecchio G, Dusi N et al (2017) Similar white matter changes in schizophrenia and bipolar disorder: a tract-based spatial statistics study. PLoS One 12(6):e0178089. https://doi.org/10.1371/journal.pone.0178089
26. Rashid B, Damaraju E, Pearlson GD, Calhoun VD (2014) Dynamic connectivity states estimated from resting fMRI identify differences among schizophrenia bipolar disorder and healthy control subjects. Front Hum Neurosci 8:897. https://doi.org/10.3389/fnhum.2014.00897
27. Mellon EA, Beesam RS, Elliott MA, Reddy R (2010) Mapping of cerebral oxidative metabolism with MRI. Proc Natl Acad Sci U S A 107(26):11787–11792. https://doi.org/10.1073/pnas.1006951107
28. Haller S, Zaharchuk G, Thomas DL, Lovblad KO, Barkhof F, Golay X (2016) Arterial spin labeling perfusion of the brain: emerging clinical applications. Radiology 281(2):337–356. https://doi.org/10.1148/radiol.2016150789
29. Bellani M, Peruzzo D, Isola M, Rambaldelli G, Perlini C, Baiano M et al (2011) Cerebellar and lobar blood flow in schizophrenia: a perfusion weighted imaging study. Psychiatry Res 193(1):46–52. https://doi.org/10.1016/j.pscychresns.2010.12.010
30. Peruzzo D, Rambaldelli G, Bertoldo A, Bellani M, Cerini R, Silvia M et al (2011) The impact of schizophrenia on frontal perfusion parameters: a DSC-MRI study. J Neural

Transm (Vienna) 118(4):563–570. https://doi.org/10.1007/s00702-010-0548-7

31. Emiliani FE, Sedlak TW, Sawa A (2014) Oxidative stress and schizophrenia: recent breakthroughs from an old story. Curr Opin Psychiatry 27(3):185–190. https://doi.org/10.1097/YCO.0000000000000054
32. Liu J, Qiu M, Constable RT, Wexler BE (2012) Does baseline cerebral blood flow affect task-related blood oxygenation level dependent response in schizophrenia? Schizophr Res 140(1–3):143–148. https://doi.org/10.1016/j.schres.2012.06.028
33. Toma S, MacIntosh BJ, Swardfager W, Goldstein BI (2018) Cerebral blood flow in bipolar disorder: a systematic review. J Affect Disord 241:505–513. https://doi.org/10.1016/j.jad.2018.08.040
34. Karthikeyan S, Fiksenbaum L, Grigorian A, Lu H, MacIntosh BJ, Goldstein BI (2019) Normal cerebral oxygen consumption despite elevated cerebral blood flow in adolescents with bipolar disorder: putative neuroimaging evidence of anomalous energy metabolism. Front Psych 10:739. https://doi.org/10.3389/fpsyt.2019.00739
35. Dai W, Chen M, Duan W, Zhao L, Bolo NR, Tamminga C et al (2020) Abnormal perfusion fluctuation and perfusion connectivity in bipolar disorder measured by dynamic arterial spin labeling. Bipolar Disord 22(4):401–410. https://doi.org/10.1111/bdi.12856
36. Squarcina L, Perlini C, Peruzzo D, Castellani U, Marinelli V, Bellani M et al (2015) The use of dynamic susceptibility contrast (DSC) MRI to automatically classify patients with first episode psychosis. Schizophr Res 165(1):38–44. https://doi.org/10.1016/j.schres.2015.03.017
37. Mauri M, Nobile M, Bellina M, Crippa A, Brambilla P (2018) Light up ADHD: I. Cortical hemodynamic responses measured by functional Near Infrared Spectroscopy (fNIRS): Special Section on "Translational and Neuroscience Studies in Affective Disorders" Section Editor Maria Nobile MD PhD. This Section of JAD focuses on the relevance of translational and neuroscience studies in providing a better understanding of the neural basis of affective disorders. The main aim is to briefly summarise relevant research findings in clinical neuroscience with particular regards to specific innovative topics in mood and anxiety disorders. J Affect Disord 234:358–364. https://doi.org/10.1016/j.jad.2017.11.087
38. Maggioni E, Molteni E, Zucca C, Reni G, Cerutti S, Triulzi FM et al (2015) Investigation of negative BOLD responses in human brain through NIRS technique. A visual stimulation study. Neuroimage 108:410–422. https://doi.org/10.1016/j.neuroimage.2014.12.074
39. Takizawa R, Fukuda M, Kawasaki S, Kasai K, Mimura M, Pu S et al (2014) Neuroimaging-aided differential diagnosis of the depressive state. NeuroImage 85(Pt 1):498–507. https://doi.org/10.1016/j.neuroimage.2013.05.126
40. Zhu Y, Quan W, Wang H, Ma Y, Yan J, Zhang H et al (2018) Prefrontal activation during a working memory task differs between patients with unipolar and bipolar depression: a preliminary exploratory study. J Affect Disord 225:64–70. https://doi.org/10.1016/j.jad.2017.07.031
41. Ehlis AC, Herrmann MJ, Plichta MM, Fallgatter AJ (2007) Cortical activation during two verbal fluency tasks in schizophrenic patients and healthy controls as assessed by multi-channel near-infrared spectroscopy. Psychiatry Res 156(1):1–13. https://doi.org/10.1016/j.pscychresns.2006.11.007
42. Kubota Y, Toichi M, Shimizu M, Mason RA, Coconcea CM, Findling RL et al (2005) Prefrontal activation during verbal fluency tests in schizophrenia--a near-infrared spectroscopy (NIRS) study. Schizophr Res 77(1):65–73. https://doi.org/10.1016/j.schres.2005.01.007
43. Shinba T, Nagano M, Kariya N, Ogawa K, Shinozaki T, Shimosato S, Hoshi Y (2004) Near-infrared spectroscopy analysis of frontal lobe dysfunction in schizophrenia. Biol Psychiatry 55(2):154–164. https://doi.org/10.1016/s0006-3223(03)00547-x
44. Hoshi Y, Shinba T, Sato C, Doi N (2006) Resting hypofrontality in schizophrenia: a study using near-infrared time-resolved spectroscopy. Schizophr Res 84(2–3):411-420. https://doi.org/10.1016/j.schres.2006.03.010
45. Lee J, Folley BS, Gore J, Park S (2008) Origins of spatial working memory deficits in schizophrenia: an event-related FMRI and near-infrared spectroscopy study. PLoS One 3(3):e1760. https://doi.org/10.1371/journal.pone.0001760
46. Coughlin JM, Horti AG, Pomper MG (2019) Opportunities in precision psychiatry using PET neuroimaging in psychosis. Neurobiol Dis 131:104428. https://doi.org/10.1016/j.nbd.2019.03.016
47. Altamura AC, Bertoldo A, Marotta G, Paoli RA, Caletti E, Dragogna F et al (2013) White matter metabolism differentiates schizophrenia and bipolar disorder: a preliminary PET study. Psychiatry Res 214(3):410–414. https://doi.org/10.1016/j.pscychresns.2013.08.011

48. Howes OD, Kapur S (2014) A neurobiological hypothesis for the classification of schizophrenia: type A (hyperdopaminergic) and type B (normodopaminergic). Br J Psychiatry 205(1):1–3. https://doi.org/10.1192/bjp.bp.113.138578
49. Borgan F, Laurikainen H, Veronese M, Marques TR, Haaparanta-Solin M, Solin O et al (2019) In vivo availability of cannabinoid 1 receptor levels in patients with first-episode psychosis. JAMA Psychiatry. https://doi.org/10.1001/jamapsychiatry.2019.1427
50. Ashok AH, Marques TR, Jauhar S, Nour MM, Goodwin GM, Young AH, Howes OD (2017) The dopamine hypothesis of bipolar affective disorder: the state of the art and implications for treatment. Mol Psychiatry 22(5):666–679. https://doi.org/10.1038/mp.2017.16
51. Da Silva T, Hafizi S, Rusjan PM, Houle S, Wilson AA, Prce I et al (2019) GABA levels and TSPO expression in people at clinical high risk for psychosis and healthy volunteers: a PET-MRS study. J Psychiatry Neurosci 44(2): 111–119. https://doi.org/10.1503/jpn.170201
52. Borgan F, O'Daly O, Veronese M, Reis Marques T, Laurikainen H, Hietala J, Howes O (2019) The neural and molecular basis of working memory function in psychosis: a multimodal PET-fMRI study. Mol Psychiatry. https://doi.org/10.1038/s41380-019-0619-6
53. Delvecchio G, Mandolini GM, Arighi A, Prunas C, Mauri CM, Pietroboni AM et al (2019) Structural and metabolic cerebral alterations between elderly bipolar disorder and behavioural variant frontotemporal dementia: a combined MRI-PET study. Aust N Z J Psychiatry 53(5):413–423. https://doi.org/10.1177/0004867418815976
54. Maggioni E, Zucca C, Reni G, Cerutti S, Triulzi FM, Bianchi AM, Arrigoni F (2016) Investigation of the electrophysiological correlates of negative BOLD response during intermittent photic stimulation: an EEG-fMRI study. Hum Brain Mapp 37(6):2247–2262. https://doi.org/10.1002/hbm.23170
55. McLoughlin G, Makeig S, Tsuang MT (2014) In search of biomarkers in psychiatry: EEG-based measures of brain function. Am J Med Genet B Neuropsychiatr Genet 165B(2): 111–121. https://doi.org/10.1002/ajmg.b.32208
56. Hasey GM, Kiang M (2013) A review of recent literature employing electroencephalographic techniques to study the pathophysiology phenomenology and treatment response of schizophrenia. Curr Psychiatry Rep 15(9):388. https://doi.org/10.1007/s11920-013-0388-x
57. Maggioni E, Bianchi AM, Altamura AC, Soares JC, Brambilla P (2017) The putative role of neuronal network synchronization as a potential biomarker for bipolar disorder: a review of EEG studies. J Affect Disord 212:167–170. https://doi.org/10.1016/j.jad.2016.12.045
58. Zangani C, Casetta C, Saunders AS, Donati F, Maggioni E, D'Agostino A (2020) Sleep abnormalities across different clinical stages of Bipolar Disorder: a review of EEG studies. Neurosci Biobehav Rev 118:247–257
59. Karch S, Thalmeier T, Lutz J, Cerovecki A, Opgen-Rhein M, Hock B et al (2010) Neural correlates (ERP/fMRI) of voluntary selection in adult ADHD patients. Eur Arch Psychiatry Clin Neurosci 260(5):427–440. https://doi.org/10.1007/s00406-009-0089-y
60. Karch S, Voelker JM, Thalmeier T, Ertl M, Leicht G, Pogarell O, Mulert C (2014) Deficits during voluntary selection in adult patients with ADHD: new insights from single-trial coupling of simultaneous EEG/fMRI. Front Psych 5:41. https://doi.org/10.3389/fpsyt.2014.00041
61. Sui J, Pearlson G, Caprihan A, Adali T, Kiehl KA, Liu J et al (2011) Discriminating schizophrenia and bipolar disorder by fusing fMRI and DTI in a multimodal CCA+ joint ICA model. NeuroImage 57(3):839–855. https://doi.org/10.1016/j.neuroimage.2011.05.055
62. Maggioni E, Arrubla J, Warbrick T, Dammers J, Bianchi AM, Reni G et al (2014) Removal of pulse artefact from EEG data recorded in MR environment at 3T. Setting of ICA parameters for marking artefactual components: application to resting-state data. PLoS One 9(11):e112147. https://doi.org/10.1371/journal.pone.0112147
63. Del Guerra A, Ahmad S, Avram M, Belcari N, Berneking A, Biagi L et al (2018) TRIMAGE: a dedicated trimodality (PET/MR/EEG) imaging tool for schizophrenia. Eur Psychiatry 50: 7–20. https://doi.org/10.1016/j.eurpsy.2017.11.007

Chapter 16

Premises of Computational Neuroscience: Machine Learning Tools and Multivariate Analyses

Ferath Kherif, Cristina Ramponi, Adeliya Latypova, and Rositsa Paunova

Abstract

This chapter provides an overview of multivariate methods and several advanced demonstrations of how these methods can be applied to machine learning applications in mental health. Multivariate methods can help identify relationships between different variables (symptoms, behavior, brain anatomy, brain function, genetics, etc.) that may influence mental health. The methods provide valuable insights into how best to target interventions for a particular brain disorder. When using these methods, it is important to know the advantages, but also the limitations, of each method and how they can be applied in different contexts. In this chapter, we define and explain the concept of multivariate analyses from the theoretical basis to practical issues related to data preparation and interpretation of results. The methods described here include factor analysis, principal component analysis, path analysis, partial least squares, and linear multivariate methods. The chapter includes examples of how multivariate methods have been used, for example, to predict treatment outcomes or identify risk factors for mental disorders. We not only discuss the many challenges of using these methods in computational neuroscience research, such as the need for large and diverse datasets, but also introduce new approaches, such as guided model-based approaches and advanced AI-based approaches, to enrich these mostly data-driven methods and obtain better insight that integrates a priori information.

Key words Multivariate brain, Mental health, Machine learning

1 Introduction

Multivariate methods [1–3] are powerful tools for understanding complex systemic psychiatric disorders [4] and are becoming increasingly popular because they can bridge the gap between mental health research and computational neuroscience by allowing the simultaneous analysis of multiple variables. Multivariate methods are clearly beneficial for deep integrating of mental health and neuroscience research as a variety of factors can be integrated, e.g., demographic data (age, gender, socioeconomic status), symptoms, and behavior in combination with many other biological data (genetic data, brain imaging, ...). By considering multiple variables

Drozdstoy Stoyanov et al. (eds.), *Computational Neuroscience*, Neuromethods, vol. 199,
https://doi.org/10.1007/978-1-0716-3230-7_16,

from these different sources of information simultaneously, neuroscience researchers can more accurately identify patterns and relationships between these different scales of observation, from biological to sociological factors. These vertical and cross-scale patterns that can be identified with multivariate methods contribute to a more comprehensive understanding of underlying causes and their effects than traditional univariate methods.

This chapter provides an introduction to multivariate methods and shows how they are applied in studies of various psychiatric disorders, including schizophrenic, unipolar, and bipolar patients. With a better understanding of multivariate methods, neuroscience researchers and clinicians can use these powerful tools to develop more effective treatments for mental health problems. Through the use of advanced algorithms and machine learning, these methods can help to better understand the complexity of mental health problems and the various factors that can influence the course of the disease. They can also be used to develop machine learning applications to help diagnose patients. For example, a personalized neural signature created with multivariate tools and a machine learning application could predict what type of therapy would be most effective for a particular patient, taking into account not only age, gender, and symptoms but also brain anatomy and function [5].

In this chapter, we briefly review the basics of multivariate methods and discuss how they can be used for machine learning applications in mental health. This chapter is divided into three parts. First, we will discuss the methodology of multivariate methods, including a description of the different types of methods, and their mathematical basis with singular value decomposition as the basic building block. To demonstrate their capabilities, we will show several applications, give a step-by-step implementation guide and show the different types of results that can be obtained. Finally, we will give an outlook on the future and present the different codes via the Multivariate Linear Model (MLM) toolbox.

2 Guide to the Methodology

2.1 Overview

Multivariate techniques [1] fall between two categories (see Fig. 1), the fully data-driven methods, which do not use a model and simply rely on the data provided to identify the key components that explain the data, captured by a global measure such as the variance of the data. Some multivariate methods are a mixture of data-driven and model-driven methods. The predefined model is used to constrain the search for the main components that explain the data, and it also allows inferences to be made about the variables included to obtain a measure of the signal-to-noise ratio.

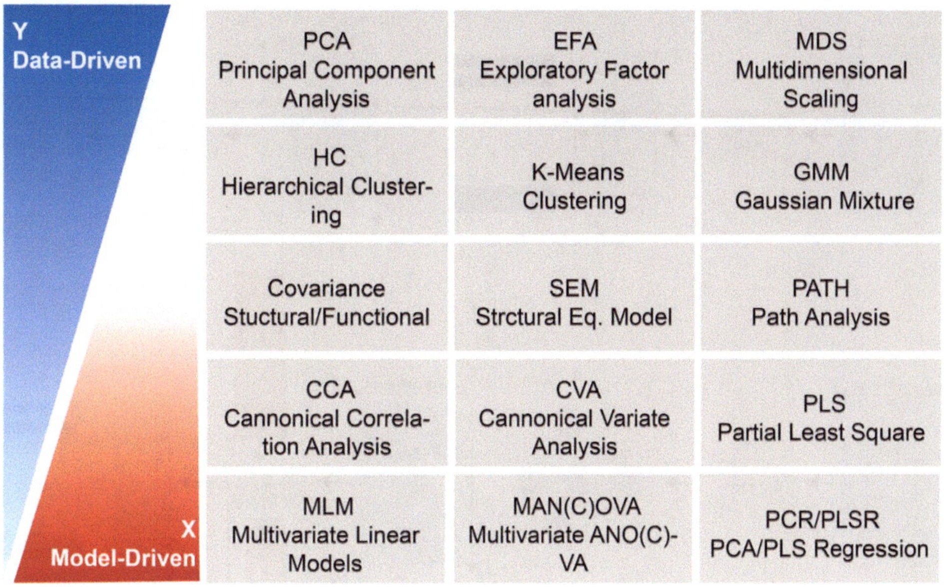

Fig. 1 Type of multivariate methods. Multivariate methods range from fully data-driven (PCA, EFA, and MDS) to model-driven approach (MLM, MANOVA, and PLSR)

- **PCA, EFA, and MDS:** The goal of multivariate analyses that are fully data-driven, such as PCA, MDS or factor analyses, is to represent the original data in a lower dimensional space without losing key information in order to discover the patterns underlying the data (see Fig. 2a).
- **Hierarchical, K-means, and GMM clustering:** Neuroscience researchers can better understand the relationships between different variables by using cluster analyses that group similar variables together (see Fig. 2b). Clustering methods can be used to classify subjects into subtypes [6] or voxels into functionally homogeneous brain regions [7].
- **Structural or functional covariance, SEM, and PATH analysis:** Neuroscience researchers can use structural equation modeling, path analysis or structural/functional covariance [8–10] to understand the causal relationships between different variables (see Fig. 2c).
- **CCA, CVA, and PLS:** Multivariate methods such as CCA, CVA [3, 11, 12], and PLS [13] are powerful tools for uncovering hidden relationships between independent (X) and dependent variables (Y) (see Fig. 2d).
- **MLM, MANOVA, and PCR/PLSR:** Multivariate linear models [14, 15] extend the previous approach and build on the variables included in the model (X) to allow for hypotheses testing of the variables contained in (Y) (see Fig. 2d).

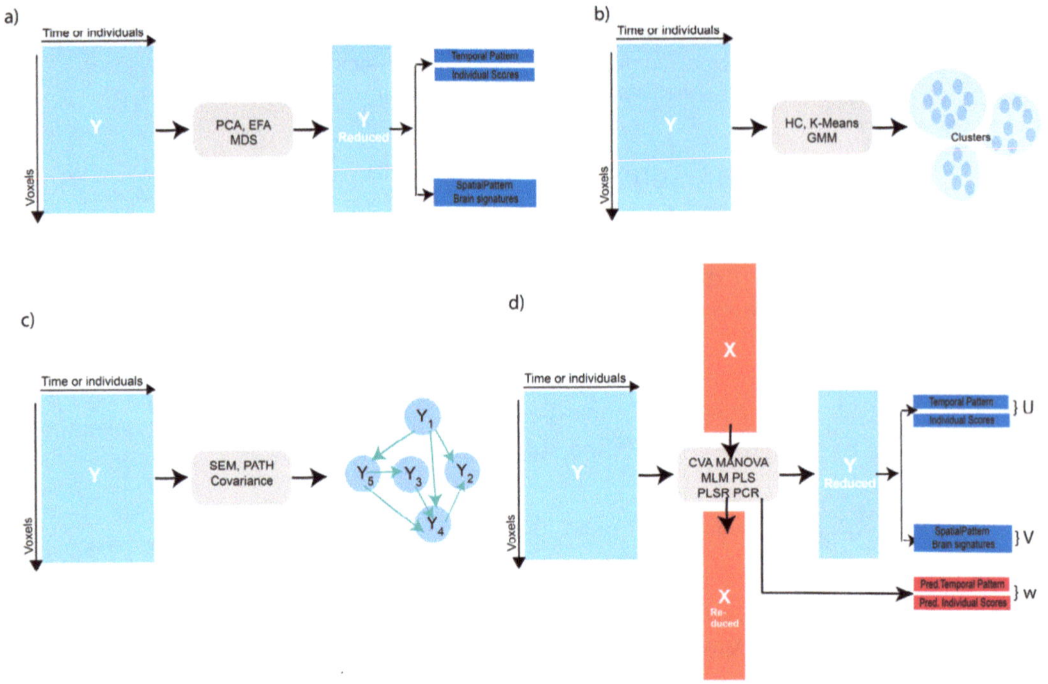

Fig. 2 Decomposition in the multivariate methods. (**a**) PCA and EFA are used to reduce the complexity of the dataset by identifying the components that best describe the variance in the data. The neuroimaging data (*Y*), such as fMRI or VBM, are transformed into orthogonal components in either the time or subject dimension and into a combined component representing the voxels, also known as eigenimages or brain signatures. (**b**) Clustering method K-Means and GMM are used to group the data points into subgroups based on their similarity. K-Means is a simple method where each data point is assigned to a single cluster, while GMM is a probabilistic method where each data point is assigned a probability of belonging to multiple clusters. (**c**) SEM and path analysis are techniques used to model the relationships between variables within a dataset. In neuroimaging, correlation-based approaches are more commonly used because of their simplicity and their ability to handle a larger number of variables or brain regions. (**d**) Bilinear methods such as PLS, CCA, and MLM are used to analyze the relationship between two groups of variables. MANOVA, PLSR, and PCR use the traditional univariate general linear model once the dimensions of *X* or *Y* or both have been reduced to a few components

Each of these methods can be used to identify potential mental health interventions and treatments. Multivariate techniques can help explain phenomena [16–18] that are difficult to understand using traditional statistical approaches, univariate statistics.

2.2 Mathematical Formulations

The Principal Component Analysis (PCA) is probably the most popular and used method for data exploration and data analysis. The other multivariate methods, such as Partial Least Squares (PLS), Multivariate Linear Model (MLM), and Canonical Variate Analyses (CVA) could be considered as PCA extensions. In this section, we would like to deep into description of these methods, the principle, and implementation.

The main theorem that underlies the multivariate methods is SVD. It states that the data matrix Y of size $n \times p$ can be decomposed in rank-1 vectors U and V and diagonal matrix S such as

$$Y = USV^T$$

The U is an orthonormal matrix of size $n \times p$, where each of the p columns represents the eigenvector of Y, $UU^T = U^T U = 1$ and $U^{-1} = U^T$. The V is an orthonormal matrix of size $p \times p$, where each of the p columns represents the eigenvector of Y^T, $VV^T = V^T V = 1$ and $V^{-1} = V^T$. And the S is a diagonal matrix of size $p \times p$, where the diagonal elements are ordered eigenvalues $\lambda_1 > \lambda_2 > \ldots > \lambda_p$ and called singular values.

$$S = \begin{bmatrix} \lambda_1 & 0 & \cdots & 0 \\ 0 & \lambda_2 & \cdots & 0 \\ \vdots & \vdots & \ddots & \vdots \\ 0 & 0 & \cdots & \lambda_p \end{bmatrix}$$

In the Principal Component Analysis, the original data is projected in a new coordinate system where the axes, which are called principal axes, correspond to directions where the data vary the most. The first axis, or the first principal axis, catches the highest variation, and the last axis catches the lowest. The eigenvectors of the data matrix Y actually represent the principle axes. The corresponding eigenvalue shows how much variance is explained by the principle axes when the data are projected on it. The sum of the eigenvalues represents the total variance of the data. Since the last components capture the lowest variance, they can be omitted. Thus, when the eigenvalues and eigenvectors are calculated, we can reconstruct data Y in lower dimensional space using only the first r principal components with largest variance explained:

$$Y_{rec} = U(:; 1 : r) diag(S_{[1 \ldots r]}) V(:, 1 : r)^T$$

The number of principal components to retain r is typically estimated through the cumulative variance explained and chosen threshold, e.g., 80%.

$$(percent) Var_i = \frac{\lambda_i}{\sum \lambda_i} * 100$$

$$(cumulative) Var_r = \sum_{1}^{r} percent Var_i$$

An alternative way to find the principal components is to calculate the eigenvectors of the covariance matrices $Y^T Y$ and YY^T. The eigenvectors of these two square symmetric matrices are the V

Table 1
Overview of multivariate methods and their implementation using SVD method. The second column represents the input information matrix that should be decomposed. We provide some shortcomings of each method. In the table, *Y* stands for data, *X* is the linear model, *R* is the orthogonal projector onto the residual space, and Σ is the temporal covariance matrix. Adapted from [14] and [15]

Method	Information matrix to decompose by SVD	Shortcomings
PCA	Y	No prior knowledge included
		Not data scale invariant
PLS	X^TY	Not model and data scale invariant
		Problem of temporal correlation
Orthonormalized PLS	$(X^TX)^{-1/2}X^TY$	Not data scale invariant
		Problem of temporal correlation
CVA	$(X^TX)^{-1/2}X^TY(Y^TRY)^{-1/2}$	Problem of $(Y^TRY)^{-1/2}$ computation
SVD-CVA	$Y_k = U_kS_kV_k(X^TX)^{-1/2}X^TY_k(Y_k^TY_k)^{-1/2}$	Problem of finding k
MLM	$(X^T\Sigma X)^{-1/2}X^TY$	Not data scale invariant

and U, respectively, that we described before. Singular values S correspond to the square roots of eigenvalues. We have the following relationships:

$$Y = USV^T$$
$$Y^TYV = VS^2 - > V = eig(Y^TY)$$
$$YY^TU = US^2 - > U = eig(YY^T)$$

Knowing the SVD theorem and PCA principle, in Table 1, we describe other multivariate methods, according to [15]. They could be considered as extension of PCA method, differing in the input information matrix that is decomposed with SVD. The advantage of extended methods is that they take into account not only data Y but also design matrix X that may include contextual, experimental, behavioral, and other information. In particular, the multivariate linear model analysis also includes in the model the temporal auto-correlation from the noise Σ and thus could be used, for example, in studies based on functional MRI.

2.3 Benefits and Limitations of Using Multivariate Methods for Mental Health

To develop clinically useful models of psychiatric and neurological disorders, neuroscience researchers need to integrate "omic" data with clinical measures of disease manifestation and progression. As there is growing evidence that not only biological but also environmental and social factors have a significant impact across the lifespan, researchers are also incorporating these other factors to gain a comprehensive understanding of how diseases manifest and progress. As a result, data from comprehensive "omic" studies (genomes, proteomes, metabolomes, microbiomes, multimodal neuroimaging) have increased in size. Thanks to the availability of vast amounts of "omic-"data (UK Biobank, ADNI, OASIS), we are now able to develop diagnostic tools for brain disorders using artificial intelligence and machine learning. Multivariate methods for mental health machine learning have not yet achieved as spectacular results in neuroscience as it has in other fields (vision, translation, etc.). There are three fundamental problems that could explain the slow development of medical neuroscience. Each method has its own challenges when used in mental health machine learning applications. The following are the major challenges, ranked from easiest to hardest:

- **Interpretability**: The first challenge is that any model used in the clinic must be interpretable. Even though multivariate methods simplify and interpret data, researchers may have difficulty interpreting the results. This is especially true when we are trying to understand the results of network or cluster analyses. Factor or PCA analysis can be helpful in reducing the number of variables in a dataset, but it can also be difficult to interpret the combination that makes up the eigenvectors.
- **Causality**: It can be difficult to interpret the latent factors and find out how they influence a mental health system.
- **Scaling, bias, and noise distribution**: The type of data used in neuroscience is more noisy than typical input for ML/AI analysis (images, sound).
- **Selection**: It can also be difficult to figure out which variables should be included in the analysis.
- **Dimensionality**: Multivariate methods are absolutely necessary to reduce dimensions without losing critical information, and however, sometimes there are too many features compared to the size of the dataset, which makes it difficult for getting a unique solution. This is a problem of high dimensionality in medicine and neuroscience that is different from other fields. The data matrices (Υ of $n \times p$) resulting from the compilation of these datasets contain many more variables (columns) per individual than there are individuals (several thousand, millions if genetic or neuroimaging variables are included).

3 Examples of Application of Multivariate Methods in Mental Health

We present here some examples of research applications where the multivariate methods have been used to analyze and discriminate psychiatric diagnoses. Fundamentally, psychiatric diagnoses are based on patients' reports (self-assessment scales), the clinician's evaluation (clinical-rating scales), and information coming from patients' relatives. This essentially leads to subjectivity in terms of the diagnostic approach. Diagnoses in psychiatry are not defined by biological signatures as in other medical specialties. As a result, we need a valid construct for their precise characteristics. In this chapter, we show how combining neuroimaging tools, such as structural and functional magnetic resonance imaging, as well as additional data including experimental, behavioral, and others with clinical assessment tools, and analyzing them with advanced statistical tools should help to bridge this gap. In the following examples, we show that multivariate methods allow us to identify specific brain signatures that can distinguish between diagnostic groups, and thus patients can then be classified accordingly.

3.1 Multivariate Methods Applied to the Classification of Schizophrenia

3.1.1 Objective

In this example, we present a study by Kawasaki et al. [19] published in 2007. This was one of the first studies to use a data-driven multivariate method to calculate an index that can later be used to classify patients into different disease categories based on the anatomical features of their brain. The authors hypothesized that regional gray matter differences between schizophrenia patients and healthy controls can be captured using structural MRI images and advanced statistical analysis. These differences can then be used to identify individuals with schizophrenia.

3.1.2 Data Used

The recruited patients and controls were scanned with a three-dimensional gradient-echo sequence (fast low-angle shot, FLASH). The images were preprocessed and analyzed with standard voxel-based morphometry (VBM) protocol [20] and partitioned into gray matter, white matter, cerebrospinal fluid, and others.

3.1.3 Method Used

As a method, the authors applied MLM analysis to voxel-based morphometry [21]. The study was conducted in two steps: first, the statistical model that classified schizophrenia patients and healthy controls (30 male/30 male) was obtained, and second, the statistical model was validated by classifying a new cohort of schizophrenia patients and healthy controls (16 males/16 males). The patterns of gray matter distributions that differ the most between the two groups were extracted with the MLM method. Since there is no temporal covariance matrix, the matrix of the methods was an orthonormalized PLS: $(X^TX)^{-1/2}X^TY$, where Y

is the data and X is the linear model. Thus, the normalized correlation between the data and a set of regressors from the design matrix was first calculated. Next, from the decomposition of the correlation matrix, the authors obtained the spatial eigenvectors, referred to as so-called "eigenimages," since the input data are the images. As the method operates on voxel-by-voxel correlation matrices, the extracted eigenimage, which best represented the variance in the correlation, reflected patterns of correlated gray matter concentrations.

3.1.4 Results

In the first step, 90% of the subjects were correctly classified by the eigenimage. In the second validation part, where the eigenimage from the first cohort was applied to the second one, the model could correctly identify 80% of the schizophrenia patients and healthy controls. These results indicate that a multivariate approach on structural MRI images can capture the main source of variation in schizophrenia and that the predictive power of the multivariate method generalizes to an independent dataset (see Fig. 3).

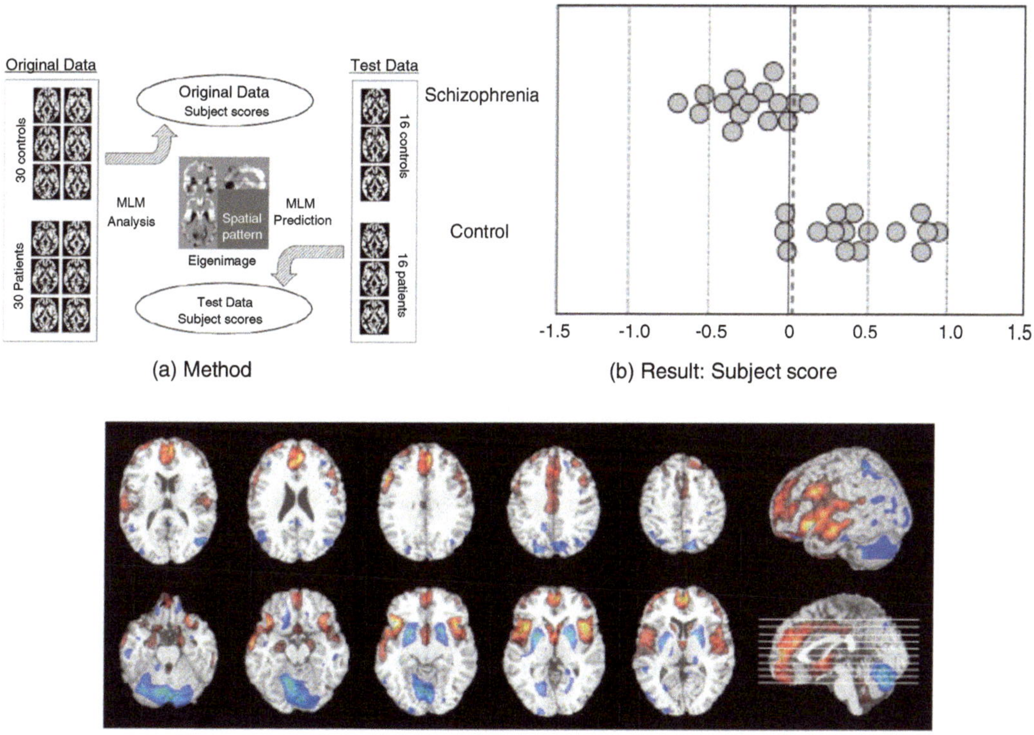

(a) Method

(b) Result: Subject score

(c) Result: Brain signature

Fig. 3 (**a**) Discriminant analysis using VBM and MLM methods. MLM analysis was used to obtain an eigenimage (**c**) and subject scores for the original data. (**b**) By using this eigenimage as a predictor in separate test data, new subject scores could be calculated and used for classification

3.2 Individual- and Group-Level Brain Signatures of Schizophrenia, Major Depressive, and Bipolar Disorders

In this section, we present the study [22]. The MLM was applied on the fMRI images of medicated depressed patients and matched healthy controls. The authors aimed to differentiate two groups by carrying out individual- and group-level multivariate analysis and final discriminant analysis.

3.2.1 Objective

3.2.2 Data

The fMRI paradigm consisted of the visual presentation of positive, negative, and neutral (PS, DS, and DN) pictures.

3.2.3 Method

MLM analysis is first performed at the individual level to determine the eigenimages for each participant that explain most of the changes in fMRI activity associated with each clinical condition (PS, DS, and DN). The paradigm of fMRI study was represented in a design matrix X. It encoded the three types of stimuli (PS, DS, and DN) and nuisance covariates that included the six rigid body motion parameters. For each subject i ($i=1, \ldots, s$), the matrix $(X_i^T \Sigma_i X_i)^{-1/2} X_i^T \Upsilon_i$ was computed and decomposed. In this case, X_i is a design matrix (time by covariates), Υ_i is a data matrix (time by voxel), and Σ_i is a temporal covariance matrix of the data. Through the correlation matrix decomposition, the authors calculate the model parameters eigenvectors U_i that are referred as clinical loadings and the spatial eigenvectors V_i that are referred as eigenimages. To consider only three active conditions (PS, DS, and DN), the authors use F-contrast encompassing these conditions. As a result, three eigenimages for each subject are obtained.

In the next step, the MLM analysis is carried out at the group level. The correlation matrix is represented as $(X_G^T \Sigma_G X_G)^{-1/2} X_G^T \Upsilon_G$, where X_G is the design matrix (subjects by covariates (diagnostic groups, age, gender)), and $\Upsilon_G = [V_1, V_2, \ldots, V_s]$ is a concatenation of eigenimages (the number of active conditions by voxel) of each subject. Through decomposition, the authors calculate eigenvectors V_G that identify the most consistent brain pattern across individuals in terms of variance explained and subject loadings U_G that represent the contribution of each subject to the main brain pattern.

In the final step, the linear discriminant analysis classifier is applied on each of the three subject loadings and used to assess power of the brain signatures to differentiate between two groups of subjects.

3.2.4 Results

With MLM, the authors could produce three brain patterns that summarized all the individual variabilities of the individual brain patterns. The discriminant analysis conducted next yielded accuracy levels for the three brain signatures ranging from 67 to 98% (see Fig. 4).

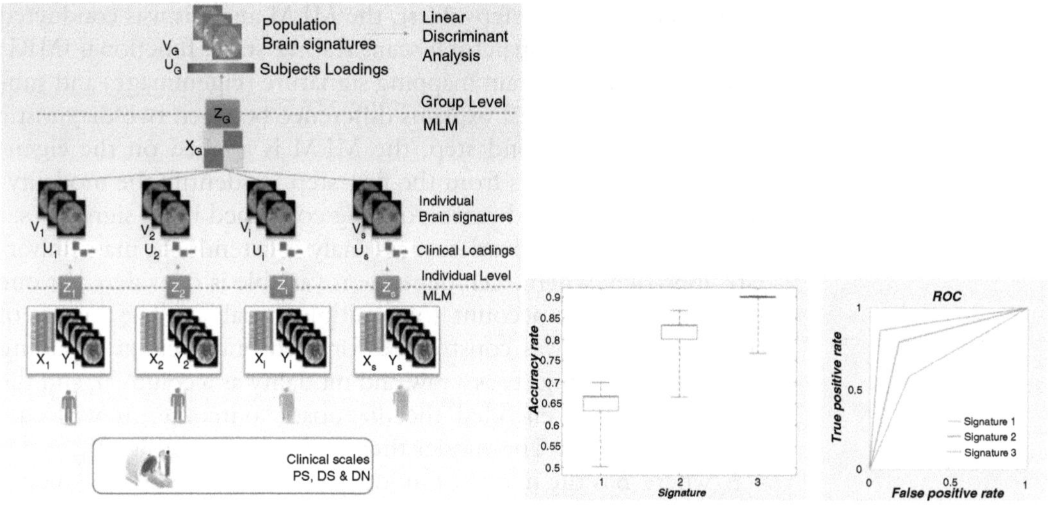

Fig. 4 (**a**) Discriminant analysis using VBM and MLM methods. MLM analysis was used to obtain an eigenimage (**c**) and subject scores for the original data. (**b**) By using this eigenimage as a predictor in separate test data, new subject scores could be calculated and used for classification

3.3 Multimodel Brain Signature with Task-fMRI, Resting State, and Morphometry in Schizophrenia and Major Depressive Disorder

3.3.1 Objective

In this section, we consider two studies: [23] and [24]. They are both based on the same dataset collected from the group of patients with schizophrenia and depression, but a different type of analysis was carried out. The study [23] investigated the contribution of the various MRI modalities to brain signatures and their ability to discriminate the two diagnostic groups. In the study [24], rather than analyzing each modality separately and determining their association with the diagnosis, the hypothesis tested was that a multivariate signature based on all of the modalities existed for each brain region.

3.3.2 Data

The input data for analysis included voxel-based morphometry, functional resting state, task-related neuroimaging, and the relevant clinical measures.

3.3.3 Methods

The participants underwent scanning with three different MRI sequences, high-resolution structural scan, resting state, and task-based functional MRI that included depression specific (DS), paranoid specific (PS), and diagnostically neutral (DN) blocks. First, the VBM analysis was carried out on the structural images to extract total intracranial volume (TIV) for each participant to account for individual differences in head size later in the model. Next, the task-related fMRI was preprocessed to obtain F-contrast for all active conditions orthogonal to the motion effect. The resting state fMRI was preprocessed to obtain individual residual mean square images. In [24], the authors parceled brain images for each modality (individual atlas) and obtained mean value for each brain region for each subject and modality. In [23], the multivariate analysis was implemented in two steps. First, the MLM analysis was conducted for each modality (structural scan, resting state, functional fMRI) to identify optimal brain mapping signature (eigenimage) and subjects loadings that best explains difference between two diagnostic groups. On the second step, the MLM is applied on the eigenimages and the scores from the first step to identify the modality-specific signatures and corresponding combined brain signatures.

In [24], the multivariate mass analysis extends the mass univariate approach, where each dependent variable is considered at one point in time, to account for multiple variables. The matrix of observation Y_{reg} was constructed for each brain region including all modalities (subjects as a raw and modality as a column), and the design matrix X encoded the diagnostic outcome, as well age, gender, and TIV. The model then was represented as $Y_{reg} = XB + E$, where B is the matrix of model parameters and E is the matrix of errors. To test the significance between diagnostic groups, the authors analyzed eigenvalues of the variance matrices.

3.3.4 Results

The study [23] demonstrates the differential contribution of the various MRI modalities as combined in principal components to brain signatures with high capacity for discrimination of the two diagnostic entities studied. The study [24] could define several brain regions with significant relevance to the diagnosis. These studies deliver evidence that a multimodal neuroimaging approach can potentially enhance the validity of psychiatric diagnoses (see Fig. 5).

4 Code and Toolbox Availability

The methods described in this chapter are available in MLM Toolbox, a toolbox developed in Matlab for use with SPM neuroimaging software. MLM is an easy-to-use open-source toolbox for SPM based on a modular implementation that allows neuroscience researchers to efficiently analyze multivariate data using different methods (PCA, PLS or MLM). MLM's wide range of input formats (nii format, gifti format or other variables coded in the design matrix), advanced analysis functions, and interactive visualization tools help researchers draw conclusions from the results (eigenimages and subject scores).

5 Conclusion

Mental health disorders are difficult to diagnose and treat, and many psychiatric disorders are diagnosed on the basis of apparent symptoms without quantitative or biological measurements such as those that can be obtained through neuroimaging. By combining

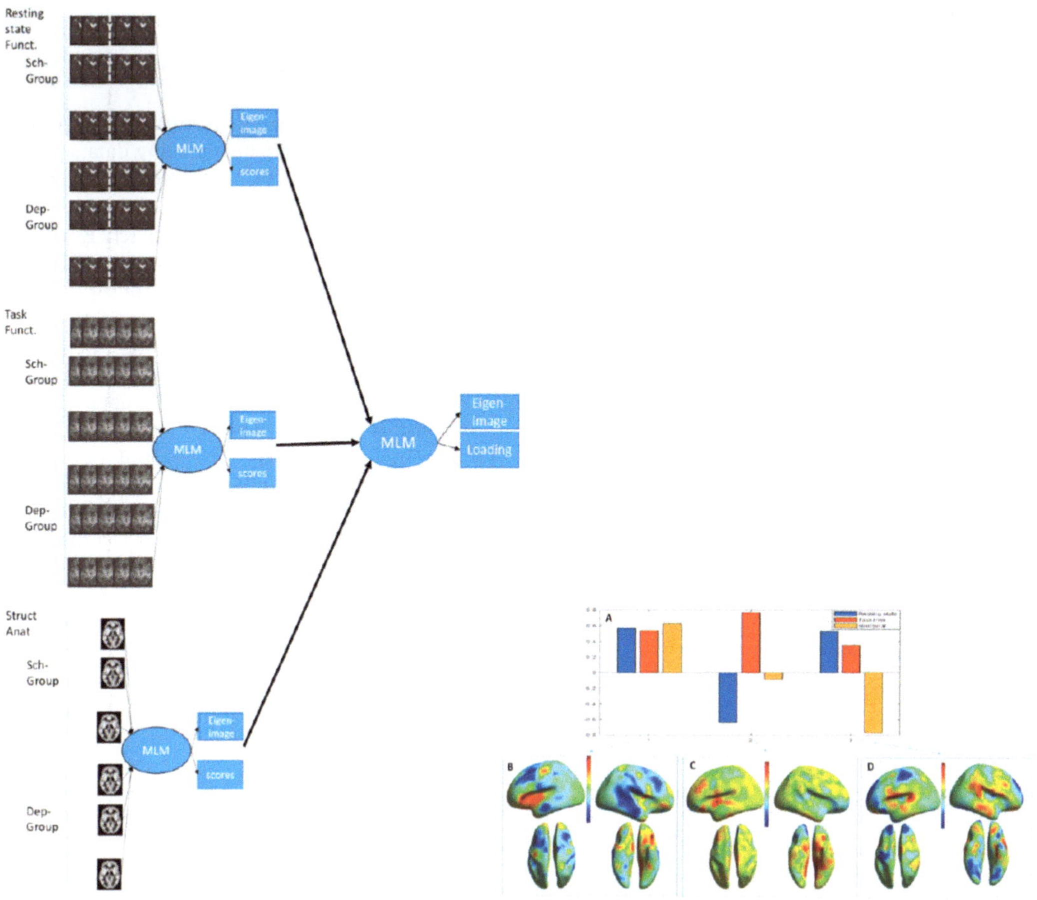

(a) Method (b) Result

Fig. 5 MLM analysis with two stages across all modalities: (**a**) Overview of the method. (**b**) Results. (A) shows bar graphs with the contribution of each modality (resting state fMRI in blue, task fMR in red, and structural MRI in yellow). For the first component, the contribution is almost the same for all modalities. (B–D) are the corresponding eigenimages computed with MLM and projected onto a 3D surface. The voxel values in the eigenimages represent the correlation of the value across all subjects at that voxel with the identified principal components

different imaging techniques, we have shown that machine learning and multivariate analysis are very powerful in determining latent factors underlying mental disorders in the form of neural signatures. In terms of predicting diagnoses and treatment outcomes, these methods could be more accurate than traditional clinical methods such as interviews and psychiatric assessments. Combined with the latest AI tools, they could pave the way for prevention strategies and personalized therapies and enable therapeutic interventions that target specific brain circuits. However, interdisciplinary research in computer science, engineering, mathematics, psychology, and psychiatry is needed to drive the development of practical applications.

References

1. Kherif F, Latypova A (2020) Principal component analysis. Machine Learning, 209–225
2. Tabachnick B, Fidell L, Ullman J (2013) Using multivariate statistics. Pearson, Boston, MA
3. Friston K, Frith C, Frackowiak R, Turner R (1995) Characterizing dynamic brain responses with fMRI: a multivariate approach. Neuroimage 2:166–172
4. Lima A, Mridha M, Das S, Kabir M, Islam M, Watanobe Y (2022) A comprehensive survey on the detection, classification, and challenges of neurological disorders. Biology:11. https://www.mdpi.com/2079-7737/11/3/469
5. Oala L, Fehr J, Gilli L, Balachandran P, Leite A, Calderon-Ramirez S, Li D, Nobis G, Alvarado E, Jaramillo-Gutierrez G, Others (2020) Ml4h auditing: From paper to practice. Mach Learn Health, 280–317
6. Kherif F, Josse G, Seghier M, Price C (2009) The main sources of intersubject variability in neuronal activation for reading aloud. J Cogn Neurosci 21:654–668
7. Simon O, Kherif F, Flandin G, Poline J, Riviere D, Mangin J, Le Bihan D, Dehaene S (2004) Automatized clustering and functional geometry of human parietofrontal networks for language, space, and number. Neuroimage 23:1192–1202
8. Melie-Garcia L, Slater D, Ruef A, Sanabria-Diaz G, Preisig M, Kherif F, Draganski B, Lutti A (2018) Networks of myelin covariance. Hum Brain Mapp 39:1532–1554
9. Sanabria-Diaz G, Melie-Garcia L, Draganski B, Demonet J, Kherif F (2021) Apolipoprotein E4 effects on topological brain network organization in mild cognitive impairment. Scientific Reports 11:845
10. Sanabria-Diaz G, Demonet J, Rodriguez-Herreros B, Draganski B, Kherif F, Melie-Garcia L (2021) Apolipoprotein E allele 4 effects on Single-Subject Gray Matter Networks in Mild Cognitive Impairment. NeuroImage Clinical 32:102799
11. Mihalik A, Chapman J, Adams R, Winter N, Ferreira F, Shawe-Taylor J, Mourão-Miranda J, Initiative A, Others (2022) Canonical correlation analysis and partial least squares for identifying brain-behaviour associations: a tutorial and a comparative study. Biol Psychiatry Cogn Neurosci Neuroimag 77(11):1055–1067
12. Wang H, Smallwood J, Mourao-Miranda J, Xia C, Satterthwaite T, Bassett D, Bzdok D (2020) Finding the needle in a high-dimensional haystack: Canonical correlation analysis for neuroscientists. NeuroImage 216: 116745
13. McIntosh A, Lobaugh N (2004) Partial least squares analysis of neuroimaging data: applications and advances. Neuroimage 23:S250–S263
14. Worsley KJ, Poline JB, Friston KJ, Evans AC (1997) Characterizing the response of PET and fMRI data using multivariate linear models. NeuroImage 6(4):305–319
15. Kherif F, Poline JB, Flandin G, Benali H, Simon O, Dehaene S, et al. (2002) Multivariate model specification for fMRI data. NeuroImage 16(4):1068–1083
16. Zufferey V, Donati A, Popp J, Meuli R, Rossier J, Frackowiak R, Draganski B, Gunten A, Kherif F (2017) Neuroticism, depression, and anxiety traits exacerbate the state of cognitive impairment and hippocampal vulnerability to Alzheimer's disease. Alzheimer's Dementia Diagnosis Assessment Disease Monitoring 7:107–114
17. Draganski B, Lutti A, Kherif F (2013) Impact of brain aging and neurodegeneration on cognition: evidence from MRI. Curr Opin Neurol 26:640–645
18. Draganski B, Kherif F, Lutti A (2014) Computational anatomy for studying use-dependant brain plasticity. Front Hum Neurosci 8:380
19. Kawasaki Y, Suzuki M, Kherif F, Takahashi T, Zhou S-Y, Nakamura K, Matsui M, Sumiyoshi T, Seto H, Kurachi M (2007) Multivariate voxel-based morphometry successfully differentiates schizophrenia patients from healthy controls. NeuroImage 34(1): 235–242. ISSN 1053-8119. https://doi.org/10.1016/j.neuroimage.2006.08.018
20. Good CD, Johnsrude IS, Ashburner J, Henson RN, Friston KJ, Frackowiak RS (2001) A voxel-based morphometric study of ageing in 465 normal adult human brains. Neuroimage 14(1):21–36
21. Ashburner J, Friston KJ (2000) Voxel-based morphometry—the methods. NeuroImage 11(6):805–821. ISSN 1053-8119. https://doi.org/10.1006/nimg.2000.0582
22. Kandilarova S, Stoyanov D, Stoeva M, et al. (2020) Functional MRI in depression—multivariate analysis of emotional task. J Med Biol Eng 40:535–544. https://doi.org/10.1007/s40846-020-00547-2
23. Stoyanov D, Kandilarova S, Aryutova K, Paunova R, Todeva-Radneva A, Latypova A, Kherif F (2021) Multivariate analysis of structural and functional neuroimaging can inform

psychiatric differential diagnosis. Diagnostics 11:19. https://doi.org/10.3390/diagnostics11010019

24. Paunova R, Kandilarova S, Todeva-Radneva A, Latypova A, Kherif F, Stoyanov D (2022) Application of mass multivariate analysis on neuroimaging data sets for precision diagnostics of depression. Diagnostics 12:469. https://doi.org/10.3390/diagnostics12020469

27. Taubert M, Roggenhofer E, Melie-Garcia L, Muller S, Lehmann N, Preisig M, Vollenweider P, Marques-Vidal P, Lutti A, Kherif F, Others (2020) Converging patterns of aging-associated brain volume loss and tissue microstructure differences. Neurobiol Aging 88:108–118

28. Marchewka A, Kherif F, Krueger G, Grabowska A, Frackowiak R, Draganski B, Initiative A (2014) Influence of magnetic field strength and image registration strategy on voxel-based morphometry in a study of Alzheimer's disease. Hum Brain Mapp 35:1865–1874

Index

Drozdstoy Stoyanov et al. (eds.), *Computational Neuroscience*, Neuromethods, vol. 199,
https://doi.org/10.1007/978-1-0716-3230-7,

MIX
Papier aus verantwortungsvollen Quellen
Paper from responsible sources
FSC® C105338

If you have any concerns about our products,
you can contact us on
ProductSafety@springernature.com

In case Publisher is established outside the EU,
the EU authorized representative is:
Springer Nature Customer Service Center GmbH
Europaplatz 3, 69115 Heidelberg, Germany

Printed by Libri Plureos GmbH
in Hamburg, Germany